# FLUID MACHINERY

## Performance, Analysis, and Design

Terry Wright

**CRC Press**

**Boca Raton   London   New York   Washington, D.C.**

**Library of Congress Cataloging-in-Publication Data**

Wright, Terry, 1938-
    Fluid machinery : performance, analysis, and design / Terry Wright.
      p.     cm.
    Includes bibliographical references and index.
    ISBN 0-8493-2015-1 (alk. paper)
    1. Turbomachines.   2. Fluid mechanics.   I. Title.
TJ267.W75   1999
621.406--dc21

98-47436
CIP

© 1999 by CRC Press LLC

No claim to original U.S. Government works
International Standard Book Number 0-8493-2015-1
Library of Congress Card Number 98-47436
Printed in the United States of America 1 2 3 4 5 6 7 8 9 0
Printed on acid-free paper

# Preface

The purpose of this book is to provide a fairly broad treatment of the fluid mechanics of turbomachinery. Emphasis is placed on the more utilitarian equipment, such as compressors, blowers, fans, and pumps that will be encountered by most mechanical engineers as they pursue careers in industry. This emphasis is intended to allow the text to serve as a useful reference or review book for the practicing engineer. Both gas and hydraulic turbines are considered for completeness, and the text inevitably includes material from the large literature on gas turbine engines. These machines traditionally have been treated as aerospace equipment and are considered at length in the literature (Bathie, 1996; Lakshiminarayana, 1996; Mattingly, 1996; Oates, 1985; Oates, 1984; and Wilson, 1984). Although recent developments in power generation for either load-peaking, distributed generation or process-cogeneration have significantly increased the chances that an engineering graduate will encounter gas turbine engines, this text will focus primarily on the more commonly encountered industrial equipment.

The performance parameters of fluid machinery are carefully developed and illustrated through extensive examples. The relationship of the inherent performance of a machine, in terms of the flow rate, head change, and sound power or noise generation through the rotating impeller, is discussed and treated as it relates to the fluid system with which the machine interacts. The dependence of machine performance on the resistance characteristics of the fluid system is emphasized throughout by examining the machine and the system simultaneously through the text. The characteristic sound pressure and sound power levels associated with a fluid machine are treated in this text as a basic performance variable—along with flow and pressure change.

The fundamental relationship between the shape and internal geometry of a turbomachine impeller and its inherent performance is treated from the beginning of the text. In the early chapters, the shape and size of a machine are related through the concepts of similarity parameters to show how head and flow combine with shape and size to yield unique relationships between the geometry and performance. The development of these "specific" speed, noise, and size relations are set out in an empirical, traditional manner as correlations of experimental data. The concepts are used to achieve a basic unification of the very broad range and variety of machine types, shapes, and sizes encountered in engineering practice.

In the later chapters, the theme of geometry and performance is continued through the approximate treatment of the flow patterns in the flow passages of the machine. The fundamental consideration of the equations for mass and angular momentum lead to the governing relations for turbomachinery flow and performance. Again, the process is related to the machine geometry, size, speed, and flow path shape. This higher level of detail is related as closely as possible to the overall considerations of size and speed developed earlier.

Following extensive examples and design exercises for a broad range of equipment, application, and constraints, the later chapters begin tightening the rigor of the calculational process. The simplifying assumptions used to develop the earlier illustrations of fundamental performance concepts are replaced with more complex and more rigorous analysis of the flow fields. The more thorough treatment of the flow analyses provides a more realistic view of the complexity and difficulties inherent in understanding, analyzing and designing turbomachinery components.

Following the development of greater rigor and greater calculational complexity, near the end of the text, some of the more vexing problems associated with turbomachinery analysis and design are introduced as advanced design topics. The influence of very low Reynolds numbers and high levels of turbulence intensity are considered as they influence the design and geometric requirements to generate specified levels of performance. Limitations on performance range and acceptable operation are introduced in these later chapters through consideration of compressibility, instability, and stalling phenomena and the inherent degradation of machine and system interaction.

Some attention is given throughout the text to the need to apply advanced analytical techniques to turbomachinery flowfields in a final design phase. However, the more approximate techniques are emphasized through most of the book. Here, the sense of a preliminary design approach is employed to promote a basic understanding of the behavior of the machinery and the relation between performance and geometry. The final chapter provides an overview of the calculational techniques that are being used to provide rigorous, detailed analysis of turbomachinery flows. Beginning with a reasonable geometry, these techniques are used to examine the influence of detailed geometric refinements and allow the designer to achieve something of an optimized layout and performance for a machine. The chapter is used to emphasize the importance of current and future computational capabilities and to point the reader toward more rigorous treatment of fluid mechanics in machine design.

Throughout most of the text, the examples and problem exercises are either partially or totally concerned with the design or selection process. They deal with system performance requirements or specifications, along with size, speed, cost, noise, and efficiency constraints on the problem solution. The purpose of this pragmatic design approach to turbomachinery applications is to expose the reader—either student or practicing engineer—to the most realistic array of difficulties and conflicting requirements possible within the confines of a textbook presentation. By using examples from a fairly large range of industrial applications, it is hoped that the reader will see the generality of the basic design approach and the common ground of the seemingly diverse areas of application.

# The Author

**Terry Wright** has acquired a wide range of experience in performance, analysis, and design of turbomachinery in both the academic and industrial
worlds. He initially joined the Westinghouse
Research Laboratories and served there for many
years as a research scientist and fellow engineer.
He studied and developed design techniques and
procedures for fans, pumps, and blowers and
became involved in design and predictive methods
for minimizing noise generation in industrial equipment. Much of his effort in this period was working
with those manufacturing divisions of the Westinghouse Corporation involved with turbomachinery.

Dr. Wright left the industrial sphere in the mid-1980s to become a professor of
mechanical engineering at the University of Alabama at Birmingham and was active
in teaching and mentoring in fluid mechanics and heat transfer, fluid measurement,
and applications in turbomachinery. During this period he conducted and directed
studies in fluid measurement and design and performance estimation, with emphasis
on prediction and minimization of turbomachinery-generated noise. While at the
university, he consulted with many industrial manufacturers and end users of turbomachinery equipment on problems with performance and noise. In addition to his
academic and research activities, Dr. Wright also served as chairman of the Department of Mechanical Engineering for several years.

He has acted as technical advisor to government and industry and has published
over 90 research and industrial papers (of limited distribution). He has also published
over 40 papers in engineering journals, pamphlets, and proceedings of the open
literature. He is a registered Professional Engineer in Mechanical Engineering and
has served on technical committees on turbomachinery and turbomachinery noise
in the American Society of Mechanical Engineers.

Dr. Wright is presently an Emeritus Professor of the University of Alabama at
Birmingham and is active in writing and society activities. He continues to interact
with the manufacturers and industrial users of turbomachinery equipment.

# Acknowledgments

I am indebted to a great many people who directly or indirectly contributed to the development of this book. My wife Mary Anne and my son Trevor have encouraged me throughout the writing process and have patiently listened to me wrestle with the wording of the text. To my parents, Roy John and Karolyn Wright, I owe the debt for their encouragement and assistance with my education and for fostering my early interest in engineering. I am particularly grateful to Professor Robin Gray of The Georgia Institute of Technology for my early training in careful, objective research, and honest, cautious reporting of results.

This book would never have been written without the patience and encouragement of hundreds of engineering students who suffered through about five versions of my turbomachinery notes and the development of the early problem sets for this book. Those who particularly stand out include Femi Agboola, Jerry Myers, and Amy Brown Wathen.

Among the many engineers, university folks, and industrial people who helped me put this text together, I want to thank Daniel Ariewitz (Westinghouse), John Beca (ITT Industrial Group), Martin Crawford (University of Alabama at Birmingham), Donna Elkins (Barron Industries), Charles Goodman (Southern Company Services), Torbjørn Jacobsen (Howden Variax), Johnny Johnson (Barron Industries), John McGill (Howden Buffalo), Ken Ramsey (Howden Sirocco), Tom Talbot (University of Alabama at Birmingham), Barbara Wilson (University of Alabama at Birmingham), and Steve Wilson (Southern Company Services).

# Table of Contents

# 1 Introduction

## 1.1 PRELIMINARY REMARKS

For convenience of review and quick reference, this introduction includes the basic fundamentals of thermodynamics and fluid mechanics needed to develop and manipulate the analytical and empirical relationships and concepts used with turbomachinery. The standard nomenclature for turbomachinery will be used where possible, and the equations of thermodynamics and fluid mechanics will be particularized to reflect practice in the industry.

## 1.2 THERMODYNAMICS AND FLUID MECHANICS

The physics and properties of the common, simple fluids used in this application include those of many gases, such as air, combustion products, dry steam, and others, and liquids, such as water, oils, petroleum products, and other Newtonian fluids transported in manufacturing processes. These fluids and the rules governing their behavior are presented here in terse fashion for only the simple working fluids, as needed for examples and problem-solving in an introductory context.

More complete coverage of complex fluids and special applications is readily available in textbooks on thermodynamics and fluid mechanics, as well as the more specialized engineering texts and journals. See, for example, White, 1994; Fox and McDonald, 1992; Black and Hartley,1986; Van Wylen and Sonntag, 1986; Baumeister, et al., 1978; the journals of the American Society of Mechanical Engineers and the American Institute of Aeronautics and Astronautics, and the *Handbook of Fluid Dynamics and Turbomachinery*, Shetz and Fuhs, 1996. Here, there will be no extensive treatment of multiphase flows, flows of mixtures such as liquid slurries or gas-entrained solids, fluids subject to electromagnetic effects, or ionized or chemically reacting gases or liquids.

## 1.3 UNITS AND NOMENCLATURE

Units will generally be confined to the International System of units (SI) and British Gravitational system (BG) fundamental units as shown in Table 1.1. Unfortunately, turbomachinery performance variables are very frequently expressed in conventional, traditional units that will require conversion to the fundamental unit systems. Units for some of these performance parameters, based on pressure change, through-flow, and input or extracted power, are given in Table 1.2.

As seen in the table, many forms exist where the units are based on instrument readings, such as manometer deflections, or electrical readings rather than the fundamental parameter of interest. As an engineer, one must deal with the nomenclature common to the particular product or industry at least some of the time. Hence, this

**TABLE 1.1**
**Fundamental Units in SI and BG Systems**

| Unit | SI | BG |
|---|---|---|
| Length | Meter | Foot |
| Mass | Kilogram | Slug |
| Time | Second | Second |
| Temperature | °C, Kelvin | °F, °Rankine |
| Force | Newton, kg-m/s² | lbf, slugs-ft/s² |
| Pressure | Pa, N/m² | lbf/ft² |
| Work | Joule, J, N-m | ft-lbf |
| Power | Watt, J/s | Ft-lbf/s, horsepower (hp) |

**TABLE 1.2**
**Traditional Performance Units in BG and SI**

| Unit | BG | SI |
|---|---|---|
| Pressure | lbf/ft²; InWG; InHg; psi; psf | N/m²; Pa; mm H₂O; mm Hg |
| Head | Foot; inches | m; mm |
| Volume flow rate | ft³/s; cfm; gpm; gph | m³/s; l/s; cc/s |
| Mass flow rate | Slug/s | kg/s |
| Weight flow rate | lbf/s; lbf/hr | N/s; N/min |
| Power | watt; kW; hp; ft-lbf/s | N-m/s; J; kJ; watts kW |

text will include the use of gpm (gallons per minute) for liquid pumps, hp (horsepower = (ft–lbf/s)/550) for shaft power, and some others as well. However, this text will revert to fundamental units for analysis and design and convert back to the traditional units if necessary or desirable.

## 1.4 THERMODYNAMIC VARIABLES AND PROPERTIES

The variables and properties frequently used include the state variables: pressure, temperature, and density ($p$, $T$, and $\rho$). They are defined respectively as: $p$ = the normal stress in the fluid; $T$ = a measure of the internal energy in the fluid; and $\rho$ = the mass per unit volume of the fluid. These are the fundamental variables that define the state of the fluid. When dealing (as one must) with work and energy inputs to the fluid, then one will be dealing with the specific energy ($e$), internal energy ($h' = u' + p/\rho$), entropy ($S$), and specific heats of the fluid ($c_p$ and $c_v$). The transport properties for friction and heat transfer will involve an additional two parameters: the viscosity ($\mu$) and the thermal conductivity ($\kappa$). In general, all of these properties are interrelated in the state variable functional form $\rho = \rho(p,T)$, $h' = h'(p,T)$, and $\mu = \mu(p,T)$. That is, they are functions of the state properties of the fluid.

The energy stated in terms of $u'$ and $e$ are measures of the thermal and internal energy of the fluid as well as potential and kinetic terms, such that $e = u' + V^2/2 + gz$.

Here, g is the gravitational constant of the flowfield, V is the local velocity, and z is the positive upward direction.

For gases, this text restricts its attention to those fluids whose behavior can be described as thermally and calorically perfect. That is, the variables are related according to $P = \rho RT$, with $R = c_p - c_v \cdot c_v$ and $c_p$ are constant properties of the particular fluid. R can also be defined in terms of the molecular weight, M, of the gas where $R = R_u/M$. $R_u$ is the universal gas constant, given as 8310 $m^2/(s^2K)$ in SI units and 49,700 $ft^2/(s^2 \,^\circ R)$ in BG units. Appendix A provides limited information on these and other fluid properties for handy reference in illustrations and problem-solving. $c_p$ and $c_v$ are related as a distinct molecular property, $\gamma = c_p/c_v = \gamma(T)$ and may be written as $c_v = R/(\gamma - 1)$ and $c_p = \gamma R/(\gamma - 1)$. For most turbomachinery air-moving applications, one can accurately assume that $c_v$ and $c_p$ are constants, although for large temperature excursions, both $c_v$ and $c_p$ increase with increasing temperature.

In addition to these fundamental variables, it is necessary to define the "real" fluid properties of dynamic and kinematic viscosity. Dynamic viscosity, $\mu$, is defined in fluid motion as the constant of proportionality between a velocity gradient and the shearing stress in the fluid. From the relationship $\tau = \mu(\partial V/\partial n)$, where n is the direction normal to the velocity V, viscosity has the units of velocity gradient $[s^{-1} = (\text{length/time})/\text{length})]$ over stress (Pa or $lbf/ft^2$). Viscosity is virtually indepen-dent of pressure in most flow-system analyses, yet it can be a fairly strong function of the fluid temperature. Typical variation in gases can be approximated in a power law form such as $\mu/\mu_0 = (T/T_0)^n$ with $\mu$ increasing with temperature. For air, the values $n = 0.7$ and $T_0 = 273K$ are used with $\mu_0 = 1.71 \times 10^{-5}$ kg/ms. Other approximations are available in the literature, and data is presented in Appendix A. The kinematic viscosity, $v = \mu/\rho$, is merely defined for convenience in forming the Reynolds number, $Re = \rho Vd/\mu = Vd/v$.

Liquid viscosities decrease with increasing temperature, and White (1994) sug-gests the form for pure water as.

$$\ln\left(\frac{\mu}{\mu_0}\right) = -1.94 - 4.80\left(\frac{273.16}{T}\right) + 6.74\left(\frac{273.16}{T}\right)^2 \qquad (1.1)$$

as being a reasonable estimate, with T in Kelvins (within perhaps 1%).

When dealing with the design and selection of liquid handling machines, the prospect of "vaporous cavitation" within the flow passages of the pump or turbine is an important constraint. The critical fluid property governing cavitation is the vapor pressure of the fluid. The familiar "boiling point" of water as 100°C is the temperature required to increase the vapor pressure to reach the value of a standard atmospheric pressure, 101.3 kPa. At a given pressure, the vapor pressure of water (or another fluid) is reduced with decreasing temperature, so that boiling or cavitation will not occur without a reduction in the pressure of the fluid. As pressure is prone to be significantly reduced in the entry (or suction) regions of a pump, if the fluid pressure becomes more or less equal to the vapor pressure of the fluid, boiling or cavitation may commence. The vapor pressure is a strongly varying function of temperature. For water, this $p_v$ ranges from nearly zero (0.611 kPa) at 0°C to 101.3

kPa at 100°C. A figure of $p_v$ vs. T is given in Appendix A. A rough estimate of this functional dependence is given as $p_v = 0.61 + 10^{-4}T^3$. The approximation (accurate to only about 6%) illustrates the strong nonlinearity that is typical of liquids. Clearly, fluids at high temperature are easy to boil with pressure reductions and cavitate readily.

The absolute fluid pressure associated with the onset of cavitation depends, in most cases, on the local barometric pressure. One recalls that this pressure varies strongly with altitude in the atmosphere and can be modeled, using elementary hydrostatics, as $p_b = p_{SL} \exp[-(g/R)\int dz/T(z)]$, integrated from sea level to the altitude z. The function T(z) is accurately approximated by the linear lapse rate model, $T = T_{SL} - Bz$, which yields $p_b = p_{SL}(1 - Bz/T_{SL})^{(g/RB)}$. Here, B = 0.0065 K/m, $p_{SL}$ = 101.3 kPa, g/RB = 5.26, and T ≅ 15K. This relation allows approximation of the absolute inlet-side pressure for pump cavitation problems with vented tanks or open supply reservoirs at a known altitude.

## 1.5   REVERSIBLE AND IRREVERSIBLE PROCESSES WITH PERFECT GASES

In turbomachinery flows, not just the state of the fluid must be known, but also the process or path between these end states is of interest in typical expansion and compression processes. The ideal process relating these end state variables is the *isentropic process*. Recalling the second law of thermodynamics,

$$TdS = dh' - \frac{dp}{\rho} \tag{1.2}$$

and assuming the usability of perfect gas relations ($c_p$ and $c_v$ constant, $dh' = c_p dT$ and $du' = c_v dT$, $R = c_p - c_v$, and $p = \rho RT$) this becomes

$$dS = \frac{c_p dT}{T} - \frac{Rdp}{p} \tag{1.3}$$

On integration between end states 1 and 2, one can write the change in entropy as

$$S_2 - S_1 = c_p \ln\left(\frac{T_2}{T_1}\right) - R \ln\left(\frac{p_2}{p_1}\right) \tag{1.4}$$

If the fluid flow process is reversible (without heat addition or friction), the process is isentropic, $S_2 - S_1 = 0$ and

$$\left(\frac{p_2}{p_1}\right) = \left(\frac{T_2}{T_1}\right)^{\gamma/(\gamma-1)} = \left(\frac{\rho_2}{\rho_1}\right)^{\gamma} \tag{1.5}$$

If the fluid process is not really isentropic, usually as a result of frictional losses in a compression or expansion process, one can use a similar form based on the efficiency of the process, $\eta$. The exponent $\gamma/(\gamma - 1)$ is replaced by $n/(n - 1) = \eta\gamma/(\gamma - 1)$ to form the polytropic relationship for a perfect gas:

$$\left(\frac{p_2}{p_1}\right) = \left(\frac{T_2}{T_1}\right)^{n/(n-1)} = \left(\frac{\rho_2}{\rho_1}\right)^n \tag{1.6}$$

Although changes in the fluid density, $\rho$, can be neglected in a great many flow processes in turbomachinery, one can rely on these polytropic relations when changes in the fluid pressure are too large to neglect density variations in the flow. The process efficiency can be approximated by the total pressure rise or head extraction efficiency of the turbomachine, developed in later chapters, without undue loss of rigor or accuracy.

## 1.6 CONSERVATION EQUATIONS

In fluid mechanics analyses, one must obey the set of natural laws that govern Newtonian physics. That is, one must satisfy: conservation of mass or $dm/dt = 0$; conservation of linear momentum or $\mathbf{F} = m(d\mathbf{V}/dt)$ (here the bold letters indicate the vector character of the terms); conservation of angular momentum or $\mathbf{M} = d\mathbf{H}/dt = d(\Sigma r \mathbf{X} \mathbf{V} \delta m)/dt$, where $\delta m$ is the mass of each term being included in the sum; and conservation of energy or $dQ'/dt - dW/dt - dE/dt = 0$. These relationships consist of two scalar equations and two vector equations (three terms each) so that one is actually dealing with eight equations. In the energy equation, the first law of thermodynamics, $Q'$ is the heat being transferred to the fluid, W is the work being done by the fluid, and E is the energy of the fluid. These equations, along with the second law and state equations mentioned above, complete the analytical framework for the fluid flow.

As with the study of fluid mechanics, these basic forms are converted to a control volume formulation using the Reynolds Transport Theorem. Conservation of mass for steady flow becomes

$$m' = \iint_{cs} \rho(\mathbf{V} \bullet \mathbf{n}) dA \tag{1.7}$$

where cs indicates integration over the complete surface of the control volume and $\mathbf{V} \bullet \mathbf{n}$ is the scalar product of the velocity into the surface unit normal vector, n (i.e., the "flux term"). The relation says simply that in order to conserve mass in this Eulerian frame of reference, what comes in to the control volume must leave it. For simple inlets and outlets, with uniform properties across each inlet or outlet, one can write

$$\sum (\rho V A)_{out} - \sum (\rho V A)_{in} = 0 \tag{1.8}$$

For incompressible flow ($\rho$ = constant), the equation reduces to

$$\sum (VA)_{out} = \sum Q_{out} = \sum (VA)in = \sum Q_{in} \qquad (1.9)$$

where Q is the volume flow rate VA and $m' = rVA = \rho Q$.

Conservation of linear momentum for steady flow becomes

$$\sum F = \iint_{cs} V\rho(V \cdot n)dA \qquad (1.10)$$

retaining the vector form shown earlier. Again, for simple inlets and outlets,

$$\sum F = \sum m'V_{out} - \sum m'V_{in} \qquad (1.11)$$

The incompressible form can be written with $m' = \rho Q$ as

$$\sum F = \rho \sum \left( QV_{out} - \sum QV_{in} \right) \qquad (1.12)$$

The Bernoulli equation for steady, incompressible, frictionless flow without heat addition or shaft work is frequently a very useful approximation to more realistic flows and can be developed from the conservation equations above and the first law. It is written as

$$\frac{p_1}{\rho} + \frac{V_1^2}{2} + gz_1 = \frac{p_2}{\rho} + \frac{V_2^2}{2} + gz_2 = \text{Constant} = p_0 \qquad (1.13)$$

where $p_0$ is the total pressure of the flowing fluid. In turbomachinery flows, where work always takes place in the flow process, this equation is not valid when the end states are located across the region of work addition or extraction, because the total pressure rises or falls. If shaft work or frictional losses of total pressure are to be included, then the Bernoulli equation must be extended to the form

$$\frac{p_1}{\rho} + \frac{V_1^2}{2} + gz_1 = \frac{p_2}{\rho} + \frac{V_2^2}{2} + gz_2 + \frac{W_s}{\rho} + \frac{W_v}{\rho} \qquad (1.14)$$

Here, $W_s$ is shaft work and $W_v$ is the viscous dissipation of total pressure. This equation is frequently rewritten in "head" form as

$$\frac{p_1}{\rho g} + \frac{V_1^2}{2g} + z_1 = \frac{p_2}{\rho g} + \frac{V_2^2}{2g} + z_2 + h_s + h_f \qquad (1.15)$$

where each term in the equation has units of length. $h_s$ and $h_f$ are the shaft head addition or extraction and the frictional loss, respectively.

These then are the basic physical relationships and variables for the flow in turbomachines and their fluid systems needed for the studies in this text. If further review or practice with these fundamentals is needed, the reader is referred to the books by White (1994), Fox and McDonald (1992), Shetz and Fuhs (1996), and Baumeister et al. (1978), or Schaums outline series (Hughes and Brighton, 1991).

## 1.7 TURBOMACHINES

The content of this material will be restricted to the study of the fluid mechanics of turbomachines. This requires that a clear definition of turbomachinery be established at the outset. Paraphrasing from earlier authors (Balje, 1981; White, 1994), turbomachines can be defined as follows:

> A turbomachine is a device in which energy is transferred to or from a continuously moving fluid by the action of a moving blade row. The blade row rotates and changes the stagnation pressure of the fluid by either doing work on the fluid (as a pump) or by having work done on the blade row by the fluid (as a turbine).

This definition precludes consideration of a large class of devices called positive displacement machines. These are not turbomachines, according to the definition, since flow does not move continuously through them. They have moving boundaries that either force the fluid to move or are forced to move by the fluid. Examples include piston pumps and compressors, piston steam engines ("turbines"), gear and screw devices, sliding vane machines, rotary lobe pumps, and flexible tube devices. They will not be further considered here, and reference is made to Balje (1981) for more detailed material. It is also pointed out that no treatment will be provided here for the very broad areas of mechanical design: dynamics of rotors, stress analysis, vibration, or other vital mechanical topics concerning turbomachinery. Reference is made to others (Rao, 1990; Beranek and Ver, 1992; Shigley and Mischke, 1989) for further study of these important topics. Because of its overriding importance in selection and siting of turbomachines, the subject of noise control and acoustics of turbomachinery will be included in the treatment of the performance and fluid mechanics.

## 1.8 CLASSIFICATIONS

Much has been written on classifying turbomachinery, and a major subdivision is implied in the definition stated above. This is the power classification, identifying whether power is added to or extracted from the fluid. Pumps, which are surely the most common turbomachines in the world, are power addition machines and include liquid pumps, fans, blowers, and compressors. They operate on such fluids as water, fuels, waste slurry, air, steam, refrigerant gases, and a very long list of others. Turbines, which are probably the oldest turbomachines, are the power extraction devices and include windmills, water wheels, modern hydroelectric turbines, the exhaust side of automotive engine turbochargers, and the power extraction end of

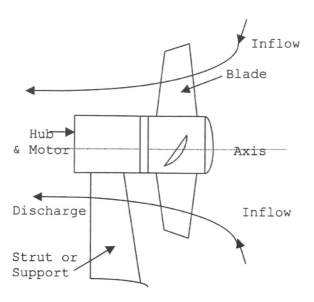

**FIGURE 1.1** Examples of an open flow turbomachine. No shroud or casing defines the limits of the flow field.

an aviation gas-turbine engine. Again, they operate on a seemingly endless list of fluids, including gases, liquids, and mixtures of the two, as well as slurries and other particulate-laden fluids.

The manner in which the fluid moves through and around a machine provides another broad means of classification. For example, some simple machines are classed as open or open flow, as illustrated in Figure 1.1. Here, there is no casing or enclosure for the rotating devices and they interact rather freely with the flowing stream—in this case, the atmosphere. Consistent with the power classification, the propeller is an open flow pumping device, and the windmill is an open flow turbine. Figure 1.2 shows examples of enclosed or encased flow devices where the interaction between the fluid and the device is carefully controlled and constrained by the casing walls. Again, these examples are power classified—in this case, as pumps.

Since all turbomachines have an axis of rotation, the predominant organization of the mass flow relative to the rotating axis can be used to further refine the classification of machines. This subdivision is referred to as flowpath or throughflow classification and deals directly with the orientation of the streamlines that define the mass flow behavior. In axial throughflow machines—pumps, fans, blowers, or turbines—the fluid moves through the machine on streamlines or surfaces approximately parallel to the axis of rotation of the device. Figure 1.3 shows an axial (or axial flow) fan characterized by flow parallel to the fan axis of rotation.

Flow that is predominantly radial in the working region of the moving blade row is illustrated in Figure 1.4, showing a cutaway or sectioned view. The device is a radial-inflow turbine and illustrates typical geometrical features of such machines. If the flow direction were reversed, the geometry would be typical of radial-discharge pumps or compressors.

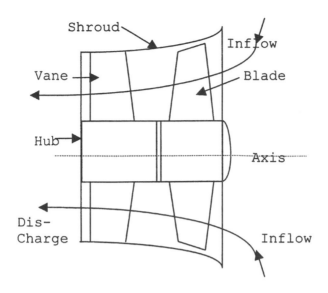

**FIGURE 1.2** Example of enclosed or encased flow with the shroud controlling the outer streamlines.

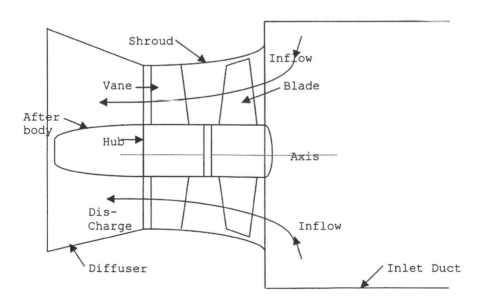

**FIGURE 1.3** Layout of an axial fan with the major components.

Since nothing is ever as simple or straightforward as one would like, there must be a remaining category for machines that fail to fit the categories of predominantly axial or predominantly radial flow. Pumps, fans, turbines, and compressors may all fall into this class and are illustrated by the mixed-flow compressor shown in Figure 1.5. Flow direction for a pump is generally from the axial path to a conical

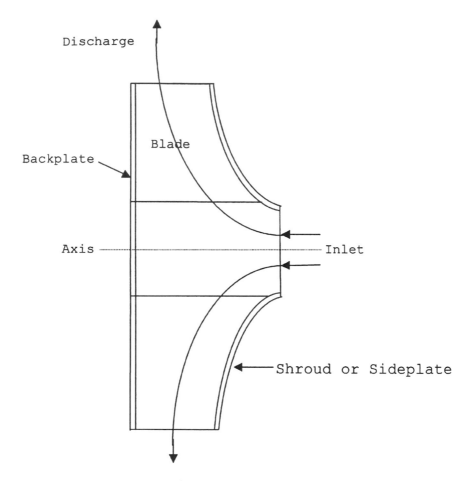

**FIGURE 1.4**   A radial throughflow turbomachine. The flow moves through the blade row in a primarily radial direction.

path moving upward at an angle between 20° and 65° (more or less). Again, reversing the direction of flow yields a path that is typical of a mixed flow turbine. In addition, the simple flow paths shown here may be modified to include more than one impeller. For radial flow machines, two impellers can be joined back-to-back so that flow enters axially from both sides and discharges radially as sketched in Figure 1.6. These machines are called double suction (for liquid pumps) or double inlet (for gas movers such as fans or blowers). The flow paths are parallel to each other, and flow is usually equal in the two sides, with equal energy addition occurring as well. Others machines might consist of two or more impellers in either axial, radial, or even mixed flow configurations. In these machines, the flow proceeds serially from one impeller to the next, with energy addition occurring at each stage. These multistage machines are illustrated in Figure 1.7.

Further breakdown of these classifications can include the compressibility of the fluid in the flow process. If the density is virtually constant in the entire flow process,

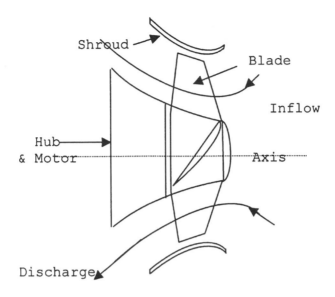

**FIGURE 1.5** A mixed throughflow machine. The flow can enter axially and exit radially or at an angle from the axis, as shown.

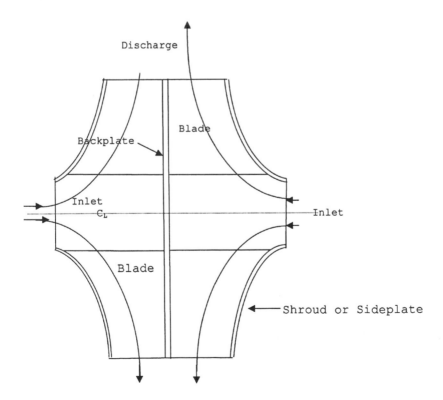

**FIGURE 1.6** A double-inlet, double-width centrifugal impeller.

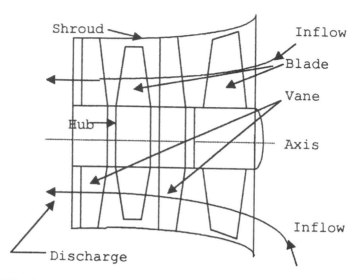

**FIGURE 1.7**   A two-stage axial fan configuration.

as in liquid pumps and turbines, the incompressible label can be added. For gas flows, if there are large absolute pressure changes or high speeds or large Mach numbers involved that lead to significant changes in density, the machines can be labeled as compressible or simply as compressors. This text will try to keep the range of turbomachines as nearly unified as possible and make distinctions concerning gas–liquid, compressible–incompressible when it is convenient or useful to do so.

A set of illustrations of turbomachines is included at the end of this chapter (Figures 1.15 through 1.19). They should help to relate the various flow paths to actual machines.

## 1.9   PERFORMANCE AND RATING

The concept of determination of the performance or "rating" of a machine can center on how to measure the performance capability, for example, in a laboratory setting. A schematic layout of a test facility is shown in Figure 1.8. Measurement is carried out with equipment and procedures specified by recognized standards to ensure accuracy, acceptance, and reproducibility of test results. As illustrated, the test stand is for pumping machines handling air, and the appropriate test standard to be used is "Laboratory Methods of Testing Fans for Rating" (AMCA, 1985) or other testing standards such as those written by the American Society of Mechanical Engineers (ASME). For example, the ASME Power Test Code PTC-10 for Exhausters and Compressors (ASME, 1965) is a more general test code that effectively includes the basically incompressible AMCA fan test code as a subcase. For liquid pumps, the appropriate guideline could be ASME PTC-8 (ASME, 1985).

Referring again to Figure 1.8, this test stand is set up to measure the performance of the gas pumping machines. Shown on the left is a small centrifugal fan. This fan draws air from the room into the unrestricted intake and discharges the air into the

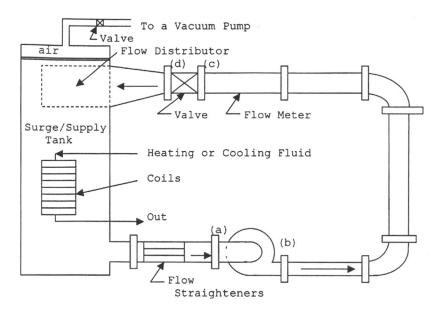

**FIGURE 1.8** Schematic of a pump performance test facility, based on the ASME pump Test Code PTC-8.2. (Adapted from ASME, 1965; Hydraulic Institute, 1983.)

carefully sealed flow box, or (more formally) plenum chamber. There are numerous pressure tappings located along the flow path through the plenum by which one determines a set of gage pressures. For a more general treatment that fully accounts for compressibility, the alternate instrumentation for determining total pressure and temperature at the fan discharge is shown as a pitot-static probe with a thermocouple for local temperature measurement. The taps are located at points 'a,' 'b,' and 'c.' At 'a,' or on the pitot-static probe, the discharge pressure of the fan is being monitored, and since the fan is doing work on the air passing through it, this $p_a$ (total or static) will be greater than the room ambient pressure seen outside the plenum or near the intake. This measurement is used to identify the pressure increase imparted by the fan to the air by calculation of the total pressure ratio across the blower, $p_{02}/p_{01}$. These values include the kinetic term associated with the fan discharge velocity and are absolute pressures, including the reference barometric pressure. If the pressure change across the machine is sufficiently small (about 1% of the barometric pressure), the incompressible variable can be used: $\Delta p_s = p_a - p_{amb}$. This value is called the *static pressure rise* of the fan. The *total pressure rise*, more nearly equivalent to the compressible pressure ratio, would include the velocity pressure of the discharge jet (with $V_j$) to yield $\Delta p_T = \Delta p_s + \rho V_j^2/2$, where $\rho$ is the ambient value.

The small jet of air being discharged by the fan is spread out (or settled) as it moves through the first chamber of the box and the row of resistive screens through which the air must pass. The amount of resistance caused by these screens is specified by the test standard to ensure that the flow is smoothly distributed across the cross-section of the plenum chamber on the downstream side of the screens. In other test arrangements such as constant area pipes or ducts, the flow settling may rely on flow straighteners such as nested tubes, honeycomb, successive perforated plates, or fine-mesh screens. A particular arrangement must yield a nearly uniform approach velocity pattern as the flow nears the inlet side of the flow metering apparatus. This meter might be a set of precise or calibrated flow measurement nozzles, as sketched (for example, ASME long radius nozzles), mounted in the center-plane of the plenum chamber (see Holman, 1989; Granger, 1988; or Beckwith et al., 1993 for details of the construction of these nozzles). The meter may also be a nozzle in the duct, a precision venturi meter, or a sharp-edged orifice plate in the pipe or duct.

The pressure tap at point 'b' supplies the value upstream of the flow nozzles and the tap at 'c' gives the downstream pressure. For compressible flow, the total pressure and temperature at the flow meter must be established to provide accurate information for calculation of the mass flow rate. The difference in pressure across the meter, perhaps read across the two legs of a simple U-tube manometer or pressure transducer, supplies the differential pressure or pressure drop through the nozzle $\Delta p_{b-c} = p_b - p_c$. This differential pressure is proportional to the square of the velocity of the air being discharged by the nozzles, so that the product of the velocity and the total nozzle area provides the volume flow rate being supplied by the fan.

Another piece of direct performance data required is a measure of the power being supplied to the test fan. This can be determined by a direct measure of the electrical watts being supplied to the fan motor, or some means may be provided to measure torque to the fan impeller along with the rotating speed of the shaft. The product of the torque and speed is, of course, the actual power supplied to the fan shaft by the driving motor. To complete the acquisition of test data in this experiment, it remains to make an accurate determination of the air density. In general, one needs to know the density at the inlet of the blower and at the flow meter. These values are needed to determine the discharge velocity from the ASME nozzles to be used in defining the mass flow rate being supplied by the machine. Nozzle velocity is calculated from the Bernoulli equation as

$$V_n = c_d \left( \frac{2\Delta p_{b-c}}{\rho} \right)^{1/2} \tag{1.16}$$

where $c_d$ is the nozzle discharge coefficient, $c_d \leq 1.0$, used to account for viscous effects in the nozzle flow. $c_d$ is a function of the diameter-based Reynolds Number for the nozzles and may be approximated by (Beckwith et al., 1993)

$$c_d = 0.9965 - 0.00653 \left( \frac{10^{16}}{Re_d} \right)^{1/2} \tag{1.17}$$

The Reynolds Number is given by $Re_d = V_n d/\nu$, and $\nu$ is the fluid kinematic viscosity. The inlet density must be accurately determined by knowing the values of barometric pressure, $p_{amb}$, ambient temperature, $T_{amb}$, and the suppression of ambient temperature achievable through evaporation of water on a temperature instrument (thermometer). This latter value, $\Delta T_{d-w} = T_d - T_w$, is used to calculate the relative humidity or water vapor content in the air. $T_d$ is measured on a dry thermometer and $T_w$ on an air-cooled thermometer inserted in a wet wick material. This non-negligible correction to the air density, $\rho$, can be carried out using a psychometric chart. (See, for example, Van Wylen and Sonntag, 1986; AMCA, 1985). The psychometric chart from the AMCA standard is reproduced here as Figure 1.9, where the weight density $\rho g$ is given as a function of the dry bulb temperature, $T_d$, the wet bulb suppression $\Delta T_{d-w}$ = $T_d - T_w$, and the barometric pressure, $p_{amb}$. All of the units are in English as shown, and $\rho g$ is in $lbf/ft^3$. The density at the flow meter can be determined from the inlet value and the total pressure ratio according to

$$\rho_{meter} = \rho_{inlet} \left( \frac{p_{02}}{p_{01}} \right) \left( \frac{T_{01}}{T_{02}} \right) \tag{1.18}$$

where the p and T values are in absolute units. For essentially incompressible test conditions, the two densities will be the same.

To complete the data gathering from the fan, the noise generated by the test fan should be measured using a microphone and suitable instrumentation to provide values of $L_w$, the sound power level in decibels. This topic will be considered in detail in Chapter 5; it is mentioned here because of the frequently overriding importance of this performance parameter.

For analysis purposes, the aerodynamic efficiency of the fan can be calculated from the main performance variables. Efficiency is the output fluid power divided by the shaft input power, $P_{sh}$. This is the traditional definition for pumps, fans, and blowers, even into the compressible regime of pressure rise. The calculation is according to $\eta_{T0} = (p_{02} - p_{01})(m/\rho_{01})/P_{sh}$. This is equivalent, for incompressible flows, to $\eta_T = \Delta p_T Q/P_{sh}$. The T subscript on $\eta$ corresponds to the use of total pressure rise in the output power calculation. Use of $\Delta p_s$ yields the commonly used static efficiency $\eta_s$. In high-speed compressors, such as those used in gas turbine engines, efficiency is defined differently (Lakshminarayana, 1996). The compressor efficiency, $\eta_c$, is characterized by the ratio of work input to an ideal isentropic compressor, to the work input to an actual compressor with the same mass flow and pressure ratio. In terms of stagnation enthalpy, $\eta_c$ becomes

$$\eta_c = \frac{\left( h_{02s} - h_{01} \right)}{\left( h_{02} - h_{01} \right)} \tag{1.19}$$

With constant values of $c_p$, this yields

$$\eta_c = \frac{\left( T_{02s} - T_{01} \right)}{\left( T_{02} - T_{01} \right)} \tag{1.20}$$

Using the isentropic relation between total temperature and total pressure, $(T_{02s}/T_{01}) = (p_{02}/p_{01})^{(\gamma-1)/\gamma}$ reduces to

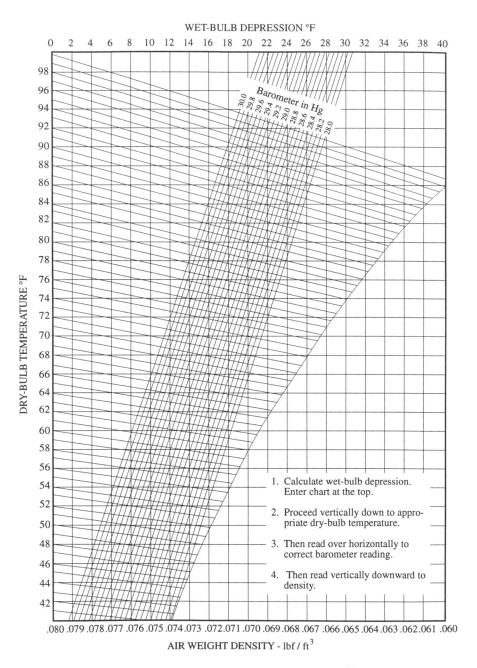

**FIGURE 1.9** A psychometric density chart. (ANSI/AMCA Standard 210-85; ANSI/ASHRAE Standard 51-1985.)

$$\eta_c = \frac{\left[\left(\dfrac{p_{02}}{p_{01}}\right)^{(\gamma-1)/\gamma} - 1\right]}{\left(\dfrac{T_{02}}{T_{01}} - 1\right)} \qquad (1.21)$$

Finally, the flow, pressure, sound power level, efficiency, and input power are the values of a performance point associated with the machine being tested. Note that this point of operation is being controlled through the effects of either a downstream throttle, an auxiliary exhaust blower, or both. Both devices allow the pressure rise or pressure ratio of the device to be varied by increasing or reducing the overall resistance imposed on the ability of the blower to pump air. Reduced resistance (a more open throttle) will allow more flow (Q or m') at a reduced pressure rise or ratio ($p_{02}/p_{01}$ or $\Delta p_s$). The throttle can be successively adjusted to provide a table of performance points that cover the full performance range of the compressor or blower. A graph of results, typical for a small centrifugal fan, is shown in Figure 1.10.

There are several important things to notice in Figure 1.10. First, there is a point of maximum efficiency or Best Efficiency Point (BEP) defined by the $\eta_s$ vs. Q curve. The corresponding values of Q, $\Delta p_s$, $P_{sh}$, and $L_w$, along with that maximum value of $\eta_s$, define the BEP. To the right of BEP (at higher flow rate), the $\Delta p_s$ decreases with increasing flow, yielding a negative slope that represents the usable, stable range of performance for the pumping machine. Slightly to the left of the BEP (at lower flow rate), the curve of $\Delta p_s$ vs. Q goes through a zero slope condition, followed by a region of positive slope where the fan operates unstably, $\Delta p_s$ falls off, $\eta_s$ declines sharply, and $L_w$ increases very strongly. In a word, everything goes wrong at once, and this region represents a virtually unusable regime of operation for the fan. This zone is called the *stalled region* and is strictly avoided in selection and operation. At the extreme left of the curve, when Q = 0.0, the fan is still producing some pressure rise but is completely sealed off. This limit condition is called the "blocked tight" or "shut-off" point. It can be used in a transient condition such as start-up, but of course the fan must then progress through the stalled region of the curve to finally reach a stable operating condition. The question of start-up procedures will be referred to other texts (e.g., Jorgensen, 1983). Here, with Q = 0.0, the efficiency $\eta_s = 0.0$ as well. At the other extreme, one has the limit of zero pressure rise at the right end of the curve. This limit is called the "free delivery," or "wide open" point, and represents a maximum flow rate for the fan. It is seen that the efficiency generally degrades away from the BEP sharply on the left and more gradually on the right. The same can be said for the sound power level, $L_w$, with a gradual increase of noise to the right of BEP and a sharp increase to the left.

Equivalent performance for a compressor with significant density change is qualitatively the same as in the foregoing discussion: simply change the axis labels to $p_{02}/p_{01}$ and m'. Figures 1.10d and 1.10e show high-speed compressor and turbine performance curves. These characteristic curves are similar to low-speed fan curves, but they also show the phenomenon of choking mass flow at high speeds and high mass flow rates. Here, the throughflow velocities have reached sonic values and the

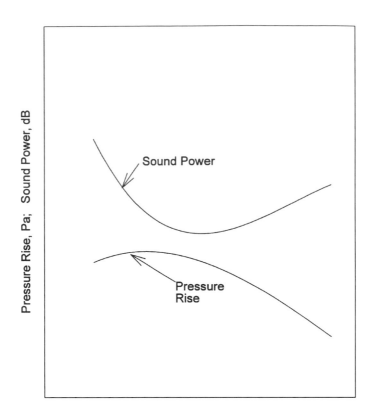

Volume Flow Rate

**FIGURE 1.10a**  Sound power performance curves.

mass flow rate is "frozen" at this level and does not continue to increase with changes in pressure. For the turbine, the pressure ratio is inverted to $p_{01}/p_{02} > 1.0$ to reflect the drop in total pressure as the turbine extracts energy from the flow stream. A low-speed hydraulic turbine performance curve is shown for completeness in Figure 1.10f. The similarities are obvious, but the head-flow curve lacks the choke effect seen in high Mach number flow of the high-speed turbine. The total head change is a drop in total head of the flow as it passes through the impeller and energy is removed from the stream.

## 1.10  PUMP AND SYSTEM

In practice or application, the fan or other pumping device would be connected to a flow 'system' rather than to a simple throttle. This system would typically consist of, say, a length of ductwork or pipe containing elbows, grills, screens, valves, heat exchangers and other resistance elements. The shape of a "resistance curve" or "system curve" can be described in terms of $\Delta p_f$ (f for friction) or in terms of head loss, $h_f$:

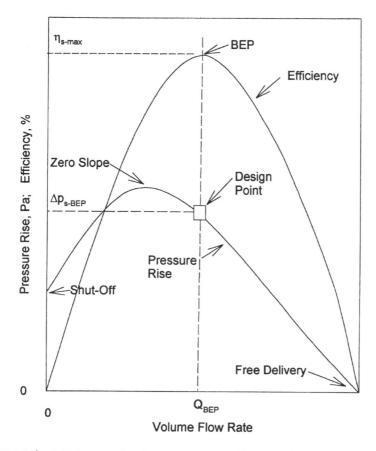

**FIGURE 1.10b**  Efficiency and performance curves with nomenclature.

$$h_f = \frac{\Delta p_f}{\rho g} = \left\{ f\left(\frac{L}{D}\right) + \sum K_m \right\}\left(\frac{V^2}{2g}\right) \qquad (1.22)$$

f is the Darcy friction factor (White, 1994); L/D is the ratio of the characteristic length and diameter of the ductwork; and $K_m$ is the loss factor for a given system element such as an elbow in the duct. $V^2$ depends on $Q^2$ as $V = Q/A_{duct}$ so that

$$h_f = f\left(\frac{L}{D}\right)\left(\frac{\left(\frac{Q}{A_{duct}}\right)^2}{2g}\right) + \sum K_M \left(\frac{\left(\frac{Q}{A_{duct}}\right)^2}{2g}\right) = \text{Constant}\left(\frac{Q}{A_{duct}}\right)^2 \qquad (1.23)$$

where the "Constant" is a fixed function of the system (although it may vary slightly with Reynolds Number). Rewritten, one obtains $\Delta p_f = \rho g h_f = CQ^2$ and observes that

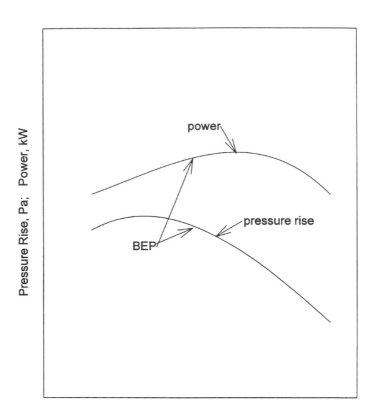

**Volume Flow Rate**

**FIGURE 1.10c**    Required power and performance curves.

the resistance must be overcome by the pressure rise of the fan, $\Delta p_s = f(Q)$. Then solve for the intersect of $\Delta p_s = \Delta p_f$ to determine a unique "operating point." Graphically, the fan characteristic curve $\Delta p_s = f(Q)$ intersects the resistance curve $\Delta p_f = CQ^2$ as shown in Figure 1.11. Note that if one adds length (L/D) or roughness (friction factor), one increases the value of C and gets a steeper parabola.

As the steeper parabola crosses the characteristic curve of the blower, the intersection occurs at the necessarily higher value of $\Delta p_s = f(Q)$, and the value of Q commensurately reduced. Similarly, reducing the pipe length, roughness, or such resistance elements as elbows or heat exchangers causes the resistance parabola to be less steep (smaller resistance constant), and the intersection with the fan curve occurs at a lower pressure and higher flow rate.

In a liquid pumping system, the resistance curve may be similarly characterized. However, it may be necessary to account for significant changes in elevation (hydrostatic head) imposed on the pump system. Here, the resistance curve can be characterized by $\Delta p_f = \rho g h_f + \rho g z = CQ^2 + \rho g z$ or $h_f = CQ^2 + z$. The additional term is

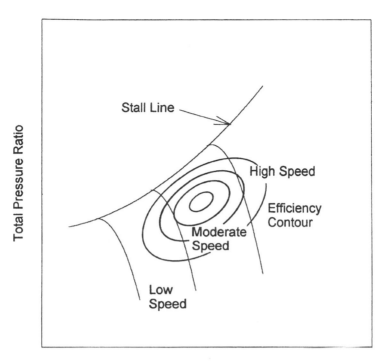

**Mass Flow Rate**

**FIGURE 1.10d**   Performance curve for a high-speed compressor.

independent of Q and results in a vertical offset of the resistance as shown in Figure 1.12.

The same comments apply to variation in resistance, as well as possible changes in the value of z, given changes in the geometry of the piping system. Recall that the hydrostatic terms are nearly always negligible in air/gas systems.

Now, one can illustrate the performance and testing concept. Consider a fan test carried out according to these procedures. The plenum chamber data gathered for a single performance point are:

$$p_a = 10 \text{ InWG}$$

$$p_b = 9.8 \text{ InWG}$$

$$p_c = 4.0 \text{ InWG}$$

$$P_m = 3448 \text{ watts (motor power measured)}$$

$$\eta_m = 0.83 \text{ (motor efficiency)}$$

$$L_w = 85 \text{ dB}$$

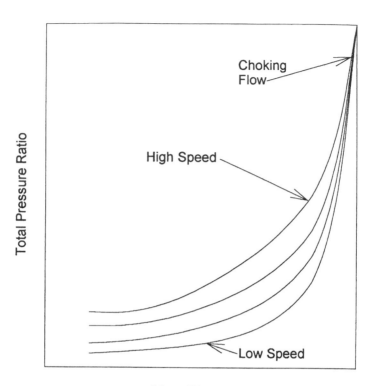

**FIGURE 1.10e**   Performance curve for a high-speed exhaust gas turbine.

Ambient air data are:

$$T_w = 60°F$$

$$T_d = 72°F$$

$$p_{amb} = 29.50 \text{ InHg}$$

Additional data for a fully compressible test must include the outlet total pressure and the total temperature at the flow meter inlet. In addition, it is known that the sum of the nozzle areas, $A_{tot}$, is based on five 6-inch diameter nozzles. The area is thus $5(\pi d_n^2/4)$. Thus, $A_{tot} = 5(\pi 0.5^2/4) \text{ ft}^2 = 0.1963 \text{ ft}^2$.

As a first step, one can calculate the ambient air density using the psychometric chart in Figure 1.9. An alternate would be to use the AMCA equations, based on pressures in InHg and temperatures in °F. $\rho$ is given by

$$\rho = 0.04119 \frac{\left(p_{amb} - 0.378 \, p_p\right)}{\left(T_d + 459.7\right)}$$

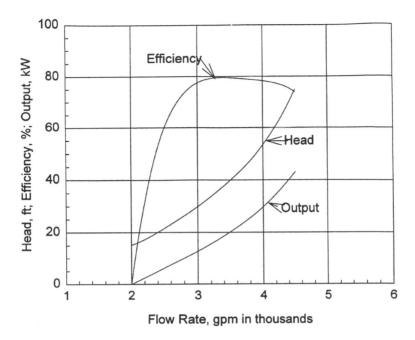

**FIGURE 1.10f**   Performance curve for a 1200 rpm hydraulic turbine.

with

$$P_p = P_e - P_{amb} \frac{(T_d - T_w)}{2700}$$

and

$$P_e = 2.96 \times 10^{-4} \, T_w^2 - 1.59 \times 10 - 2T_w + 0.41$$

The wet-bulb suppression is

$$\Delta T_{d-w} = T_d - T_w = (72 - 60)°F = 12°F$$

At the top of the psychometric chart, one enters with 12° suppression and drops vertically to an intersect with the downward sloping line for $T_d = 72°F$. Then move horizontally to the left to intersect the upward sloping line for $p_{amb} = 29.5$ InHg. Finally, drop down vertically to the abscissa and read the value for the weight density, $\rho g = 0.0731$, so that the mass density is

$$\rho = \frac{(\rho g)}{g} = \frac{0.0731}{32.17} = 0.00227 \ \text{slugs}/\text{ft}^3$$

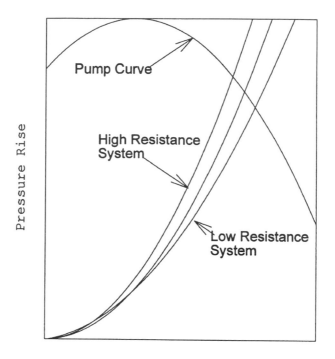

FIGURE 1.11   Pump performance with system curves of increasing resistance.

Once $\rho$ is obtained, one can look up the dynamic viscosity, $\mu$, and calculate $\nu$. At $T_d = 72°F$,

$$\mu = 3.68 \times 10^{-7} \text{ slugs/ft/s}$$

so

$$\nu = \frac{\mu}{\rho} = 1.621 \times 10^{-4} \text{ ft}^2/\text{s}$$

Next, one can calculate the nozzle pressure drop as

$$\Delta p_{b-c} = p_b - p_c = (9.8 - 4.0) \text{ InWG} = 5.8 \text{ InWG}$$

Convert this value back to basic units as

$$\Delta p_{b-c} = 5.8 \text{ InWG} \times \left( \frac{5.204 \left( \text{lbf/ft}^2 \right)}{\text{InWG}} \right) = 30.18 \text{ lbf/ft}^2$$

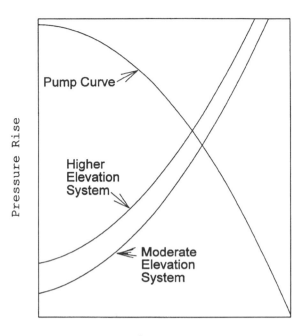

Flow Rate

**FIGURE 1.12**   The influence of elevation change on pump performance.

Now insert the data into the equation for nozzle velocity, noting that there is no Reynolds Number as yet to calculate $c_d$ (since velocity is unknown). That does not present a problem here because one can assume that $c_d \cong 1.0$ and make an iterative correction using an updated $V_n$. A first guess is then

$$V_n = c_d \left( \frac{2 \Delta p_{b-c}}{\rho} \right)^{1/2}$$

$$= 1.0 \times \left( 2 \times 30.18 \ \text{lbf/ft}^2 / 0.00227 \ \text{slugs/ft}^3 \right)^{1/2}$$

$$= 163.1 \ \text{ft/s}$$

Now, one can calculate a Reynolds Number by

$$Re_d = \frac{V_n d}{\nu} = \frac{(163.1 \times 0.5)}{\left( 1.621 \times 10^{-4} \right)}$$

so

$$Re_d = 5.09 \times 10^5$$

Use this value to estimate $c_d$ by

$$c_d = 0.9965 - 0.00653\left(\frac{10^6}{Re_d}\right)^{1/2}$$

so

$$c_d = 0.9965 - 0.00653\left(\frac{10^6}{Re_d}\right)^{1/2}$$

$$= 0.9965 - 0.00653\left(\frac{10^6}{5.09 \times 10^5}\right)^{1/2}$$

and

$$= 0.9874$$

Adjust the velocity for $c_d < 1.0$ by multiplying the first guess by 0.987 to get $V_n = 161.0$ ft/s. A new Reynolds Number is calculated as

$$Re_d = 0.987 \times 5.09 \times 10^5 = 5.02 \times 10^5$$

Then, a new $c_d$ value is calculated as 0.9873. No effective change is seen in $c_d$, so accept the value of $V_n = 161.0$ ft/s. The volume flow rate becomes

$$Q = V_n \times A_{tot}$$

$$= 161.0 \text{ ft/s} \times 0.1963 \text{ ft}^2$$

$$= 31.57 \text{ ft}^3/\text{s}$$

The traditional unit commonly used for Q is cubic feet per minute (ft³/min), or cfm. That is, Q = 1896.3 cfm.

The remaining piece of data is the wattage measured on the fan drive motor, which one can use as the fan shaft power by multiplying by the given motor efficiency. Fan shaft power is then

$$P_{sh} = \eta_m \times P_m \tag{1.24}$$

so

$$P_{sh} = 0.83 \times 3448 \text{ watts} = 2862 \text{ watts}$$

$$= 2110 \text{ lbf - ft/s} = 3.84 \text{ hp}$$

Finally, calculate the fan static efficiency using flow rate, static pressure rise, and fan shaft power. That is:

$$\eta_s = \frac{Q\Delta p_s}{P_{sh}} \tag{1.25}$$

$$= 31.57 \times \frac{52.0}{2110} = 0.778$$

Summarizing the performance, one obtains

$$Q \quad = \quad 31.57 \ \text{ft}^3/\text{s} = 1896 \ \text{cfm}$$

$$\Delta p_s \quad = \quad 52.0 \ \text{lbf}/\text{ft}^2 = 10.0 \ \text{InWG}$$

$$P_{sh} \quad = \quad 2110 \ \text{lbf-ft}/\text{s} = 3.84 \ \text{hp}$$

$$L_w \quad = \quad 85 \ \text{dB}$$

$$\eta_s \quad = \quad 0.778$$

These numbers make up a characteristic point on the performance curves of the fan. The plot of $\Delta p_s$ and $P_{sh}$ are traditionally used to form the characteristic curves, but one should include the efficiency and sound power level as useful information.

In order to calculate total efficiency (incompressibly) or to correct for compressibility in either the fan efficiency or a compressor efficiency, one must include the additional measurement of total pressure in the discharge jet and the total temperature there. The values are

$$\Delta p_T = 11.5 \ \text{InWG} \ \text{and} \ T_{02} = 76.6°\text{F}$$

The incompressible total would simply utilize the $\Delta p_T$ value to yield

$$\eta_T = \frac{\Delta p_T Q}{P_{sh}} = 0.895$$

In absolute, compressible units, $p_{02} = 2147.8 \ \text{lbf}/\text{ft}^2$, $p_{01} = 2088.0 \ \text{lbf}/\text{ft}^2$; and with density recalculated for the nozzle as $0.00231 \ \text{slugs}/\text{ft}^3$ yielding $Q = 31.3 \ \text{ft}^3/\text{s}$:

$$\eta_T = \frac{p_{01}Q\left(\dfrac{p_{02}}{p_{01}} - 1\right)}{P_{sh}} = 0.887$$

Using the compressor efficiency method with

$$p_{02} = 102.70 \text{ kPa}; \qquad p_{01} = 99.96 \text{ kPa}$$

$$T_{02} = 297.8\text{K}; \qquad T_{01} = 295.2\text{K}$$

gives

$$\eta_c = \frac{\left( \left( \dfrac{p_{02}}{p_{01}} \right)^{(\gamma-1)/\gamma} - 1 \right)}{\left( \dfrac{T_{02}}{T_{01}} - 1 \right)} \qquad (1.26)$$

$$= 0.909$$

These total property-based values are within a few percent for a fan whose pressure rise is only marginally compressible.

Another type of performance loop that is in common use for liquid pump testing is illustrated in Figure 1.13. Here, the test pump is connected in a piping network forming a closed loop with a pressure- and temperature-controlled reservoir. The primary performance information is based on pressure readings upstream and downstream of the pump, and another set of pressure readings taken upstream and downstream of a sharp-edged orifice plate, positioned downstream of the pump itself. In addition, power measuring instrumentation is required for the pump torque and speed or to monitor the wattage input to a calibrated drive motor. Usually, the reservoir temperature and pressure and the temperature and pressure at the suction eye of the pump will also be monitored. This data is useful in determining the cavitation characteristics of the pump (as will be discussed in Chapter 4).

An important part of the flow loop is the downstream valve used to control the system resistance imposed on the pump. This valve will usually be a "non-cavitating" type to avoid instabilities and noise in the pump flow (Karassik).

As shown in Figure 1.13, the pressures are read from taps as $p_a$, $p_b$, and $p_c$ on gages or mercury manometers and converted to pressure differentials in Pa units. For illustration, one can read the power input to the motor with a wattmeter, as $P_m$, if one also knows the motor efficiency, $\eta_m$. Across the pump, $(p_b - p_a)$ provides the differential pressure to calculate the head rise. Calculating the fluid liquid density from the temperature $T_{in}$, °C, one can calculate the head rise as

$$H = \frac{(p_b - p_a)}{\rho g} \qquad (1.27)$$

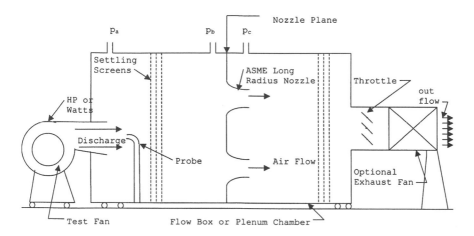

**FIGURE 1.13**   A flow test rating facility.

The pressure differential across the sharp-edged orifice is used to infer the volume flow rate of the pump in the same manner used to determine the air flow through the long radius ASME nozzles used in the fan performance test setup. The necessary characteristics for the orifice flow are the Reynolds Number and the value of $\beta$ for the plate. $\beta$ is defined as $\beta = d/D$, the ratio of pipe inner diameter to the open diameter of the orifice. The discharge coefficient for the orifice is then given as

$$c_d = f\left(Re_d, \beta\right) \tag{1.28}$$

where the Reynolds Number is defined for the throat of the orifice as

$$Re_d = \frac{V_t d}{\nu} \tag{1.29}$$

$V_t$ is the throat velocity, d is the throat diameter, and $\nu$ is the kinematic viscosity of the liquid. The flow is calculated as

$$Q = c_d A_t \left(\frac{2\left(p_c - p_b\right)}{\rho}\right)^{1/2} \tag{1.30}$$

$c_d$ is given (ISO-ASME, 1981) for standard taps as

$$c_d = f(\beta) + 91.71\beta^{2.5} \, Re_D^{-0.75} \tag{1.31}$$

$$f(\beta) = 0.5959 + 0.0312\ \beta^{2.1} - 0.184\ \beta^8 \qquad (1.32)$$

$$A_t = \left(\frac{\pi}{4}\right) d^2 \qquad (1.33)$$

An example will help to illustrate the procedures for data acquisition and analysis. Sample data required to define a single point on the characteristic performance curve for the pump are:

$$T_{inlet} = 25°C \qquad\qquad p_a = 2.5\ kPa \qquad\qquad P_m = 894\ watts$$

$$p_b = 155.0\ kPa \qquad p_c = 149.0\ kPa \qquad \eta_m = 0.890$$

$$D = 10\ cm \qquad\qquad d = 5\ cm$$

The inlet temperature yields a water density of 997 kg/m³ and a kinematic viscosity of $0.904 \times 10^{-6}$ m²/s. As usual, one cannot calculate an *a priori* Reynolds number to establish $c_d$, so that one must estimate an initial value of $c_d$ and iterate a final result. If one assumes that $Re_D$ is very large, then

$$c_d \cong 0.5959 + 0.0312\ (0.5)^{2.1} - 0.184\ (0.5)^8 = 0.6025$$

The first estimate for volume flow rate is given by

$$Q = c_d A_t \left(\frac{2(p_b - p_c)}{\rho}\right)^{1/2}$$

$$= (0.6025)(0.001963\ m^2)\left[\frac{(6000\ N/m^2)}{(997\ kg/m^3)}\right]^{1/2}$$

or

$$Q = 0.004103\ m^3/s$$

so that $V_t = 2.09$ m/s. As the first refinement, one calculates the Reynolds Number as

$$Re_D = \frac{Re_d}{\beta} = \frac{\left(\dfrac{V_t d}{\nu}\right)}{\beta} = \frac{\left(\dfrac{2.09 \times 0.05}{0.904 \times 10^{-6}}\right)}{0.5} = 2.304 \times 10^5$$

Recalculating $c_d$ yields:

$$c_d = 0.6025 + \frac{91.7\,(0.5)^{2.5}}{\left(2.304 \times 10^5\right)^{0.75}}$$

$$= 0.6025 + 0.00154 = 0.604$$

This 0.2% change is negligible, so accept the Q as

$$Q = 0.00410 \ \text{m}^3/\text{s} = 4.10 \ l/\text{s}$$

Volume flow rate in liters per second, while not in fundamental SI units, is common practice in metric usage.

The head rise for the pump can be taken directly from the pressure differential measured across the pumps so that

$$H = \frac{\left(p_b - p_a\right)}{\rho g} = \frac{(155,000 - 2,500)\left(\text{N}/\text{m}^2\right)}{\left(997 \times 9.81 \ \text{N}/\text{m}^3\right)} \tag{1.34}$$

$$H = 15.59 \ \text{m}$$

The fluid power, or the power input to the water, is

$$P_{fl} = \rho g Q H = \left(9.81 \ \text{m}/\text{s}^2\right)\left(997 \ \text{kg}/\text{m}^3\right)\left(0.004289 \ \text{m}^3/\text{s}\right)(15.59)$$

$$P_{fl} = 654.1 \ \text{watts} \tag{1.35}$$

Converting the input wattage to the motor to shaft power as

$$P_{sh} = P_m \eta_m = 796 \ \text{watts}$$

allows calculation of the pump efficiency according to

$$\eta = \frac{P_{fl}}{P_{sh}} = \frac{654.1 \ \text{watts}}{796 \ \text{watts}} = 0.822 \tag{1.36}$$

In summary, the test point for the pump is defined by

$$Q = 0.004289 \ \text{m}^3/\text{s}$$

$$H = 15.59 \ \text{m}$$

$$P_{sh} = 796 \ \text{watts}$$

$$\eta = 0.822$$

As in the fan test, the valve downstream of the pump can be opened or closed to decrease or increase the flow resistance and provide a large range and number of points to create the performance characteristic for the pump. Extensive reviews of pump test standards and procedures are available in the pump literature (e.g., Karassic et al., 1986, Chapter 13).

## 1.11  SUMMARY

This chapter introduces the fundamental concepts of Fluid Machinery. Following a review of basic fluid mechanics and thermodynamics, a discussion of the conservation equations for fluid flow is presented.

The definition of turbomachinery or rotodynamic machinery is given, and the distinctions drawn between turbomachines and positive displacement devices are delineated. Within the broad definition of turbomachines, the concepts of classification by work done on or by the fluid moving through the machine are used to differentiate the turbine family (work-extracting machines) from the pumping family (work-adding machines). Further subclassifications of the various kinds of fluid machinery are based on the path taken by the fluid through the interior of a machine. These are described in terms of axial flow, radial flow, and the in-between mixed flow paths in pumps and turbines.

Finally, the concepts of performance capability and rating of fluid machinery are introduced and developed. The techniques used to test the performance of turbomachinery are introduced to illustrate the performance variables. The complete dependence of the blower or pump on the characteristics of the system it operates in is illustrated and emphasized in this discussion as well.

We close this chapter with a group of figures which illustrate the kind of equipment the text deals with primarily. Figure 1.14 shows a packaged fan assembly with accessories. Figure 1.15 shows a typical compressor impeller, and Figure 1.16 shows several axial fan impellers. Figure 1.17 shows a large fan installation in a power plant, while Figures 1.18, 1.19, 1.20, and 1.21 show a variety of pump configurations, installations and arrangements.

## 1.12  PROBLEMS

1.1. If the pump in Figure P1.1 is operated at 1750 rpm and is connected to a 2-inch i.d. galvanized iron pipe 175 ft long, what will be the flow rate and bhp?

1.2. The pump of Figure P1.1 is installed in the system of Figure P1.2 as sketched. What flow rate, power, and efficiency result? All pipe is 2-inch i.d. commercial steel.

1.3. An Allis-Chalmers pump has the performance data given by Figure P1.1. At the pump's best efficiency point, what values of Q, H, and bhp occur? Estimate maximum or free-delivery flow rate.

**FIGURE 1.14** An SWSI centrifugal showing drive-motor arrangement at left, volute and discharge, and the inlet box at the right end.

**FIGURE 1.15** Centrifugal compressor impeller showing blade layout details. (Barron Industries, Birmingham, AL.)

**FIGURE 1.16** Axial fans showing impellers of different types. (New York Blower Company, Chicago, IL.)

1.4. The pump of Figure P1.1 is used to pump water 800 ft uphill through 1000 ft of 2-inch i.d. steel pipe. You can do this if you connect a number of these pumps in series. How many pumps will it take? (HINT: fix the pump flow at BEP.)

1.5. Two pumps of the type of Figure P1.1 can be used to pump flows with a vertical head rise of 130 ft.
   (a) Should they be in series or parallel?
   (b) What size steel pipe would you use to keep both pumps operating at BEP?

1.6. A fuel transfer pump delivers 400 gpm of 20°C gasoline. The pump has an efficiency of 80% and is driven by a 20-bhp motor. Calculate the pressure rise and head rise.

**FIGURE 1.17**   Fan installation in a power plant. (Barron Industries) (Note the dual-inlet ducts to supply flow to both sides of the DWDI Fan.)

1.7. If the pump and pipe arrangement of Problem 1.1 includes an elevation increase of 30 ft, what flow will result?

1.8. A small centrifugal fan is connected to a system of ductwork. The fan characteristic curve can be approximated by the equation $\Delta p_s = a + bQ$, where $a = 30$ lbf/ft$^2$ and $b = -0.40$ (lbf/ft$^2$)/(ft$^3$/s). Duct resistance is given by $h_f = (fL/d + \Sigma Km)(V2/2g)$. The resistance factor is $(fL/d + \Sigma K_m) = 10.0$, and the duct cross-sectional area is 1 ft$^2$.

(a) Calculate the volume flow rate of the fan-duct system.

(b) What is the static pressure of the fan?

(c) Estimate the horsepower required to run the fan (assume the static efficiency of the fan is $\eta_s = 0.725$).

1.9. A slurry pump operates with a mixture of sand and water, which has a density of 1200 kg/m$^3$ and an equivalent kinematic viscosity of $5 \times 10^{-5}$ m$^2$/s. The pump, when tested in pure water, generated a head rise of 15 meters at a flow rate of 3 m$^3$/s. Estimate the head rise in the sand slurry with a flow rate of 3 m$^3$/s.

**FIGURE 1.18**   Centrifugal pumps showing casing layouts and flow paths for single-suction and double-suction. (ITT Industrial Pump Group, Cincinnati, OH.)

**FIGURE 1.19**   An end-suction pump with a cutaway view of the flow path, impeller, and mechanical drive components. (ITT Industrial Pump Group, Cincinnati, OH.)

**FIGURE 1.20**   View of a pump installation with two pumps, showing the drive motor arrangemants. (ITT Industrial Pump Group, Cincinnati, OH.)

**FIGURE 1.21**  An installation with six double-suction pumps operating in parallel. (ITT Industrial Pump Group, Cincinnati, OH.)

1.10.  A three-stage axial fan has a diameter of 0.5 m and runs at 1485 rpm. When operating in a system consisting of a 300-m long duct with a 0.5-m diameter (f = 0.040), the fan generates a flow rate of 5.75 m³/s. Estimate the static pressure rise required for each of the three stages.

1.11.  In order to supply a very high rate of flow, 375 gpm, a group of pumps identical to the pump of Figure P1.1 are operated in parallel. How many pumps are needed to operate the group at Q = 375 gpm at the BEP point of operation? Develop a graph for the combined performance of this group of pumps.

1.12.  A double-width centrifugal blower supplies flow to a system whose resistance is dominated by the pressure drop through a heat exchanger. The heat exchanger has a K-factor of 20 and a cross-sectional area of 1 m². If the blower is generating a pressure rise of 1 kPa, what is the volume flow rate (m³/s) for each side of the double-sided impeller?

1.13.  A medium-sized pump delivers 1.50 m³/s of SAE 50 engine oil (SG = 0.9351). The pump has a total efficiency of 0.679 and is direct-connected to a motor that can supply a shaft power of 25 kW. What head rise can the pump generate at this flow and power?

1.14.  The centrifugal fan described in Problem 1.8 must be used to meet a flow-pressure rise specification in metric units. Convert the equation for its pumping performance into SI units and
   (a)  Estimate the flow rate achievable through a 15-cm diameter duct for which the resistance is characterized by $K_{res} = (fL/D + \Sigma K_m) = 1.0$.
   (b)  Develop a curve of Q vs. diameter for 15 cm < D < 45 cm with $K_{res} = 1.0$.
   (c)  Develop a curve of Q vs. $K_{res}$ with a diameter of 10 cm and $1 < K_{res} < 5$.

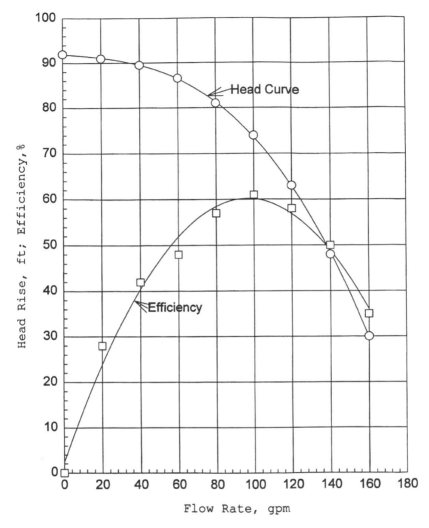

**FIGURE P1.1** Pump performance with a 9-inch impeller running at 1750 rpm.

1.15. Develop a set of performance curves in SI units for the pump data given in
    Figure P1.1 (Q in m³/hr, H in m, and P in kW). Use these curves to predict
    flow rate, head rise, and required power if two of these pumps are connected
    in series to a filter system with a resistance coefficient of $K_m = 50.0$. Assume
    the diameter of the flow system is 5 cm and neglect other resistances.

1.16. Estimate the pressure rise of the slurry pump of Problem 1.9 at the stated
    conditions. Compare this result to the pressure rise with water as the
    working fluid.

1.17. Use the pump performance curves developed in Problem 1.15 to estimate
    the BEP power and efficiency for the SI pump. Estimate the shut-off and
    free-delivery performance.

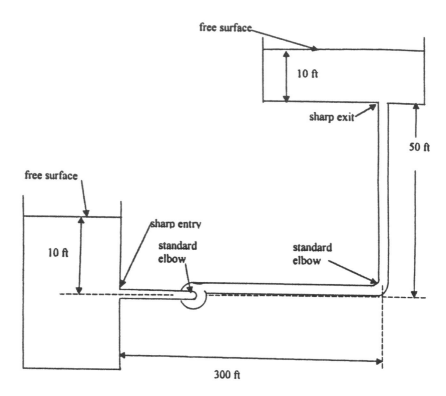

**FIGURE P1.2**   Pump and system arrangement.

1.18. Estimate the flow delivery of the fan of Problem 1.12 if the fan is a single-width version of the given fan.

1.19. What flow rate will the pump of Problem 1.15 generate through 100 m of a pipe that is 5 cm in diameter with a friction factor of f = 0.030?

1.20. One can arrange the three stages of the axial fan of Problem 1.10 to operate in a parallel, side-by-side arrangement, delivering their individual flow rates to a plenum that is connected to a duct. If this duct is to be 110 m long (f = 0.04), what size must the duct be to ensure a total flow delivery of 17.25 m³/s? (Neglect any plenum losses.)

# 2 Dimensional Analysis In Turbomachinery

## 2.1 DIMENSIONALITY

Recall the study of dimensional analysis in fluid mechanics [see any good under-graduate text (White, 1994; Fox and McDonald, 1992) and review as necessary]. The concept is linked to the principle of dimensional homogeneity and is absolutely central to the ability to correctly design experiments and to extract the maximum possible information from a mass of test data. When the concept is applied to turbomachines, it will provide a greater depth of understanding of performance, classification of machines, and the fundamental reasons behind the wide configurational differences in machines. The principle is paraphrased here as:

If an equation correctly represents a physical relationship between variables in a physical process, it will be homogeneous in dimensions. That is, each term will add to the other terms only if they are all of like dimensions.

This is a basic "Truth Test" or screen to determine the fundamental validity of any expression. For example, consider the familiar Bernoulli equation for steady, incompressible, irrotational (or frictionless) flow:

$$p_0 = p + \frac{1}{2}\rho V^2 + \rho g z \qquad (2.1)$$

In SI units, each term must be in N/m² or Pa (lbf/ft² in BG units). For review, the relation states:

$$\begin{aligned}
\left(\text{Total or Stagnation pressure}\right) &= \left(\text{Static or Thermodynamic pressure}\right) \\
&+ \left(\text{Velocity or Dynamic pressure}\right) \qquad (2.2) \\
&+ \left(\text{Hydrostatic pressure}\right)
\end{aligned}$$

where the Total Pressure is the invariant sum of the three terms on the right-hand side. In a typical flow situation, velocity—and hence velocity pressure—may increase at the expense of a decrease in static or hydrostatic pressure. Typically, in gas flows (such as the fan test), hydrostatic changes are negligible in comparison to the velocity pressure and static pressure variations. That is, one can approximate (or neglect) the $\rho g z$ term, so

$$p_o = p_{\text{Total}} = p_{\text{Static}} + \frac{1}{2}\rho V^2 \qquad (2.3)$$

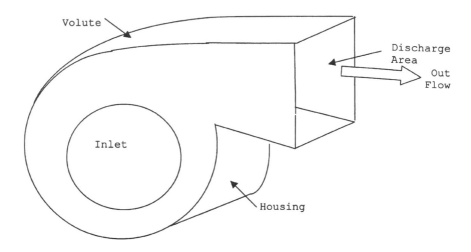

**FIGURE 2.1**   Discharge velocity averaged on the outlet area.

In the process of adding or extracting fluid energy, a turbomachine changes the total pressure of the fluid stream as the fluid passes through the machine. This process is the mechanism whereby work is done either on or by the fluid. This change is designated as $\Delta p$, either $\Delta p_T$ or $\Delta p_S$ (for Total or Static). Thus, changes in p can be interpreted as changes in or through a turbomachine:

$$\Delta p_S = \Delta p_T - \frac{1}{2}\rho V^2 \tag{2.4}$$

or

$$\Delta p_T = \Delta p_S + \frac{1}{2}\rho V^2 \tag{2.5}$$

What velocity is used? Traditionally, it is the value given by $V_d = Q/A_{discharge}$. This assumes a uniform exit velocity profile, as shown in Figure 2.1.

## 2.2   SIMILITUDE

The accepted definitions of similarity are defined in terms of geometric, kinematic, and dynamic properties. Specifically,

> A model and prototype are geometrically similar if all physical or body dimensions in all three axes (Cartesian) have the same linear ratio.

For our purposes, this means that all angles, flow directions, orientations, and even surface roughness properties must "scale."

The motions of two systems are kinematically similar if similar elements lie at similar locations at similar times.

By including time similarity, one introduces such elements as velocity and acceleration—as in a fluid flow problem. Note that for similitude in a kinematic sense, geometric similarity is presupposed. In a fluid mechanics sense, kinematic similarity leads to requirements in Reynolds Number, Mach Number, and perhaps Froude Number.

Dynamic similarity requires that the additional effects of force-scale or mass-scale be maintained between a model and a prototype, presupposing length-scale and time-scale.

For most purposes, the maintenance of Reynolds, Mach, and Froude numbers, and perhaps Weber and Cavitation numbers, will be required. Much is said of the requirement for maintaining precise similarity when comparing the behavior of bodies or machines of different "scale" (size, speed, fluid properties). The real goal in comparing is to be able to accurately relate or predict the performance or behavior of one body (or machine, or test) from another set of conditions or from another situation. To do this, strictly speaking, similitude must be precisely maintained. In practice, it is frequently impossible or impractical to achieve these ideal conditions. In such cases, it may be necessary to extrapolate or to attempt to achieve or maintain the same "regime." Maintenance of the "regime" might mean "nearly" the same Reynolds number or roughness is "acceptably" small (perhaps hydraulically smooth). Figure 2.2 (from Emery et al., 1958) illustrates compressor blade lift or fluid turning angle, $\theta_{fl}$, versus Reynolds Number with "limited" data. Although this data is well-behaved, clearly judgment, experience, and risk are key issues in trying to push such information beyond the range of the original data.

## 2.3 Π-PRODUCTS

Dimensional analysis is usually performed using either "power products" or the Π-product method. We will concentrate on Π-products. Recall the basic rules. If one has a set of performance variables based on a number of fundamental dimensions, the object is to group the variables into products that are dimensionless, thus forming new variables or coefficients which will be fewer in number than the original variable list. The steps in the procedure are as follows:

1. List and count the variables ($N_v$).
2. List and count the fundamental units ($N_u$).
3. Select a number of variables as "primary" (usually, this group consists of $N_u$ of the problem variables).
4. Form Π-products from the primaries plus a remaining variable. Once one obtains ($N_v - N_u$) new dimensionless variables, the variable list is reduced by $N_u$.

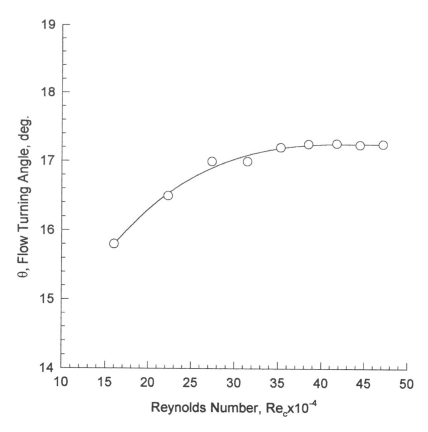

**FIGURE 2.2**   Variation of blade fluid turning angle with Reynolds Number.

As an example, consider the lift force on an airfoil. The variables are

$F_L$—lift force (N)
$\rho$—air density (kg/m³)
c, b—chord and span of the airfoil (m) *or*
A—airfoil area: A = c·b (m²)
$\mu$—viscosity of the air (kg/m·s)
V—air velocity (m/s)

Following the process outlined above, the steps are

Step 1.   Variables: $F_L$, $\rho$, A, $\mu$, V ($N_v$ = 5)
Step 2.   Units: kg·m/s², kg/m³, m², kg/m s, m/s M,L,T ($N_u$ = 3)
Step 3.   Choose Primaries (3) NOTE: These three must not be able to form a dimensionless product by themselves. Choose V and A (since V contains time and A does not) and choose $\rho$ to go with them because it contains mass while V and A do not.
Step 4.   Form Π-products: $N_v - N_u$ = 2; should develop 2 products.

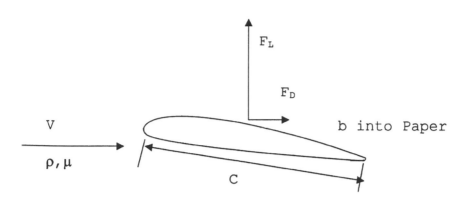

**FIGURE 2.3** Lift on an airfoil.

$$\Pi_1 = \rho^a V^b A^c F_L; \quad \Pi_2 = \rho^a V^b A^c \mu$$

Solve for the exponents according to the basic rule that $\Pi$ is dimensionless; therefore, the right-hand side must also be dimensionless, e.g., $\Pi_1 = \rho^a V^b A^c F_L$ yields the unit equation:

$$M^0 L^0 T^0 = \left(\frac{M}{L^3}\right)^a \left(\frac{L}{T}\right)^b (L^2)^c \left(\frac{ML}{T^2}\right)$$

Time: $T^0 = T^{-b} \times T^{-2}; \quad 0 = -b - 2 \quad \text{or} \quad b = -2$

Mass: $M^0 = M^a \times M^1; \quad 0 = 1 + a \quad \text{or} \quad a = -1$

Length: $L^0 = L^{-3a} \times L^b \times L^{2c} \times L^1; \quad 0 = -3a + b + 2c + 1 \quad \text{or} \quad c = -1$

Therefore,

$$\Pi_1 = \rho^a V^b A^c F_L = \rho^{-1} V^{-2} A^{-1} F_L$$

This is usually written as the lift coefficient $C_L$ as

$$C_L = \frac{F_L}{\left(\frac{1}{2}\rho V^2 A\right)} \tag{2.6}$$

where the 1/2 is "conventional" in terms of $1/2\rho V^2$ as the velocity pressure in Bernoulli's equation.

The second $\Pi$-product is

$$\Pi_2 = \rho^a V^b A^c \mu$$

The unit equation becomes

$$M^0 L^0 T^0 = M^a L^{-3a} L^b T^{-b} L^{2c} M^1 L^{-1} T^{-1}$$

$$\text{Mass:} \ \ 0 = a + 1; \quad a = -1$$

$$\text{Time:} \ \ 0 = -b - 1; \quad b = -1$$

$$\text{Length:} \ \ 0 = -3a + b + 2c - 1; \quad c = -\frac{1}{2}$$

Therefore,

$$\Pi_2 = \rho^{-1} V^{-1} A^{-1/2} \mu$$

This is usually written as a Reynolds Number, where

$$Re = \frac{1}{\Pi_2} = \frac{\rho V A^{1/2}}{\mu} = \frac{V A^{1/2}}{\nu} \tag{2.7}$$

More conventionally, $A^{1/2}$ is replaced by C and $\nu = \mu/\rho$ is the kinematic viscosity.

## 2.4 DIMENSIONLESS PERFORMANCE

One can now move directly to performance of turbomachines. There are many ways to choose the variables (e.g., the head H, pressure rise $\Delta p_T$, or frequently gH to replace H) so that going through the analysis will not result in a unique set of dimensionless variables.

To start, one can use

gH—head variable, $m^2/s^2$
Q—volume flow rate, $m^3/s$
P—power, Watts $W = kg \cdot m^2/s^3$
N—speed, $s^{-1}$ (radians/s)
D—diameter, m
ρ—fluid density, $kg/m^3$
μ—fluid viscosity, $kg/m \cdot s$
d/D—dimensionless diameter ratio

d/D is included as a reminder that geometric similitude is imposed to eliminate considerations of shape or proportion.

There are three basic units involved: mass, length, and time (M, L, T as before), so one can expect to reduce the seven variables shown above to four. Choosing ρ, N, and D as primaries, the following Π-products are generated:

$$\Pi_1 = \rho^a N^b D^c Q$$

$$\Pi_2 = \rho^a N^b D^c \mu$$

$$\Pi_3 = \rho^a N^b D^c gH$$

$$\Pi_4 = \rho^a N^b D^c P$$

Look at $\Pi_1 = \rho^a N^b D^c Q$

The unit equation is $M^o L^o T^o = M^a L^{-3a} T^{-b} L^c L^3 T^{-1}$

Mass : $0 = a$

Time : $0 = -b - 1; \quad b = -1$

Length : $0 = -3a + c + 3; \quad c = -3$

$$\Pi_1 = N^{-1} D^{-3} Q = \frac{Q}{ND^3}$$

This is usually written as a "flow coefficient" $\phi$:

$$\phi = \frac{Q}{ND^3} \tag{2.8}$$

Now, look at the second $\Pi$-product.

$$\Pi_2 = \rho^a N^b D^c \mu$$

The unit equation is

$$M^o L^o T^o = M^a L^{-3a} T^{-b} L^c M^1 L^{-1} T^{-1}$$

Mass : $0 = a + 1; \quad a = -1$

Time : $0 = -b - 1; \quad b = -1$

Length : $0 = -3a + c - 1; \quad c = -2$

and one can write

$$\Pi_2 = \rho^{-1} N^{-1} D^{-2} \mu = \frac{\mu}{ND^2 \rho}$$

This is usually written as

$$Re = \frac{ND^2\rho}{\mu} = \frac{ND^2}{\nu} \tag{2.9}$$

$\Pi_3$ is analyzed in the same manner and is found to be

$$\Pi_3 = \frac{gH}{N^2D^2}$$

usually written as a "head coefficient" $\psi$:

$$\psi = \frac{gH}{N^2D^2} = \frac{\Delta p_T}{\rho N^2D^2} \tag{2.10}$$

$\Pi_4$ becomes

$$\Pi_4 = \frac{P}{\rho N^3D^5}$$

usually written as a "power coefficient" $\xi$:

$$\xi = \frac{P}{\rho N^3D^5} \tag{2.11}$$

This gives four new variables:

$$\xi = \frac{P}{\rho N^3D^5}$$

$$Re = \frac{ND^2\rho}{\mu}$$

$$\psi = \frac{gH}{N^2D^2} \tag{2.12}$$

$$\phi = \frac{Q}{ND^3}$$

These can also be written in function groups as

$$\xi = f_1\left(\phi,\ \text{Re},\ \frac{L}{D}\right)$$

$$\psi = f_2\left(\phi,\ \text{Re},\ \frac{L}{D}\right) \tag{2.13}$$

$$\phi = f_3\left(\phi,\ \text{Re},\ \frac{L}{D}\right)$$

where L/D is included as a reminder of geometric similitude. If one hypothesizes a "weak" dependence on Re (e.g., look at f vs. Re, for large Re with moderate relative roughness, on a Moody diagram), one can write, approximately:

$$\psi = g_1(\phi)$$

$$\xi = g_2(\phi) \tag{2.14}$$

as a description of turbomachine performance. This is really compact. For convenience, recall the concept of the efficiency defined as:

$$\eta_T = \left(\frac{\text{Net power input to the fluid}}{\text{Power input to the shaft}}\right)$$

$$= \frac{\rho Q g H}{P} = \frac{Q \Delta p_T}{P} \tag{2.15}$$

Then,

$$\eta_T = \frac{\rho Q g H}{P} = \frac{\left[\left(\dfrac{Q}{ND^3}\right)\left(\dfrac{gH}{N^2D^2}\right)\right]}{\left(\dfrac{P}{(\rho N^3 D^5)}\right)} \tag{2.16}$$

or

$$\eta_T = \frac{\phi \psi}{\xi} \tag{2.17}$$

Thus, a more complete description is

$$\psi = g_1(\phi)$$

$$\xi = g_2(\phi) \tag{2.18}$$

$$\eta_T = g_3(\phi)$$

Having developed these relationships, what can one do with them? Go back to the characteristic curves generated in the discussion of testing in Chapter 1. P and $\Delta p_T$ were plotted as functions of Q. Take the known values of $\rho$, N, and D and convert the data tables of P, $\Delta p_T$, and Q to $\phi$, $\psi$, $\xi$, and $\eta_T$. If one plots these, the dimensionless representation of the fan's performance is obtained. The advantage is that the $\psi$–$\phi$, $\xi$–$\phi$, and $\eta_T$–$\phi$ curves can be used to generate an infinite number of gH–Q, P–Q, and $\eta_T$–Q curves by choosing any values of $\rho$, N, and D we want as long as we maintain geometric similitude. The major restriction lies in the assumption of weak Reynolds Number dependence (and Mach Number dependence, which will be discussed later).

Consider an example. We test a fan such that

$$0 < Q < 100 \ \text{m}^3/\text{s}$$

$$0 < \Delta p_T < 1500 \ \text{kPa}$$

$$0 < P < 100 \ \text{kW}$$

with design point performance of Q = 80 m³/s; $\Delta p_T$ = 1000 Pa; P = 90 kW, where

$$D = 1.2 \ \text{m}; \quad N = 103 \ \text{s}^{-1}; \quad \rho = 1.2 \ \text{kg/m}^3$$

At the design point (or system intersect/operating point if correctly matched),

$$Q = 80 \ \text{m}^3/\text{s}, \quad D = 1.2 \ \text{m}$$

$$\Delta p_T = 1000 \ \text{Pa}, \quad N = 103 \ \text{s}^{-1}$$

$$P = 90 \ \text{kW}, \quad \rho = 1.2 \ \text{kg/m}^3$$

The normalizing factors to be used with the dimensionless variables are formed as

$$ND^3 = 177$$

$$\rho N^2 D^2 = 18,181$$

$$\rho N^3 D^5 = 3.270 \times 10^6$$

and $\phi$, $\psi$, $\xi$, and $\eta_T$ are calculated as

$$\phi = \frac{80}{177} = 0.452$$

$$\psi = 0.055$$

$$\xi = 0.0279$$

$$\eta_T = \frac{\phi\psi}{\xi} = 0.890$$

Now, what would this fan do at a different size and speed? For example:

$$D = 30 \text{ inches} = 2.5 \text{ ft}$$

$$N = 1800 \text{ rpm} = 188.5 \text{ s}^{-1}$$

$$\rho g = 0.074 \text{ lbf/ft}^3$$

$$\rho = 0.0023 \text{ slugs/ft}^3$$

Calculate

$$ND^3 = 2945 \text{ ft}^3/\text{s}$$

$$\rho N^2 D^2 = 510.8 \text{ lbf/ft}^3$$

$$\rho N^3 D^5 = 1.504 \times 10^6 \text{ lbf - ft/s}$$

Now use $\phi$, $\psi$, and $\xi$ to calculate $\Delta p_T$, Q, P.

$$Q = \phi(ND^3) = 1325 \text{ ft}^3/\text{s} = 79,252 \text{ cfm}$$

$$\Delta p_T = \psi(\rho N^2 D^2) = 28.1 \text{ lbf/ft}^2 = 0.20 \text{ psi} = 5.4 \text{ InWG}$$

$$P = \xi(\rho N^3 D^5) = 41,950 \text{ lbf - ft/s} = 76.3 \text{ HP} = 56 \text{ kW}$$

$$\eta_T = \frac{Q\Delta p_T}{P} = \frac{(1325 \text{ ft}^3/\text{s})(28.1 \text{ lbf/ft}^2)}{41950 \text{ lbf - ft/s}} = 0.890$$

According to the scaling rules, $\eta_T$ should be conserved in the resizing process, so this last calculation serves as a check on the other results.

As an additional check, one should examine the relation between the model and prototype Reynolds numbers. Using a standard value for $v$ ($v = 0.000161$ ft²/s), one

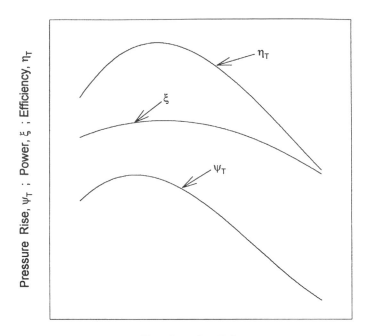

**FIGURE 2.4**   Dimensionless performance curves.

obtains $Re_D = 0.73 \times 10^7$. The value for the original fan was $1 \times 10^7$. Are these sufficiently large or sufficiently close together to justify neglecting differences in viscous effects? One can use the Darcy friction factor to estimate this effect. Using (Haaland, 1983)

$$\frac{1}{f^{1/2}} \cong -1.8 \ \log_{10}\left[\left(\frac{6.9}{Re_D}\right)+\left(\left(\frac{\varepsilon}{D}\right)\frac{}{3.7}\right)^{1.11}\right] \qquad (2.19)$$

and assuming a very small roughness as a worst case, one obtains $f \cong 0.0085$ for the prototype and $f \cong 0.008\,13$ for the model. For this hydraulically smooth case, the frictional losses should be slightly less for the prototype (4%) and might increase the efficiency by perhaps 1/2 of 1%, if all of the inefficiency were attributable to friction ($0.04 \times (1 - \eta) = 0.0044$). A later chapter will deal more rigorously with this Reynolds Number influence, but as a general rule, changes in values of $Re_D$ above $1 \times 10^7$ have little influence on efficiency.

## 2.5   COMPRESSIBLE SIMILARITY

When significant changes in density are likely to occur through the turbomachine, the use of volume flow rate, Q, and the head rise, H, become inappropriate. Work instead with the mass flow rate, m', in kg/s. The head rise must be replaced by the change in stagnation enthalpy $\Delta h_{0s}$ in J/kg or a related variable, as discussed in Chapter 1. If attention is restricted to adiabatic flow in a perfect gas, one can relate $\Delta h_{0s}$ to the change in stagnation temperature through the machine as energy is added to the flow; that is,

$$\Delta h_{0s} = c_p \Delta T_0 \tag{2.20}$$

In addition, one can consider the total energy of the fluid in relation to other variables, for example, the total pressure, to include other thermodynamic variables, depending on the convention for a particular kind of equipment.

A modified variable list may now include the gas constant and specific heats: R, $c_p$, $c_v$, or $\gamma = c_p/c_v$. They can be written in the function form as

$$\Delta h_{0s} = f_1\left(N, D, m', T_0, \rho_0, \mu, \gamma, \frac{d}{D}\right)$$

$$P = f_2\left(N, D, m', T_0, \rho_0, \mu, \gamma, \frac{d}{D}\right) \tag{2.21}$$

$$\eta = f_3\left(N, D, m', T_0, \rho_0, \mu, \gamma, \frac{d}{D}\right)$$

As before, d/D and $\gamma$ need not be included in a dimensional analysis because they are already non-dimensional parameters of the problem. They can of course be considered in the underlying constraints on maintaining similarity and can be used in combination with any dimensionless parameters resulting from a dimensional analysis, perhaps to form new parameters.

The new list is

R,$c_p$: with dimensions (m²/s²K)
$\Delta h_{0s} = c_p T_0$: with dimensions (m²/s²)
m': with dimensions (kg/s)
$\rho_{01} = p_{01}/RT_{01}$: with dimensions (kg/m³)
$a_{01} = (\gamma RT_{01})^{1/2}$: with dimensions (m/s)

along with part of the previous set

$\mu$: with dimensions (kg/m·s)
N: with dimensions (radians/s or s⁻¹)
D: with dimensions (m)

Note that $T_{01}$ could have been used directly in the analysis rather than the derived variable $a_{01}$. Thus, there are eight variables requiring the three dimensions length (m or L), time (s or T), and mass (kg or M). One must choose three primary variables to form five $\Pi$-products. The same three variables used before are appropriate: $\rho_{01}$, N, and D.

The only change is to stipulate the total or stagnation inlet density because of the variability of density expected through the machine.

Form the $\Pi$s as

$$\Pi_1 = \rho^a\ N^b\ D^c\ Q$$
$$\Pi_2 = \rho^a\ N^b\ D^c\ P$$
$$\Pi_3 = \rho^a\ N^b\ D^c\ \Delta h_{0s}$$
$$\Pi_4 = \rho^a\ N^b\ D^c\ a_{01}$$
$$\Pi_5 = \rho^a\ N^b\ D^c\ \mu$$

The routine examination of the non-dimensionality of each $\Pi$ product gives the results as

$$\Pi_1 = \frac{m'}{\left(\rho_{01}ND^3\right)} \qquad \text{mass flow coefficient}$$

$$\Pi_2 = \frac{P}{\left(\rho_{01}N^3D^5\right)} \qquad \text{power coefficient}$$

$$\Pi_3 = \frac{\Delta h_{0s}}{\left(N^2D^2\right)} \qquad \text{work or head coefficient}$$

$$\Pi_4 = \frac{ND}{a_{01}} \qquad \text{Mach Number}$$

$$\Pi_5 = \frac{\rho_{01}ND^2}{\mu} \qquad \text{Reynolds Number}$$

These can be used as they stand or they can be rearranged using isentropic flow and perfect gas relations to form the more conventional variables used for high-speed compressors and turbines. One can use $p/\rho^\gamma = $ constant for isentropic flow for a perfect gas with $p = \rho RT$. Note that $a_0^2 = \gamma RT_0$, so that

$$\Delta h_{0s} = c_p\Delta T_0 = c_p\left(T_{02} - T_{01}\right)$$

One can combine $\Pi_3$ and $\Pi_4$ with this result to form

$$\psi = \Pi_3 \times \Pi_4 = \frac{c_p\left(T_{02} - T_{01}\right)}{\left(\gamma RT_{01}\right)}$$

(2.22)

$$= \left(\frac{c_p}{\gamma R}\right)\left(\frac{T_{02}}{T_{01} - 1}\right)$$

Note that $c_p/\gamma R = \gamma/(\gamma - 1)$, which is a non-dimensional parameter already. Dropping the $\gamma$ expression and the '1' from the expression yields

$$\psi = \frac{T_{02}}{T_{01}} \qquad (2.23)$$

Further, since $(T_{02}/T_{01}) = (p_{02}/p_{01})^{(\gamma-1)/\gamma}$ for isentropic flow, one can write $\psi$ as

$$\psi = \frac{p_{02}}{p_{01}} \qquad (2.24)$$

This is the conventional "head" variable for high-speed compressors.

For the flow rate variable, again combine parameters so that

$$\phi = \frac{m'}{\left(\rho_{01} ND(D^2)\right)\left(\dfrac{ND}{a_{01}}\right)} = \frac{m'}{\left(\rho_{01} a_{01} D^2\right)} \qquad (2.25)$$

to form a new parameter. Introducing $a_{01} = (\gamma RT_{01})^{1/2}$, $\rho_{01} = p_{01}/RT_{01}$, one can write

$$\phi = \frac{m'}{\left(\rho_{10}(\gamma RT_{01})^{1/2} D^2\right)} = \frac{\left(m'RT_{01}\right)}{\left(p_{01}D^2(\gamma RT_{01})^{1/2}\right)} \qquad (2.26)$$

or in final form, having dropped the $\gamma$ out of the parameter

$$\phi = \frac{m'(RT_{01})^{1/2}}{p_{01}D^2} \qquad (2.27)$$

This process yields the traditional forms of the dimensionless performance variables for high speed gas compressors and turbines as

$$\frac{p_{02}}{p_{01}} = f_1\left[\left(\frac{m'(RT_{01})^{1/2}}{p_{01}D^2}\right), \left(\frac{ND}{(RT_{01})^{1/2}}\right), \mathrm{Re}, \gamma\right] \qquad (2.28)$$

$$\eta = f_2\left[\left(\frac{m'(RT_{01})^{1/2}}{p_{01}D^2}\right), \left(\frac{ND}{(RT_{01})^{1/2}}\right), \mathrm{Re}, \gamma\right] \qquad (2.29)$$

Traditional use of these compressible parameters frequently involves a simplified form that rests on the assumptions that (1) there is negligible influence of Reynolds Number, (2) the impeller diameter is held constant, and (3) there is no change in the gas being used (e.g., air as a perfect gas). Under these assumptions, R, $\gamma$, Re, and D are dropped from the expressions, leaving a set of dimensional scaling parameters

$$\frac{P_{02}}{P_{01}} = f_1 \left[ \left( \frac{m'(T_{01})^{1/2}}{P_{01}} \right), \left( \frac{N}{(T_{01})^{1/2}} \right) \right] \tag{2.30}$$

$$\eta = f_2 \left[ \left( \frac{m'(T_{01})^{1/2}}{P_{01}} \right), \left( \frac{N}{(T_{01})^{1/2}} \right) \right] \tag{2.31}$$

The form for the head-flow characteristic curves for these compressible flows is very similar to that introduced for the low-speed, low-pressure machines. Figures 2.5 and 2.6 show the $\Psi$ vs. $\phi$ curves of a compressor and a high-speed turbine, using the Mach Number to differentiate the curves relating to different values of fixed speed. Qualitatively, the most notable difference is seen in the very sharp steepening of the characteristics at the highest flow rates. Here, the compressible flow has reached a local sonic velocity in the machine so that the mass flow becomes choked at Ma = 1.0 (Anderson, 1984).

## 2.6   SPECIFIC SPEED AND DIAMETER: CORDIER ANALYSIS

The scaling rules developed earlier, whether the conventional low-speed rules or the compressible similarity laws, can be used rather formally to examine the influence of size change and speed change on machine performance. Several examples were produced so that these changes could be examined in a somewhat isolated fashion. What seemed to be missing from these studies was a systematic way to vary more than one parameter at a time in order to change both head and flow performance a specified amount. Consider a constrained change in "size" such that, using a scaled up version of a fan, one can generate specified values for pressure rise and volume flow rate. As in a previous example, where $\phi = 0.453$ and $\psi = 0.055$, one can require 20,000 cfm at 1 InWG pressure rise in a 30-inch duct with the fan sized to fit the duct, e.g., D = 30 inches. Then,

$$Q = ND^3\phi = N(2.5 \text{ ft})^3 (0.452) = 333.3 \text{ ft}^3/\text{s}$$

Solving for N, one obtains

$$N = 47.2 \text{ s}^{-1} = 451 \text{ rpm}$$

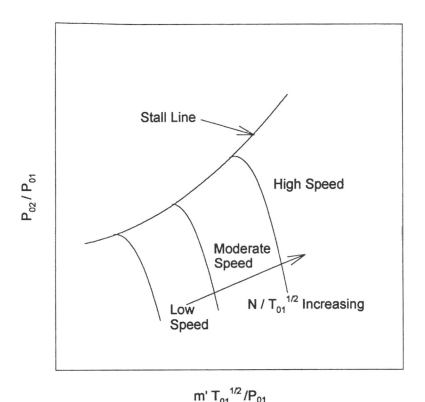

$$m' \, T_{01}^{1/2} / P_{01}$$

**FIGURE 2.5** Compressor characteristic curves.

What is the pressure rise? ($\rho = 0.0023$ slugs/ft$^3$)

$$\Delta p_T = \rho N^2 D^2 \psi = 32.0 \text{ lbf/ft}^2 (0.055) = 0.340 \text{ InH}_2\text{O}$$

and

$$P = \rho N^3 D^5 \xi = 2.36 \times 10^4 (0.0279) \text{ lbf-ft/s} = 1.2 \text{ hp}$$

$$\eta_T = 0.89$$

Previously, it was required that $\Delta p_T = 1$ InWG. One has the wrong size and speed! It is now necessary to constrain $\Delta p_T$ and Q, so impose both conditions, which results in two equations in two unknowns:

$$\Delta p_T = \rho N^2 D^2 \psi \quad \text{or} \quad N = \left\{ \frac{(\Delta p_T)}{(\rho D^2 \psi)} \right\}^{1/2} \tag{2.32}$$

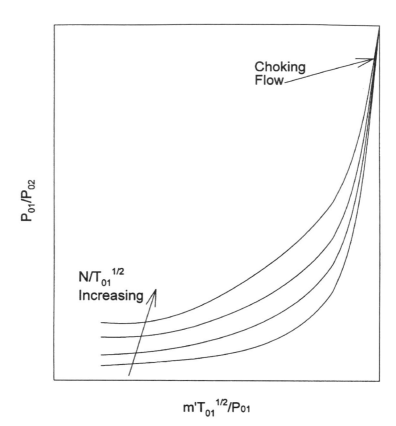

**FIGURE 2.6**   Turbine characteristic curves.

$$Q = ND^3\phi \quad \text{or} \quad N = \left\{ \frac{Q}{(D^3\phi)} \right\} \tag{2.33}$$

by equating through N:

$$N = \left\{ \frac{\Delta p_T}{\rho D^2 \psi} \right\}^{1/2} = \frac{Q}{(D^3\phi)} \tag{2.34}$$

This yields

$$D^2 = \frac{\left(\dfrac{Q}{\phi}\right)}{\left(\dfrac{\Delta p_T}{\rho\psi}\right)^{1/2}}$$

or

$$D = \frac{Q^{1/2}}{\left(\dfrac{\Delta p_T}{\rho}\right)^{1/4}} \cdot \frac{\psi^{1/4}}{\phi^{1/2}} \qquad (2.35)$$

Go back to $N = Q/D^3\phi$ so that

$$N = \frac{\left(\dfrac{\Delta p_T}{\rho}\right)^{3/4}}{Q^{1/2}} \cdot \frac{\phi^{1/2}}{\psi^{3/4}} \qquad (2.36)$$

This states that if one wants a specified $\Delta p_T$ and Q with a known $\rho$ from a fan with a given $\phi$, $\psi$, and $\eta_T$, one can use these equations to specify compatible values of N and D. For the example:

$$Q = 20,000 \text{ cfm}$$

$$\Delta p_T = 1 \text{ InH}_2\text{O} = 5.204 \text{ lbf/ft}^2$$

with $\rho g = 0.074$ lbf/ft³, one obtains

$$N = 1025 \text{ rpm}$$

$$D = 1.91 \text{ ft}$$

What has been done is to derive two new dimensionless variables:

$$N_s = \frac{NQ^{1/2}}{\left(\dfrac{\Delta p_T}{\rho}\right)^{3/4}} = \frac{\phi^{1/2}}{\psi^{3/4}} \qquad (2.37)$$

$$D_s = \frac{D\left(\dfrac{\Delta p_T}{\rho}\right)^{1/4}}{Q^{1/2}} = \frac{\psi^{1/4}}{\phi^{1/2}} \qquad (2.38)$$

Most importantly, these two variables are usually specified at best efficiency. Selecting as done above preserves the operation at BEP so it is an optimized "size" change. That is,

1. Scale to a Performance Specification: $\Delta p_T$, Q, $\rho$
2. Use $N_s$ and $D_s$ at $\eta_{T\ MAX}$
3. Arrive at the optimum N and D

This requires geometric similarity and reasonable Re and Ma.

## 2.7   THE CORDIER DIAGRAM

In the 1950s, Cordier (1955) carried out an intensive empirical analysis of "good" turbomachines using extensive experimental data. He attempted to correlate the data in terms of $N_s$, $D_s$, and $\eta_T$ using $\Delta p_T$ rather than $\Delta p_s$ in forming the dimensionless parameters. He found (subject to uncertainty and scatter in the data) that turbomachines, which for their type had good to excellent efficiencies, tended to group along a definable curve when plotted with their values of $N_s$ vs. $D_s$. He further found that the efficiencies of these machines grouped into a definable, if rough, curve as a function of $D_s$. Machines whose efficiencies could be classified as poor for a particular type of device were found to scatter away from the locus of excellent machines in $D_s$–$N_s$ coordinates.

Cordier correctly reasoned that 75 years of intense, competitive development of turbomachinery design had led to a set of "rules" of acceptable design practice, and that his curve was, in effect, a practical guideline to both design layout and effective selection of good machinery for a specified purpose. This concept was further developed by Balje (1981). The specified purpose is, of course, stated in terms of $\Delta p_T$ , Q, and $\rho$. The Cordier concept, with modifications and amplifications, is illustrated in Figure 2.7a. Cordier's original $N_s$–$D_s$ and $\eta_T$–$D_s$ curves are shown along with the definitions involved. In addition, two "boxes" covering the good data range (taken from Balje, 1981) are roughly superimposed on the curve, to indicate the type of accuracy or margin associated with the curve. Finally, the two straight lines (in log–log form) that lie within the bounds of accuracy of the original form are chosen as an approximation to the Cordier curve.

There are standardized techniques available in many spreadsheet and math-graphics software packages and a brief note on the equations or curve fits shown is in order (see, for example, Holman, 1989). If one assumes that the data is reasonably fittable by an equation in the form

$$y = ax^b$$

then in log-log form, one can rewrite as

$$\log y = \log a + b \log x \quad (\text{or, } N_s = aD_s^{\ b})$$

One can then fit this form to Balje's box of data.

$$\text{At} \quad x = D_s = 1, \quad y = N_s = 9.0; \quad a = 9.0$$

$$\text{At} \quad x = 8.5, \quad y = 0.1; \quad b = \frac{\left(\log(0.1) - \log(9)\right)}{\log(8.5)} = -2.103$$

So that, in the upper box,

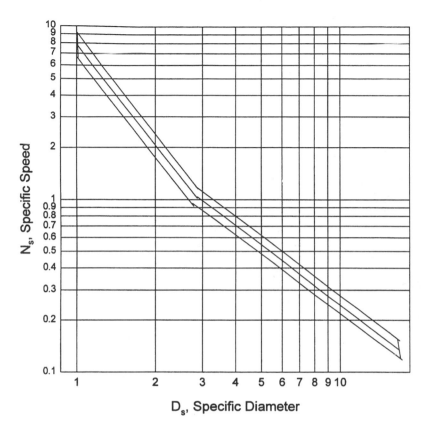

**FIGURE 2.7a** Cordier's correlation of $N_s$ and $D_s$ for high efficiency turbomachines.

$$N_s \cong 9.0 \, D_s^{-2.103} \qquad (\text{for } D_s \leq 2.8) \qquad \qquad (2.39)$$

The lower box is similarly fitted to yield

$$N_s \cong 3.25 \, D_s^{-1.126} \qquad (\text{for } D_s \geq 2.8) \qquad \qquad (2.40)$$

The efficiency band shown on Figure 2.7b represents the best total efficiency that can reasonably be expected. One can curve-fit this band to a piecewise set of simple equations and define $\eta_{T-C}$ as an upper limit value, according to

$$\eta_{T-C} = 0.149 + 0.625 \, D_s - 0.125 \, D_s^2, \qquad \text{for } D_s \leq 2.5$$

$$\eta_{T-C} = 0.864 + 0.0531 \, D_s - 0.0106 \, D_s^2, \qquad \text{for } 2.5 \leq D_s \leq 5 \qquad (2.41)$$

$$\eta_{T-C} = 1.1285 - 0.0529 \, D_s, \qquad \text{for } 5 \leq D_s \leq 20$$

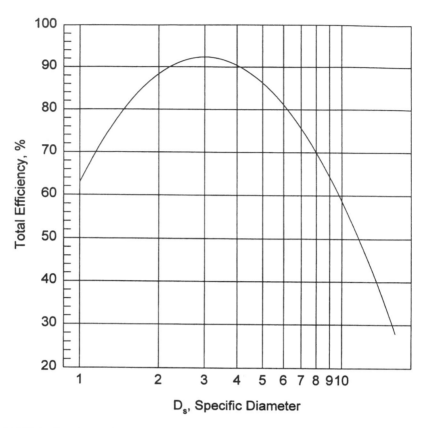

**FIGURE 2.7b**  Correlation of total efficiency with specific diameter.

As these efficiencies are, for a given type of turbomachine, rather high, they may be modified to reflect the approximate moderating influence of "size effect" (low Reynolds Number) and "cost effect" (a larger than ideal clearance gap, for example, due to-out-of-roundness or poor concentricity) by de-rate or reduction factors. One can assume that the losses scale roughly with Re = $ND^2/\nu$ and that Cordier's best machine values were based on quite high Re values ($10^7$). Then losses can be scaled with the factor $(10^7/Re)^{0.17}$ (see Koch and Smith, 1976). Further assume an approximate scaling of loss with radial clearance (tip gap for axial machines and shroud gap for centrifugals as shown in Figures 2.8 and 2.9). (See, for example, Wright, 1974; Wright, 1984c.) Thus, one can model losses as scaling with $g(\delta_o)$, $\delta_o = \delta/0.001$ where $\delta = \varepsilon/D$. These two effects are included in loss estimation according to

$$(1-\eta_T) = (1-\eta_{Tc})f(Re)g(\delta_o) \tag{2.42}$$

$$f(Re) = \left(\frac{10^7}{Re}\right)^{0.17}, \quad Re = \frac{ND^2}{\nu} \tag{2.43}$$

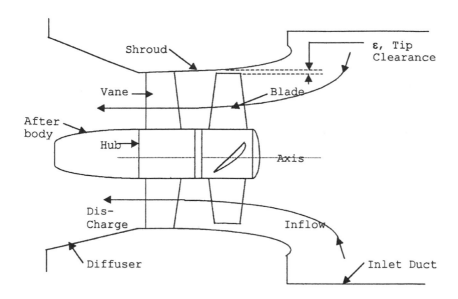

**FIGURE 2.8**   Axial fan assembly with clearance detail.

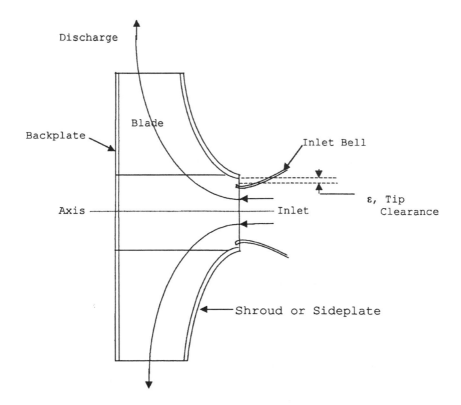

**FIGURE 2.9**   Centrifugal fan impeller with clearance detail.

$$g(\delta_o) = 1 + 2.5 \tanh(0.3(\delta_0 - 1)), \quad \delta_o = \frac{\delta}{0.001}, \quad \delta = \frac{\varepsilon}{D} \qquad (2.44)$$

The underlying assumptions are the "excellent" Re value of $10^7$ and the "excellent" clearance ratio of $\delta = 0.001$. Small size will influence both parameters to decrease efficiency, and low-cost equipment will inevitably require large relative clearances.

These approximate relations will allow us to make simple but reasonable estimates of performance. Understand that these curves and the concept itself are not the results of a rigorous analysis but merely represent a way of unifying turbomachinery design and performance into a very useful and revealing form.

To summarize at this point, one now has a set of rules for scaling the performance of a turbomachine under the restrictions of geometric similitude, subject to "acceptable" changes in the Reynolds number and the Mach number. They are

$$\phi = \frac{Q}{ND^3} = \phi(Re, Ma)$$

$$\psi = \frac{gH}{N^2D^2} = \frac{\left(\dfrac{\Delta p_T}{\rho}\right)}{(N^2D^2)} = \psi(\phi, Re, Ma)$$

$$\xi = \frac{P}{(\rho N^3 D^5)} = \xi(\phi, Re, Ma)$$

$$\eta = \frac{(\rho g H Q)}{P} = \frac{\Delta p_T Q}{P} = \eta(\phi, Re, Ma) = \frac{\phi\psi}{\xi}$$

$$(2.45)$$

or in the common compressible forms

$$\frac{p_{02}}{p_{01}} = f_1\left[\left(\frac{m'(T_{01})^{1/2}}{p_{01}}\right), \left(\frac{N}{(T_{01})^{1/2}}\right)\right]$$

$$v = f_2\left[\left(\frac{m'(T_{01})^{1/2}}{p_{01}}\right), \left(\frac{N}{(T_{01})^{1/2}}\right)\right]$$

$$(2.46)$$

## 2.8   SUMMARY

This chapter has introduced the concepts of dimensionality and reviewed the rule of homogeneity of units. The basic tenets of similitude were re-examined to define the rules of similarity. The need for geometric, kinematic, and dynamic similarity was examined through examples, and the basic rules were discussed in terms of fluid flow. The method of Π-products was reviewed and discussed. Several examples illustrated the techniques and rules involved and some of the potential difficulties were discussed.

Additionally, this chapter restructured the fundamental geometric and performance variables for turbomachines using the method of Π-products. The dimensionless flow, head rise, power, and efficiency parameters were developed as $\phi$, $\psi$, $\xi$, and $\eta_T$. These were used to generalize the performance characteristics of a fluid machine to illustrate the power and value of the dimensionless representation. These results were extended to include the traditional dimensionless variables for high-speed machines operating with significantly compressible flows.

The concepts and definitions of specific diameter and specific speed, $N_s$ and $D_s$, were developed as a natural outgrowth of constrained scaling of machine performance. Finally $N_s$, $D_s$, and $\eta_T$ were combined into the classic Cordier diagram, and algorithms interrelating these variables were introduced.

## 2.9 PROBLEMS

2.1. Repeat Problem 1.1 with the pump running at 1200 rpm. Calculate the specific speed and diameter at BEP.

2.2. Using the pump data from Figure P1.1, what speed would provide a shutoff head of 360 ft? For what speed would the bhp at best efficiency be 0.55 hp? What speed would yield a flow rate at peak efficiency of 150 gpm?

2.3. A pump that is geometrically similar to the pump of Problem 2.1 has an impeller diameter of 24 inches. What will the flow rate, head, and bhp be at the point of best efficiency?

2.4 A variable-speed wind tunnel fan provides a flow rate of 150,000 cfm against a resistance of 200 lbf/ft² when it is running at 1500 rpm. What flow and pressure rise will the fan develop at 2000 rpm? At 3000 rpm?

2.5. A centrifugal air compressor delivers 32,850 cfm of air with a pressure change of 4 psi. The compressor is driven by an 800 hp motor at 3550 rpm.
   (a) What is the overall efficiency?
   (b) What will the flow and pressure rise be at 3000 rpm?
   (c) Estimate the impeller diameter.

2.6. A commercially available fan has BEP performance of Q = 350 ft³/s, $\Delta p_s$ = 100 lbf/ft², and $\eta_s$ = 0.86. Density is $\rho$ = 0.00233 slugs/ft³, and speed and size are given as N = 485 s⁻¹ and D = 2.25 ft, respectively.
   (a) At what speed will the fan generate 20 lbf/ft² at BEP?
   (b) What will the flow rate be at that speed?
   (c) Estimate the fan efficiency at that speed.
   (d) Calculate the required motor horsepower.

2.7. Assume that the power required to drive a pump is a function of the fluid density ($\rho$), the rotational speed (N), the impeller diameter (D), and the volume flow rate (Q). Use dimensional analysis to show that the power coefficient is a function of the flow coefficient $P' = f(\phi)$ and $\phi = Q/ND^3$.

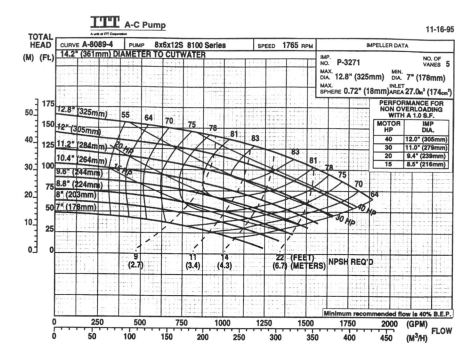

**FIGURE P2.12** An Allis-Chalmers double-suction pump. (ITT Industrial Group, Cincinnati, OH.)

2.8. If the pressure rise of an axial fan is a function of density ($\rho$), speed (N), and diameter (D), use dimensional analysis to show that a pressure rise coefficient, $\psi$, can be written as

$$\psi = \frac{\Delta p_T}{\rho N^2 D^2}$$

2.9. An axial fan with a diameter of 6 ft runs at 1400 rpm with an inlet air velocity of 40 ft/s. If one runs a one-quarter scale model at 4200 rpm, what axial inlet air velocity should be maintained for similarity? Neglect Re effects.

2.10. Use the method of $\Pi$-products to analyze centripetal acceleration. With a particle moving on a circular path of radius R at a speed of V, show that the acceleration (a) is a function of the group $V^2/R$.

2.11. Repeat Problem 2.10 with angular rotation N (radians/s) substituted for V.

2.12. Use the simple scaling laws to relate the performance of the 12.1- and 7.0-inch diameter impellers (Figure P2.12). What is wrong? Can you find the appropriate exponent for D (not 3) in the flow scaling formula that correctly relates the two impellers? (HINT: Try some cross plots at points related by simple head scaling, and look for an integer exponent other than 3 for the diameter).

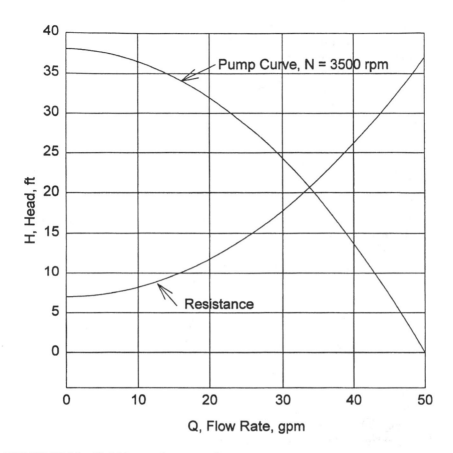

**FIGURE P2.15**   Variable-speed pump performance curve.

2.13. Explain and discuss the results of Problem 2.12 in terms of geometric features. Use dimensional analysis to support your discussion.

2.14. A pump is driven by a two-speed motor having speeds of 1750 rpm and 1185 rpm. At 1750 rpm, the flow is $Q = 45$ gpm, the head is $H = 90$ ft, and the total efficiency is $\eta = 0.60$. The pump impeller has a diameter of 10 inches.
   (a) What values of Q, H, and $\eta$ are obtained if the pump runs at 1185 rpm?
   (b) Find the specific speed and the specific diameter of the pump (at 1750 rpm).

2.15. A pump curve and a system curve are show in Figure P2.15. The pump is operating at $N = 3550$ rpm. If one reduces the pump speed gradually, at what speed will the volume flow rate go to zero?

2.16. A pump running on a two-speed motor delivers 250 gpm at 114 ft of head pumping water at 1750 rpm at 10 hp. The low-speed mode operates at 1150 rpm. Calculate flow rate, head rise, and power at the low speed (*Note:* $\eta = 0.72$ at 1750 rpm).

2.17. Using the information in Figure P2.17, construct the performance curves of a vane-axial fan with D = 2.5 m, N = 870 rpm, and ρ = 1.20 kg/m³.

(a) If this fan is supplying air to a steam condenser heat exchanger with a face area of 10 m2 and an aerodynamic loss factor of K = 12, find the operating point of the fan in terms of Q (m³/s), $\Delta p_T$ (kPa), and P (kW).

(b) Can this fan be used with two such heat exchangers in series? With three?

2.18. We want to use the fan represented in Figure P2.17 as an electronic cooling blower with a personal computer. Space available dictates a choice of fan size less than 15 cm. Scale a 15-cm fan to the proper speed to deliver 0.15 m³/s of air at $\eta_T \geq 0.55$.

2.19. A 0.325-m diameter version of the fan of Figure P2.17 runs on a variable-frequency speed-control motor that operates continuously between 600 and 3600 rpm. If total efficiencies between 0.60 and 0.40 are acceptable, define the region of flow-pressure performance capability of this fan as a zone or area on a $\Delta p_T$ vs. Q plot. What maximum power capability must the motor provide?

2.20. A ship-board ventilation fan must supply 150 kg/s mass flow rate of air to the engine room with ρ = 1.22 kg/m³ and a static pressure rise of 1 kPa. Select the smallest size for this fan for which the required power is less than 225 kW. Describe the fan in terms of the design parameters.

2.21. A three-speed vane axial fan, when operating at its highest speed of 3550 rpm, supplies a flow rate of 5000 cfm and a total pressure of 30 lbf/ft² with ρ = 0.00233 slugs/ft³ (lower speeds of 1775 and 1175 rpm are also available).

(a) Estimate the size and efficiency of the fan at 3550 rpm.

(b) Estimate the flow rate and pressure rise at the lower speed setting of 1175 rpm.

(c) Use the Reynolds numbers for the fan at the three speeds to estimate the efficiency at the lower two speeds.

2.22. Show that $\psi_s = \psi_T - (8/\pi^2)\phi^2$ and state your assumptions.

2.23. Use the non-dimensional performance curves given in Figure P2.17 for a vane axial fan to generate new curves for the static pressure and static efficiency performance of this fan. Determine the operating point for maximum static efficiency and estimate the free delivery flow coefficient for the fan.

2.24. Use the curves of Figure P2.17 to specify a fan that can supply 10 m³/s of air with a static pressure rise of 500 Pa (ρ = 1.01 kg/m³). Estimate the required power and total pressure rise at this performance.

2.25. The pump of Problem 1.1 (Figure P1.1) can be used in a side-by-side arrangement with other identical pumps (with suitable plumbing) to provide an equivalent pump of much higher specific speed.

(a) Generate a table of H, P, and η for a set of five pumps in parallel, and calculate the specific speed and diameter of this equivalent "pump."

(b) Sketch the "plumbing" arrangement for the layout.

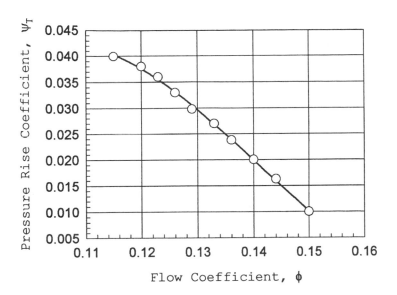

**FIGURE P2.17a** Dimensionless performance of a vane-axial fan, fan characteristic.

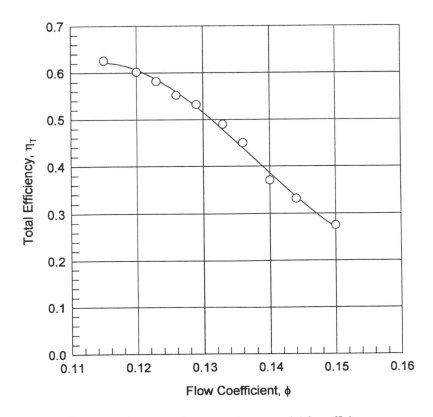

**FIGURE P2.17b** Dimensionless performance of a vane-axial fan, efficiency curve.

2.26. For the fan developed in Problem 2.17, if the condenser has a rectangular face 3 m by 5 m and a K-factor of 20, size two fans to operate in parallel to cool the condenser. State size, speed, pressure rise, flow rate per fan, and total power required.

2.27. A small centrifugal pump is tested at $N = 2875$ rpm in water. It delivers 0.15 m³/s at 42 m of head at its best efficiency point ($\eta = 0.86$).

(a) Determine the specific speed of the pump.

(b) Compute the required input power.

(c) Estimate the impeller diameter.

(d) Calculate the Reynolds Number and estimate the efficiency according to Cordier and the Re correction (assume the clearance is small).

2.28. A pump is working with a liquid slurry of sand and water with a density of 1250 kg/m³ and a kinematic viscosity of $4.85 \times 10^{-5}$. The pump was tested in pure water with a BEP performance pressure rise of 150 kPa at 2.8 m³/s with $\eta = 0.75$. Estimate the pressure rise and efficiency with a slurry flow rate of 2.8 m³/s.

2.29. A pump running on a two-speed motor delivers 60 m³/hr at a head of 35 m of water at 1485 rpm and 8.0 kW. The pump efficiency is $\eta = 0.72$ at the high speed. At the low-speed setting, the pump operates at 960 rpm. Estimate flow rate, head rise, and power for low-speed operation.

# 3 Scaling Laws, Limitations, and Cavitation

## 3.1 SCALING OF PERFORMANCE

The examples and development thus far have shown how to use $\phi$ and $\psi$ with $\xi$ to change the level of performance of a given design through changes in speed, diameter, and density. This was done by first calculating $\phi$, $\psi$, and $\xi$ from a known machine, followed by recalculating $\Delta p_T$, Q, and P by choosing new values of $\rho$, N, and D.

A more common procedure for scaling involves rearranging the low-speed non-dimensional parameters given above into ratios of the physical parameters by remembering that, with a geometrically similar change, $\phi$, $\psi$, and $\xi$ will be invariant (subject to restrictions or changes in Re and Ma, which will be discussed later). For example, for a model pump of known dimensions tested with performance curves in a given fluid, one has a complete description of the model in terms of $\rho_m$, $N_m$, $D_m$, $Q_m$, $P_m$, $\Delta p_{Tm}$, and $\eta_m$.

The "m" subscript identifies the variables of the model. To predict the performance of a full-size prototype pump that may be significantly larger than the model, one can characterize the prototype by $\rho_p$, $N_p$, $D_p$ and $\Delta p_p$, $Q_p$, and $P_p$. One can calculate

$$\phi = \frac{Q_m}{\left(N_m D_m^{\ 3}\right)} = \frac{Q_p}{\left(N_p D_p^{\ 3}\right)} \tag{3.1}$$

because $\phi$ does not change if geometric similitude is maintained. The equation can be rearranged as

$$\frac{Q_p}{Q_m} = \frac{\left(N_p D_p^{\ 3}\right)}{\left(N_m D_m^{\ 3}\right)} = \left(\frac{N_p}{N_m}\right)\left(\frac{D_p}{D_m}\right)^3 \tag{3.2}$$

or

$$Q_p = Q_m \left(\frac{N_p}{N_m}\right)\left(\frac{D_p}{D_m}\right)^3 \tag{3.3}$$

This states that Q changes linearly with speed ratio and with the third power of the size ratio (diameter ratio). This relation is called a scaling rule or a "fan law" or "pump law" and can be used directly for quick calculation of scaling effects on performance. For example, if one doubles speed, one doubles the flow. If one halves the speed, one halves the flow. Double the diameter, and obtain 8 times the flow; halve the diameter, and obtain 1/8 of the flow. The same thing can be done with pressure or $\psi$ by writing (with $\psi$ invariant when geometric similitude is maintained):

$$\psi = \frac{\Delta p_{Tm}}{\left(\rho_m N_m{}^2 D_m{}^2\right)} = \frac{\Delta p_{Tp}}{\left(\rho_p N_p{}^2 D_p{}^2\right)} \tag{3.4}$$

or

$$\frac{\Delta p_{Tp}}{\Delta p_{Tm}} = \frac{\left(\rho_p N_p{}^2 D_p{}^2\right)}{\left(\rho_m N_m{}^2 D_m{}^2\right)} \tag{3.5}$$

$$\Delta p_{Tp} = \Delta p_{Tm} \frac{\left(\rho_p N_p{}^2 D_p{}^2\right)}{\left(\rho_m N_m{}^2 D_m{}^2\right)} \tag{3.6}$$

This states that pressure rise varies linearly with density and parabolically with speed and diameter. For the power:

$$P_p = P_m \left(\frac{\rho_p}{\rho_m}\right)\left(\frac{N_p}{N_m}\right)^3\left(\frac{D_p}{D_m}\right)^5 \tag{3.7}$$

Recalling that $\eta_T$ is invariant with the scaling assumption, one adds, for completeness

$$Q_p = Q_m \left(\frac{N_p}{N_m}\right)\left(\frac{D_p}{D_m}\right)^3 \tag{3.8}$$

$$\Delta p_{Tp} = \Delta p_{Tm} \frac{\left(\rho_p N_p{}^2 D_p{}^2\right)}{\left(\rho_m N_m{}^2 D_m{}^2\right)} \tag{3.9}$$

$$P_p = P_m \left(\frac{\rho_p}{\rho_m}\right)\left(\frac{N_p}{N_m}\right)^3\left(\frac{D_p}{D_m}\right)^5 \tag{3.10}$$

$$\eta_{Tp} = \eta_{Tm} \tag{3.11}$$

These last relations are the rules or laws of scaling for low-speed fans or liquid pumps.

As an example, consider a turbine with a head drop of 1.5 m, an efficiency of 0.696, and a volume flow rate of 55 m³/s. The full-scale turbine is intended to operate in warm sea water ($\rho$ = 1030 kg/m³) while moored in the gulf stream south of Florida's east coast. It runs at a speed of 98 rpm and has a full-scale diameter of 4 m. To devise a scale model test with a power extraction of 10 kW and a model running speed of 1000 rpm in fresh water ($\rho$ = 998 kg/m³), determine the proper model diameter and calculate the required head and flow to the model turbine to maintain similarity.

The turbine can be scaled to model size using the similarity rule for power (Equation (3.10))

$$P_p = P_m \left(\frac{\rho_p}{\rho_m}\right)\left(\frac{N_p}{N_m}\right)^3\left(\frac{D_p}{D_m}\right)^5$$

since the densities and speeds are known. That is,

$$\frac{D_m}{D_p} = \left(\left(\frac{\rho_p}{\rho_m}\right)\left(\frac{P_m}{P_p}\right)\left(\frac{N_p}{N_m}\right)^3\right)^{1/5}$$

where the original full-scale power, $P_p$, is given by

$$P_p = \rho g H Q \eta_T = 1030 \times 9.81 \times 55 \times 1.5 = 579 \text{ kW}$$

Then the diameter ratio for the model is

$$\frac{D_m}{D_p} = \left(\frac{1030}{998}\right)^{1/5}\left(\frac{98}{1000}\right)^{3/5}\left(\left(\frac{10}{579}\right)\right)^{1/5} = 0.111$$

Therefore, the proper diameter is $D_m = 0.444$ m. Note that the $R_e$ values are both very large, with $R_e = 1 \times 10^9$ for the prototype and $R_e = 1 \times 10^8$ for the model. The assumption of equal efficiencies for the two is a very reasonable one in this instance. It remains to calculate the head and flow requirements for the model. These are estimated directly from the scaling laws as

$$H_m = H_p \left(\frac{N_m}{N_p}\right)^2\left(\frac{D_m}{D_p}\right)^2 = 1.5 \text{ m}\left(\frac{1000}{99}\right)^2\left(\frac{0.446}{4.0}\right)^2 = 1.9 \text{ m}$$

and

$$Q_m = Q_p \left(\frac{N_m}{N_p}\right)\left(\frac{D_m}{D_p}\right)^3 = 55 \text{ m}^3/\text{s}\left(\frac{1000}{99}\right)\left(\frac{0.446}{4.0}\right)^3 = 0.760 \text{ m}^3/\text{s}$$

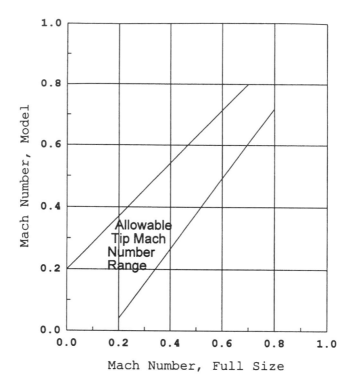

FIGURE 3.1 Mach Number limitations on testing and scaling.

## 3.2 LIMITATIONS ON SCALING

What about Reynolds Number and Mach Number limitations? The AMCA (Air Movement and Control Association) and ASHRAE (the American Society of Heating, Refrigeration, and Air conditioning Engineers) state in their jointly issued testing standard (ANSI/AMCA/ASHRAE) Standard 210-85 (AMCA, 1985), "Laboratory Methods of Testing Fans for Rating", and AMCA Publication 802-82, "Establishing Performance Using Laboratory Models" (AMCA, 1982), that the influence of Re is to be neglected. That is, use the fan laws with $\eta_p = \eta_m$, subject to the restrictions. The Reynolds Number for the model must be greater than $3 \times 10^6$ and the model must be at least 35 inches or 0.900 m in diameter (or 1/5 the size of the full-scale prototype, whichever is larger). For smaller full-scale equipment, the model is the prototype. Fan Reynolds Number is $Re_D = ND^2/\nu$.

In the same publications, Mach Number restrictions are based on a comparison of the model and prototype "tip speed Mach numbers", $M_t$, where $M_t = \pi ND/(\gamma RT)^{1/2}$; where N is in $s^{-1}$ units; $\gamma$ is the ratio of specific heat; R is the gas constant; T is the absolute temperature at the fan inlet; $(\gamma RT)^{1/2}$ is the speed of sound; and $\pi ND/60$ is the "tip speed" of the fan (if N is in rpm). The required relation between $M_{tm}$ and $M_{tp}$ is given graphically in Figure 3.1. That is, the model $M_t$ must lie between values as given by:

$$M_{tm} = 0.200 + 0.857 \, M_{tp} \qquad \text{(upper limit)} \qquad (3.12)$$

$$M_{tm} = 1.167 - 0.213 \, M_{tp}; \quad M_{tp} \geq 0.183 \qquad \text{(lower limit)} \qquad (3.13)$$

or

$$M_{tm} \geq 0, \quad M_{tp} < 0.183 \qquad (3.14)$$

In addition, corrections for compressibility effects on flow, power, and pressure rise (but not efficiency) are permitted and specified by the standards. These will be discussed in a later section.

There are other standards for compressors [see, for example, ASME Power Test Code PTC-10 (ASME, 1965)] where efficiency can be scaled with Reynolds Number according to:

$$\frac{\left(1 - \eta_p\right)}{\left(1 - \eta_m\right)} = \left(\frac{Re_m}{Re_p}\right)^{0.1} \qquad \text{(centrifugal)} \qquad (3.15)$$

Provided $Re_m \geq 10^5$ (using tip speed, $Re = ND^2/v$).

For axial flow machines, scaling is done according to

$$\frac{\left(1 - \eta_p\right)}{\left(1 - \eta_m\right)} = \left(\frac{Re_m}{Re_p}\right)^{0.2}, \qquad \text{for } Re \geq 10^5 \qquad (3.16)$$

For both types of compressors, $\psi$ can be adjusted according to

$$\frac{\psi_p}{\psi_m} = \frac{\eta_p}{\eta_m} \qquad (3.17)$$

with no correction on $\phi$ or $\xi$. A review and analysis on compressor and fan efficiency scaling is available (Wright, 1989). In that work, a method is suggested for efficiency scaling according to

$$\frac{\left(1 - \eta_p\right)}{\left(1 - \eta_m\right)} = 0.3 + 0.7\left(\frac{f_p}{f_m}\right) \qquad (3.18)$$

where f is the Darcy friction factor. The influence of both Reynolds Number and surface relative roughness ($\varepsilon/D$) are included in estimating the losses of a prototype versus a model machine. The method has been tested against centrifugal compressor data, but it has not yet been evaluated for axial machines. For centrifugal machines, the appropriate Re can be based on the axial outlet width, $w_2$, as the length scale: $Re_w = \rho N D w_2 / \mu$.

This method is very similar to the European ICAAMC (International Compressed Air and Allied Machinery Committee) method (Casey, 1985) given by:

$$\frac{\left(1-\eta_p\right)}{\left(1-\eta_m\right)} = \frac{\left(0.3 + \dfrac{0.7 f_p}{f_{fr}}\right)}{\left(0.3 + \dfrac{0.7 f_m}{f_{fr}}\right)} \tag{3.19}$$

The subscript "fr" refers to "fully rough" conditions at high Reynolds Number for a given relative roughness. The ICAAMC method is very empirical and is not clearly related to the physics, particularly in the use of $f_{fr}$. The ICAAMC also recommends adjustments in $\phi$ and $\psi$ according to:

$$\frac{\psi_p}{\psi_m} = 0.5 + 0.5\left(\frac{\eta_p}{\eta_m}\right) \tag{3.20}$$

$$\frac{\phi_p}{\phi_m} = \left(\frac{\psi_p}{\psi_m}\right)^{1/2} \tag{3.21}$$

These relationships are heavily evaluated against experimental compressor data (Strub et al., 1988; Casey et al., 1987). The ICAAMC equations are also limited to centrifugal machines but show considerable promise that they can be generalized to cover axial fans, compressors, and pumps, as well as centrifugal fans and pumps through the choice of an appropriate form of Reynolds Number.

Finally, it is important to remember that the scaling of performance, particularly efficiency, according to changes in Reynolds Number is a subject of ongoing controversy and is not fully agreed upon in the international turbomachinery community at this time.

A scaling rule for pump efficiency that has universal acceptance is also lacking. At this time, the "best" or most commonly used rule is one developed by Moody (Moody, 1925)

$$\frac{\left(1-\eta_p\right)}{\left(1-\eta_m\right)} = \left(\frac{D_m}{D_p}\right)^{1/2} \tag{3.22}$$

dating to about 1925. The inference from the work of Kittredge (1967) is that what is really implied is the form

$$\frac{\left(1-\eta_p\right)}{\left(1-\eta_m\right)} = \left(\frac{Re_m}{Re_p}\right)^{0.25} \tag{3.23}$$

which is very similar to the PTC-10 compressor scaling rules. Uncertainty about the exponent, however, is the major problem, as Kittridge quotes some 17 different scaling rules with exponents ranging from approximately 0.1 to 0.45. The methods date back as far as 1909.

ASME PTC-8 (ASME, 1965) states that "results extrapolated from model tests... are beyond the scope of the code."

A few examples of efficiency scaling are helpful in understanding what the controversy is all about. Suppose one tests a pump in model size that has a Reynolds Number of $10^5$. The full-scale machine will operate at a Reynolds Number of $10^6$, and one wants to estimate the increase in efficiency for the full scale from the measured model efficiency of $\eta_m = 0.80$. Using the Moody Rule as

$$\frac{\left(1-\eta_p\right)}{\left(1-\eta_m\right)} = \left(\frac{Re_m}{Re_p}\right)^n \tag{3.24}$$

if n = 0.25, $\eta_p = 0.887$, which is a very substantial improvement. On the other hand, if one uses a conservative scaling exponent, n = 0.1, efficiency is predicted as $\eta_p = 0.841$. Using n = 0.45, $\eta_p = 0.929$. The results indicate potential gains in efficiency from about 4% up to nearly 13%. This broad range of results is unacceptable, particularly since there is great temptation to take commercial advantage of using the larger exponents to predict really superior performance. A long history of commercial manipulation of efficiency scaling or "size effect" has led to a widespread feeling of healthy skepticism about scaling and to the adoption of excessively restrictive rules, including the AMCA statement on total restriction or no scaling at all.

## 3.3   COMPRESSIBILITY CORRECTIONS

When working with fans and compressors, one is working with a fluid that is compressible (typically air, with a ratio of specific heat $\gamma = 1.4$). Thus, one must worry about changes in density as the flow passes through the turbomachine. Following the AMCA standards (210-85 and 802-82), consider all test data and scaled performance to be incompressible as long as $\Delta p_T = 4$ InWG or about 20.8 lbf/ft². This is about 1% of typical atmospheric pressure, and an isothermal or isentropic calculation indicates a negligible change in density.

Above the 4 InWG level, however, density change can begin to have a significant effect on measurement of Q, $\Delta p_T$, and P. What the standards do is to provide an "approved" method of both correcting test data to equivalent incompressible model performance, and predicting full-scale prototype performance from the incompressible data. The procedure uses a polytropic compression analysis as outlined in Chapter 1 and derived in the standards mentioned earlier (as well as in any good thermodynamics text). In general, a compressibility coefficient, $k_p$, is calculated, and test data can be converted to equivalent incompressible values according to

$$Q_i = k_p Q \tag{3.25}$$

$$\Delta p_{Ti} = k_p \Delta p_T \tag{3.26}$$

where the "i" subscript identifies the equivalent incompressible variable. $k_p$ is given by the following equation:

$$k_p = \left(\frac{n}{n-1}\right) \frac{\left[\left(\frac{p_b + p_2}{p_b + p_1}\right)^{\left(\frac{n-1}{n}\right)} - 1\right]}{\left[\left(\frac{p_b + p_2}{p_b + p_1}\right) - 1\right]} \tag{3.27}$$

Here, $p_b$ is the barometric pressure and $p_2 - p_1 = \Delta p_T$, where these gage pressures measured are converted to absolute values for the thermodynamic calculation (usually $p_{1gage}$ is zero). n is the polytropic exponent related to the isentropic pressure exponent $\gamma$, the ratio of specific heats ($\gamma = 1.4$ for air). n is established by:

$$\frac{n}{n-1} = \eta_T \left(\frac{\gamma}{\gamma-1}\right) \tag{3.28}$$

as noted in Chapter 1.

When scaling model performance directly from test data up to the conditions for a prototype, one can recalculate the compressibility coefficient. Use the "m" subscript for models and the "p" subscript for the full-scale prototype as before and the $k_p$ values and the "as-tested" model performance. Then,

$$Q^P = Q_m \left(\frac{N_p}{N_m}\right) \left(\frac{D_p}{D_m}\right)^3 \left(\frac{k_{pm}}{k_{pp}}\right) \tag{3.29}$$

$$\Delta p_{Tp} = \Delta p_{Tm} \left(\frac{\rho_p}{\rho_m}\right) \left(\frac{N_p}{N_m}\right)^2 \left(\frac{D_p}{D_m}\right)^2 \left(\frac{k_{pm}}{k_{pp}}\right) \tag{3.30}$$

$$P_p = P_m \left(\frac{\rho_p}{\rho_m}\right) \left(\frac{N_p}{N_m}\right)^3 \left(\frac{D_p}{D_m}\right)^5 \left(\frac{k_{pm}}{k_{pp}}\right)^2 \tag{3.31}$$

Here, $\eta_{Ti} = \eta_{Tm}$ is used in both $k_p$ calculations. Note that for fully incompressible model data, $k_{pm} = 1.0$ simplifies these relations so that Q, $\Delta p_T$, and P are proportional to $1.0/k_{pp} \geq 1.0$. That is, if the prototype is going to operate at high Mach Number or high pressure/high speed, then output (Q, $\Delta p_T$) will be significantly higher than expected from the simple incompressible scaling. Typically, a smaller than expected fan might do the job.

Consider the example of a model test such that

$$\gamma = 1.4, \quad \rho_m = 0.0023 \text{ slugs/ft}^3, \quad P_m = 37.1 \text{ hp}$$

$$Q_m = 10,000 \text{ cfm}, \quad \eta_{Tm} = 0.85$$

$$\Delta p_{Tm} = 20 \text{ InWG}, \quad p_b = 2116 \text{ lbf/ft}^2$$

Here, the data is scaled to a prototype for which

$$\rho_p = 0.0023 \text{ slugs/ft}^3, \quad \frac{D_m}{D_p} = 2$$

$$p_b = 2116 \text{ lbf/ft}^2, \quad \frac{N_p}{N_m} = 2$$

Since $\Delta p_{Tm} > 20$ InWG, look at the equivalent incompressible model performance, assuming $p_1$ is zero. Then,

$$\frac{n}{n-1} = 0.85\left(\frac{1.4}{1.4-1}\right) = 2.975; \quad \frac{n-1}{n} = 0.3361$$

$$k_p = 2.975 \frac{\left[\left(\dfrac{2116+20\cdot5.204}{2116}\right)^{0.3361} - 1\right]}{\left[\left(\dfrac{2116+20\cdot5.204}{2116}\right) - 1\right]} = 0.9840 = k_{pm}$$

and

$$Q_i = 0.9840 \times 10,000 \text{ cfm} = 9840 \text{ cfm}$$

$$\Delta p_{Ti} = 0.9840 \times 20 \text{ InWG} = 19.68 \text{ InWG}$$

$$P_i = 0.9840^2 \times 37.1 \text{ hp} = 35.9 \text{ hp}$$

Now, a new value of $k_p$ is needed for the prototype: $k_{pp}$. Use $\eta_{Tmi} = \eta_{Tp}$ and an incompressibly scaled $\Delta p_{Tp}$ to calculate $k_{pm}$ as follows:

$$\left(\frac{n}{n-1}\right) = 0.850\left(\frac{1.4}{1.4-1}\right) = 2.975; \quad \left(\frac{n-1}{n}\right) = 0.3361$$

Then, estimate

$$\Delta p_{Tp} = \Delta p_{Tm} \left(\frac{\rho_p}{\rho_m}\right) \left(\frac{D_p}{D_m}\right)^2 \left(\frac{N_p}{N_m}\right)^2 \qquad (3.32)$$

$$\Delta p_{Tp} = 20 \ \text{InWG}(1)(2)^2(2)^2 = 320 \ \text{InWG}$$

Now,

$$p_b + p_2 = (2116 + 1665) \ \text{lbf/ft}^2 = 3781.3 \ \text{lbf/ft}^2$$

$$p_b + p_1 = 2116 \ \text{lbf/ft}^2 \qquad (\text{assuming } p_1 = 0)$$

Then,

$$k_{pp} = \frac{2.975(0.2154)}{(0.78700)} = 0.8145$$

Finally:

$$Q_p = 10,000 \ \text{cfm} \times (2)(2)^3 \left(\frac{0.9840}{0.8145}\right) = 193,297 \ \text{cfm}$$

$$\Delta p_{Tp} = 20 \ \text{InWG} \times (1)(2)^2(2)^2 \left(\frac{0.9840}{0.8145}\right) = 386.6 \ \text{InWG}$$

$$P_p = 37.1 \ \text{hp} \times (1)(2)^3(2)^5 \left(\frac{0.9840}{0.8145}\right)^2 = 13,837 \ \text{hp}$$

and, of course,

$$\eta_{Tp} = 0.85$$

With no compressibility corrections, the predicted values are:

|  | (Value) | (Error) |
|---|---|---|
| $Q = 160,000 \ \text{cfm}$ | | 21% |
| $\Delta p_T = 320 \ \text{InWG}$ | | 21% |
| $P_p = 9498 \ \text{hp}$ | | 21% |
| $\eta_T = 0.85$ | | |

The difference is clearly significant, but of course a very strong example was used. As an exercise, calculate the errors if, for a different prototype, $\rho_p/\rho_m = 1$, $D_p/D_m = 1$, and $N_p/N_m = 1.6$.

## 3.4  CAVITATION ALLOWANCE IN PUMPS

When working with liquids in turbomachines, one does not need to be concerned with Mach Number effects or the influence of compressibility on performance. However, one does have to worry about another equally important problem called *cavitation*. Pure cavitation is defined as a localized boiling and recondensation of the fluid as it occurs inside the impeller passages of a liquid pump. Vaporization or boiling occurs as a function of the vapor pressure of the fluid, which primarily depends on the fluid temperature. Water, for example, has a vapor pressure at 32°F of 12.49 lbf/ft²; at 100°F of 100.5 lbf/ft²; and at 212°F of 2116 lbf/ft². At 212°F, the water boils when $p_{atm} = 2116$ lbf/ft². At lower temperatures, the fluid boils if the local pressure drops to the level of the vapor pressure at that particular temperature. From the graph of Figure 3.2, one can see that if a pump is generating 30 m of head with 6 m of suction and 24 m of downstream resistance, then the water in the eye or suction line of the pump will boil when the temperature is allowed to reach approximately 75.4°C. That is, when the vapor pressure matches the absolute pressure in the suction line, one obtains

$$p_v = p_b + p_{suction} = (101,300 - 6 \times 9790)Pa = 42,560 \text{ Pa}$$

Using the algorithm of Figure 3.2,

$$\frac{p_v}{p_b} = \left(\frac{101,300}{42,560}\right) = 0.420 = \left(\frac{T}{100}\right)^{3.068} \tag{3.33}$$

Solve for T as $T = (0.420)^{1/3.068} \times 100°C = 75.4°C$. One can also find the suction line pressure for which the water will boil at "room temperature," $T = 20°C$, by a more direct calculation. That is,

$$p_v = p_b\left(\frac{T}{100}\right)^{3.068} = 101,300 \text{ Pa}\left(\frac{20}{100}\right)^{3.068} = 726 \text{ Pa}$$

so that $p_{suction} = p_v - p_b = -100,574$ Pa or $h_{suction} = -10.3$ m with 19.7 m of downstream resistance.

Such inlet mass boiling is referred to as a massive suction line cavitation and leads to a "vapor lock" of the pump, where the impeller is trying to operate with a fluid whose density is several orders of magnitude lower than the intended value. The result is a virtually complete loss of head generation, which may lead to a backflow surge of water into the pump. This surge of liquid into the impeller may well destroy the pump and should be avoided. More routinely, if the pump is selected or installed

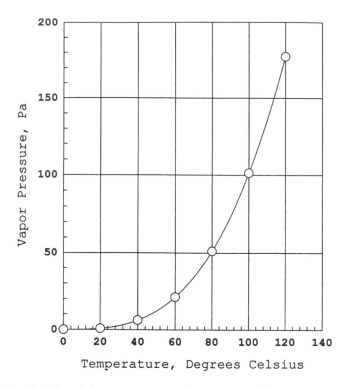

**FIGURE 3.2**   Variation of the vapor pressure of pure water with temperature.

to avoid this massive cavitation, there may still be low-pressure regions (or regions of depressed static pressure) in the impeller where the local velocities reach maximum value. These regions typically occur near the entry of the pump—at the inlet rim of the pump shroud or along the leading edges of the blades (sometimes along the juncture of the hub and blade or shroud and blade). These regions may be fairly small, imposing a short residence time of the flowing liquid in the low-pressure zone, after which the liquid passes on into a zone of much higher pressure. If the fluid resides in the low-pressure zone long enough to vaporize, the bubbles or pockets of vapor will be imbedded in the fluid, usually very close to the impeller surfaces. As the bubble moves into a zone of higher pressure, the vapor will condense, leading to sudden collapse of the bubble. The event is repeated more or less continuously as more fluid passes along the impeller into and out of a zone of critically low pressure.

As bubbles collapse, a phenomenon similar to an intense water hammer occurs locally and at high frequency. Typically, at frequencies reaching 25 kHz, shock waves induce temperatures in the range of 500 to 800°C and pressures of 4000 atmospheres. The subsequent intense hammering of impeller surfaces in the collapse region leads to local fracturing and breakdown of the surface structure, followed by progressive pitting and erosion of the impeller. Prolonged exposure to even very limited cavitation can lead to structural failure of the pump, preceded by progressive deterioration of efficiency and performance. Clearly, good design or selection and installation of a pump requires that the machine operate as far away as possible from conditions that will lead to cavitation.

Conditions leading to cavitation can be expressed in terms of the absolute pressure or head in the inlet of a pump (or the outlet region of a hydraulic turbine). If the absolute pressure at the suction side of the machine is expressed as the sum of hydrostatic, dynamic, and vapor pressures, one can write:

$$h_{sv} = h_p + h_z + h_q - h_{vp} \qquad (3.34)$$

where $h_{sv}$ is the absolute pressure in head form (frequently, if awkwardly, written as NPSHA), $(h_p + h_q)$ is the total head $h_T$, $h_z$ is the hydrostatic head, $h_{vp}$ is the head form of vapor pressure, and $h_q$ is the dynamic head, given by $V^2/(2g)$.

Note that hz can be positive or negative, depending on suction line conditions. hsv is also called the NPSHA or net positive suction head available at the eye of the machine. Following Shepherd's development (Shepherd, 1974), he argues that conditions in the impeller itself are worse due to inlet friction losses ($h_f$). Local dynamic pressure reduction is also a factor as $((1 + \lambda)V^2/2g)$, where $\lambda$ is an amplification factor for blade relative velocity. Thus, one can write the local surface pressure $h_{surf}$ as:

$$h_{surf} = h_z - h_f - \frac{(1+\lambda)V^2}{2g} + h_{atm} \qquad (3.35)$$

If local cavitation first occurs when local static head plus atmospheric is just equal to the vapor pressure, one obtains

$$h_{vp} = h_{atm} + h_s$$

such that

$$h_{sv} = h_f + \frac{(1+\lambda)V^2}{2g} \qquad (3.36)$$

Both sides of the equation are proportional to $N^2$ for a given pump, as is H or gH of the pump itself. This allows one to look at the ratio $h_{sv}/H$ as the characteristic number for a pump's propensity to cavitation. This is the Thoma cavitation index (Thoma, 1932):

$$\sigma_v = \frac{h_{sv}}{H} \qquad (3.37)$$

What is normally done is to establish the values of $h_{sv}$ and H or gH at which the pump begins to cavitate by experiment. The $\sigma_v$ characterizes the pump in a dimensionless fashion and is subject to scaling. In fact, $h_{sv}$ can be used to form its own specific speed according to:

$$S = \frac{NQ^{1/2}}{\left(gH_{sv}\right)^{3/4}} = \frac{N_s\left(gH\right)^{3/4}}{\left(\sigma_v gH\right)^{3/4}} = \frac{N_s}{\sigma_v^{3/4}} \tag{3.38}$$

Typical values of $\sigma_v$, based on collected experimental data (Shepherd, 1964; Stepanoff, 1948) are presented as a curve fit:

$$\sigma_v = 0.241\, N_s^{4/3} \qquad \text{(single - suction pumps)} \tag{3.39}$$

$$\sigma_v = 0.153\, N_s^{4/3} \qquad \text{(double - suction pumps)} \tag{3.40}$$

As an example for using these relationships, consider a centrifugal pump that will run at $H = 100$ ft in water at 150°F. One wants to estimate the lowest inlet head one can operate at without cavitation. Assume a specific speed of $N_s = 0.75$. First, estimate $\sigma_v$ from $\sigma_v = 0.241\, N_s^{4/3} = 0.164$. Then, from $h_{sv} = h_a + h_z - h_{vp}$, with $h_{sv}/H = \sigma_v$, one obtains $h_{sv} = 0.164 \times 100$ ft $= 16.4$ ft. Solve for $h_z$; $h_z = h_{sv} + h_{vp} - h_a$. Atmospheric head will be about 34 ft; and at 150°F, $h_{vp} = 8$ ft. This yields

$$h_z = (16.4 + 8 - 34) \text{ ft} = -9.6 \text{ ft}$$

or 9.6 ft of suction.

At a water temperature of 100°F, $h_{vp} = 1.6$ ft, so allowable suction would be 16.2 feet (negative head). On the other hand, a water temperature of 200°F gives $h_{vp} = 25.6$ ft, so a positive head of 8 ft would be required (no suction or lift allowed). $h_z$ is the required hydrostatic head or pressure head at the inlet to avoid cavitation.

As a practical matter, the manufacturer of a pump frequently includes requirements on NPSH (referred to as NPSHR for required) along with the H–Q performance data for a specific pump. If such information is available, it should be used directly instead of calculating $\sigma_v$ as in the example.

An example of a pump performance chart showing values of NPSHR is shown in Figure 3.3. Note that the NPSHR is a function of Q in the figure and that $\sigma_v$ is also a function of Q through $N_s$. Taking the 394-mm diameter pump at 140 m³/hr, one gets $H = 46$ m and $\eta_T = 0.67$. $Q = 0.0389$ m³/s, $N = 153.8$ s⁻¹, and $(\Delta p/\rho) = 451.3$ m²/s²; then $N_s = 0.301$. Note that $D_s = 9.21$, and Cordier gives $D_s = 8.6$ at $\eta_T = 0.69$, which is reasonably close and within Balje's data band.

Now, since the pump is single-suction:

$$\sigma_v = 0.241(0.301)^{4/3} = 0.051$$

$$\text{NPSHR} = \sigma_v H = 2.3 \text{ m}$$

The curves specify about 2.30 m, so agreement is very good.

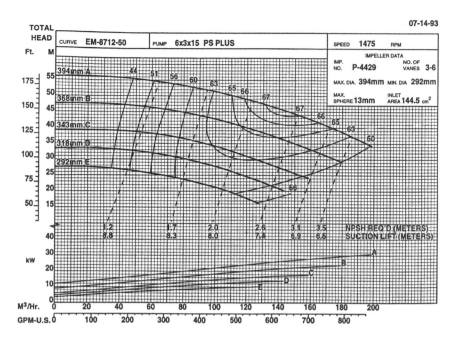

**FIGURE 3.3**  Performance curves for a single-suction Allis-Chalmers pump  (ITT Industrial Pump Group, Cincinnati, OH.)

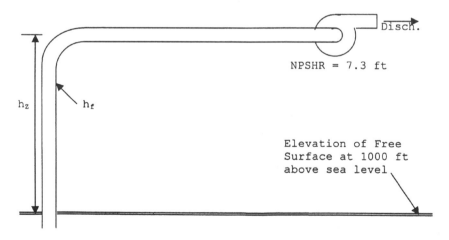

**FIGURE 3.4**  A pump operating at NPSHR = 7.3 ft.

One can boil the whole problem down (substituting NPSHA for $h_{sv}$) if one knows or can estimate the NPSHR (from a pump curve or from $\sigma_v H$). It can be stated that

$$NPSHA = h_T + h_z - h_{vp} - h_f \geq NPSHR \qquad (3.41)$$

where $h_T$ is the atmospheric or supply tank pressure. Take a look at some examples.

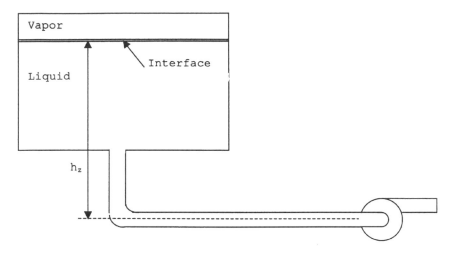

**FIGURE 3.5**  A closed tank pumping problem with a volatile fluid.

First, examine a pump that requires NPSHR = 7.3 ft for a given flow rate. Install the pump at an altitude of 1000 ft and pump the water at 85°F (see Figure 3.4). Find the maximum "lift" of the pump from an open pond or tank. 85°F water gives $h_{vp}$ = 1.38 ft; 1000 ft of altitude gives $h_T$ = 32.8 ft. There is no frictional loss to consider, so $h_f$ = 0, and

$$NPSHA = 32.8 + h_z - 1.38 - 0 \geq 7.3 \text{ ft} \geq NPSHR$$

$$h_z \geq 7.3 + 1.38 - 32.8 = -24.1 \text{ ft (lift)}$$

If there was a friction loss in the inlet pipe of 3 ft, then

$$h_z \geq 7.3 + 1.38 - 32.8 + 3 = -21.1 \text{ ft}$$

or one could only lift the water 21 ft instead of 24 ft.

Consider another example where there is a high-temperature fluid in a closed, evacuated tank so that the "cover gas" pressure is simply the vapor pressure, $h_T = h_{vp}$. One wants to move the fluid with a pump requiring NPSHR = 9.1 ft. If there is an inlet side friction drop of 1 ft, calculate the height of the tank above the pump inlet required to avoid cavitation (see Figure 3.5). Begin with Equation (3.41).

This reduces to:

$$NPSHA = h_{vp} + h_z - h_{vp} - h_f = h_z - h_f \geq NPSHR$$

$$h_z \geq NPSHR + h_f = 9.1 \text{ ft} + 1 \text{ ft} = 10.1 \text{ ft}$$

The tank must have an elevation above the pump centerline of at least 10.1 ft (say 11 ft for conservatism).

Now look at a pump that must move gasoline (S.G. = 0.73). At its temperature, the gasoline has a vapor pressure of $\rho g h_{vp}$ = 11.5 psi. The gasoline is in a vented tank at sea level and the free surface is 2 ft above the pump inlet; $h_f$ = 3 ft. Will the gasoline cavitate if the pump has NPSHR = 9.1 ft?

Use Equation (3.41):

$$NPSHA = h_T + h_z - h_{vp} - h_f \geq NPSHR$$

where

$$h_T = \frac{p_b}{(\rho g)} = \frac{h_b}{S.G.} = \frac{34 \text{ ft}}{0.73} = 46.6 \text{ ft} \qquad \text{(of gasoline)}$$

The given value of vapor pressure must be converted to head as

$$h_{vp} = \frac{11.5 \text{ psi}}{(\rho g)}$$

$$= \frac{11.5 \times 144 \text{ lbf/ft}^2}{(0.73 \times 62.4 \text{ lbf/ft}^3)}$$

$$h_{vp} = 36.4 \text{ ft}$$

So,

$$NPSHA = (46.6 + 2 - 36.4 - 3) \text{ ft} = 9.2 \text{ ft} \geq 9.1 \text{ ft}$$

The pump works in the system without cavitation, although a margin of only 0.1 ft is not considered to be very good design practice. Workers in the field suggest a 10% margin, or about 0.91 ft. If $h_f$ were equal to 4 ft, one would have to raise the tank (or lower the pump).

## 3.5  SUMMARY

The concepts of scaling turbomachinery performance according to a fixed set of rules was introduced. The previously developed rules of similitude were reformulated to generate a set of fan laws or pump affinity rules for machinery with essentially incompressible flows, and examples were presented for illustration.

The major restrictions on scaling of performance with changes in speed, size, and fluid properties were identified in terms of the matching of Reynolds and Mach numbers between a model and a prototype. This precise condition for similitude and permissible scaling was extended to embrace the idea of a "regime" of Reynolds and Mach numbers for which scaling will be sufficiently accurate. Several algorithms by which the simple scaling rules could be extended to account for the influence of significant changes in Reynolds Number on efficiency and performance of fans pumps and compressors were presented and illustrated with examples. The influence of Mach Number and allowable disparities between values for a model and a prototype was reviewed for fans and compressors, with clear limitations governing the accuracy of scaled performance. Corrections for compressibilty effects on scaling between model and prototype were developed and illustrated with examples. The scaling rules for fully compressible flow machines were developed and presented in the dimensional form in traditional use for high-speed compressors and turbines.

For liquid pumps and turbines, the phenomenon of boiling cavitation was introduced in terms of pressure in the flowfield and the temperature-dependent vapor pressure of the fluid. The influence of low-pressure regions near the suction or entry side of a pump (discharge side of a turbine) was related to the onset of conditions favorable to cavitation or local boiling and recondensation of the flowing fluid. The governing parameters of net positive suction head (NPSH or NPSHA) at the pump impeller inlet and the value of NPSH needed to preclude the development of cavitation (NPSHR) were presented. The semi-empirical prediction of NPSHR (Thoma, 1944) for a pump, in terms of the specific speed of the pump, was presented and illustrated with a range of examples to provide a uniform approach to analysis of cavitation problems. Examples for design and selection of pumping equipment were also provided.

## 3.6   PROBLEMS

3.1. The pump of Figure P1.1 is to be sized requiring geometric similarity such that at its best efficiency point, it will deliver 300 gpm at a speed of 850 rpm. Show that the diameter should be 16.7 inches and the head will be 58.8 feet.

3.2. Calculate the specific speed and specific diameter for the pump of Problem 3.1 and compare the results to the Cordier diagram. Comment on the comparison of efficiency.

3.3. The characteristic curve of a forward-curved bladed furnace fan can be approximately described as (see Figure P3.3)

$$\Delta p_s = \frac{\Delta p_{s\,BEP}\left(1 - \dfrac{Q^2}{Q^2_{max}}\right)}{\left(1 - \dfrac{Q^2_{bep}}{Q^2_{max}}\right)}$$

(a) Derive the above equation from the sketch by assuming a parabolic curve fit of the form

$$\Delta p_s = a + bQ^2$$

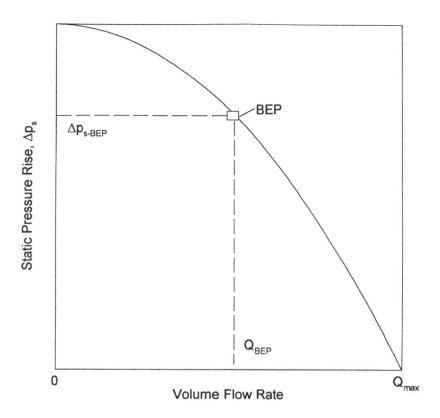

**FIGURE P3.3**   Characteristic curve for a forward-curved blade furnace fan.

(b) For a particular fan, $\Delta p_{sBEP} = 1$ InWG, $Q_{BEP} = 2000$ cfm, and $Q_{max} = 3500$ cfm. The fan supplies air ($\rho g = 0.075$ lbf/ft$^3$) to a duct-work system with D = 1 ft, L = 100 ft, and f = 0.025. What flow rate will the fan provide?

(c) If the duct diameter in part (b) were D = 1.5 ft, what flow rate would result?

3.4. If the fan of Problem 3.3 is pumping air at 250°F and 2116 lbf/ft$^2$ inlet conditions, what volume flow rate and static pressure rise will it move through the same duct?

3.5. A model fan is tested at 1200 rpm in a 36.5-inch diameter size to meet AMCA rating standards. This fan delivers 13,223 cfm at a total pressure rise of 2.67 InWg (standard air density) with a total efficiency of 89%.

(a) What type of fan is it?

(b) If the model is scaled to prototype speed and diameter of 600 rpm and 110 inches, respectively, what will the flow rate and pressure rise be (standard density)?

(c) If both model and prototype are run in standard air, what is the estimated efficiency for the prototype fan?

3.6. The fan described in Problem 3.5 is to be modified and used in a low-noise, reduced-cost application. The diameter is to be increased slightly to 40 inches, and the speed is to be reduced to 885 rpm. The tip clearance of the fan is to be increased to 0.125 inches. Recalculate the performance and, based on the scaling laws, include the effects of tip clearance and Reynolds Number on efficiency and required horsepower.

3.7. In a large irrigation system, an individual pump is required to supply 200 ft³/s of water at a total head of 450 ft. Determine the type of pump required if the impeller runs at 450 rpm. Determine the impeller diameter. Estimate the required motor horsepower.

3.8. Verify in detail the curve fits used for the Cordier $N_s$–$D_s$ relations. Generate a curve fit for the $\eta_T$–$D_s$ curve.

3.9. A centrifugal pump must deliver 250 ft³/s at 410 ft of head running at 350 rpm. A laboratory model is to be tested; it is limited to 5 ft³/s and 300 hp (water with $\rho g = 62.4$ lbf/ft²). Assuming identical model and full-scale efficiency, find the model speed and size ratio.

3.10. A hydraulic turbine must yield 30,000 hp with 125 ft of head at 80 rpm. The lab allows 50-hp models at a head of 20 ft. Determine the model speed, the size ratio, and the model flow rate.

3.11. A pump that was designed for agricultural irrigation is to be scaled down in size for use in recirculation of water in a small aquarium for pet fish. The irrigation pump supplies 600 gpm at 100 ft of head and is a well-designed pump, with an efficiency of 75% at 1200 rpm. The aquarium pump must deliver 6 gpm at 5 ft of head.
(a) Find the proper speed and diameter for the aquarium pump.
(b) Estimate the efficiency of the aquarium pump.
(c) How do you expect the small pump to deviate from the simple scaling laws?

3.12. If the fan of Problem 3.5 is scaled to a prototype speed and diameter of 1800 rpm and 72.0 inches, what will the flow and total pressure rise be? Estimate $\eta_T$.

3.13. Shown in Figure P3.13 are the performance curves of an Allis-Chalmers single-suction pump.
(a) Estimate the head-flow curve for a 10-inch diameter impeller and plot the result on the curve.
(b) Calculate the $N_s$ and $D_s$ values for the BEP of the 9-inch, 9.6-inch, and 10-inch curves. Compare the experimental efficiencies to the Cordier values (based on the $\eta_{TC}$ parabolic algorithms—the Alternate Efficiency equations).
(c) Can you use a 10-inch impeller with the 5-hp motor? Support your answer with suitable calculations.

3.14. A small centrifugal pump is tested at $N = 2875$ rpm in water. It delivers 252 gpm at 138 ft of head at its best efficiency point ($\eta = 0.76$).
(a) Determine the specific speed of the pump.
(b) Compute the required input power.
(c) Estimate the impeller diameter.
(d) Sketch two views of the impeller shape.

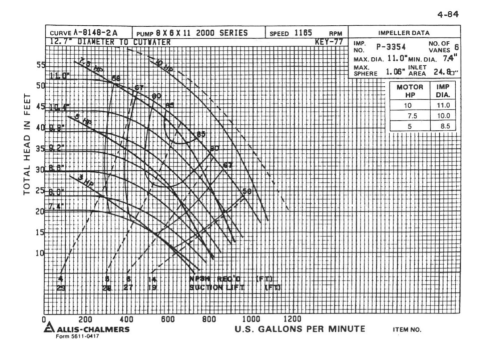

**FIGURE P3.13**   Performance curves for a single-suction Allis-Chalmers pump. (ITT Industrial Pump Group, Cincinnati, OH.)

3.15. Two low-speed, mixed-flow blowers are installed to operate in parallel flow, pumping chilled air (5°C) through a refrigerated truck trailer filled with red, seedless grapes in plastic bags. The resistance to flow through the fairly tightly packed cargo yields a K-factor of 150.0 with an effective cross-sectional area of 5 m².

   (a) If the required flow rate for initial cool-down is 10.0 m³/s, calculate the net pressure rise required.

   (b) After initial cool-down, the flow rate requirement can be supplied by only one of the two fans. What flow and pressure rise will result?

   (c) Select a suitable fan size with N = 650 rpm for this installation.

3.16. If one uses the pump of Problem 2.16 to pump jet engine fuel, one would expect a reduction in efficiency due to higher viscosity or reduced Reynolds Number. Assuming a specific gravity for JP-4 at 25°F of 0.78 and a kinematic viscosity 1.4 times that of water (see Appendix A), determine if the 10-hp motor will be adequate to drive the pump at 1750 rpm and 250 gpm, within a standard overload factor of 15%.

3.17. If one can run the model test of Problem 2.9 in a closed-loop, variable-pressure test duct, what pressure should be maintained (at the same temperature) to achieve complete dynamic similitude?

3.18. A model pump delivers 20 gpm at 40 ft of head running at 3600 rpm in 60°F water. If the pump is to be run at 1800 rpm, determine the proper

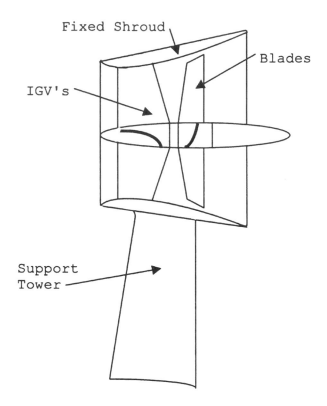

**FIGURE P3.19**   A shrouded wind turbine.

water temperature to maintain dynamic similarity. Find the flow rate and head under these conditions.

3.19. A shrouded wind turbine operates in a 30 mph wind at standard air density. The shroud allows the 30-ft diameter turbine to capture a volume flow rate of twice its area times the freestream velocity, $V_¥$. According to the Betz limit, the optimum head recovered by the turbine should be 59% of the freestream velocity pressure.

   (a) Calculate the fluid power captured by the turbine in kilowatts.

   (b) With the turbine running at 90 rpm, estimate its efficiency and net power output (kW).

3.20. For the model and prototype fans of Problem 3.5, use the ASME, ICAAMC and Wright models to estimate efficiency for the full-scale fan. Assume that the model efficiency of $\eta_T = 0.89$ is accurate. Compare these methods to the alternate efficiency model used to modify the Cordier value of total efficiency, assuming that $\delta = 1.0$. Compare and discuss the various results for the full-scale prototype fan. (Hint: Assume hydraulically smooth surfaces.)

3.21. A 40-inch diameter axial fan operating at 965 rpm has a tip clearance of 0.10 inches. The fan produces 15,000 cfm at 2.0 InWG. Estimate the total efficiency of the fan for a range of surfaces roughness values from $\varepsilon/D = 0.001$ to $0.01$. Develop a graph of $\eta_T$ and $\eta_s$ vs. $\varepsilon/D$. Can you achieve a value of $\eta_T = 0.80$? Select a combination of parameters to achieve this efficiency and defend your choice on a design decision basis. (Hint: Begin your efficiency calculations based on the Cordier efficiency and modify using clearance and Re, assuming initially that roughness is negligibly small.)

3.22. The very large pump described in Problem 3.7 is to be tested in air to verify the manufacturer's performance claims. Predict the static pressure rise (in InWG), the flow rate (in cfm), and the static efficiency running in standard air (68°F).

3.23. A pump is designed to provide 223 gpm at 98 ft of head in 85°F water. The 9.7-inch impeller operates at 1750 rpm. Estimate the NPSHR and maximum suction lift assuming no inlet losses.

3.24. A double-suction version of the pump in Problem 3.23, providing 450 gpm at 95 ft of head, has upstream head losses of 5 ft. Estimate the NPSHR and maximum suction lift for 85°F water. Repeat for 185°F water.

3.25. The pump of Problem 3.23 must pump gasoline (S.G. = 0.73, $h_{vp}$ = 11.5 psi) from an evacuated tank. Find the NPSHR and maximum suction lift.

3.26. The influence of water temperature on cavitation in pumps is related to the behavior of $h_{vp}$, the vapor pressure head, with temperature. NPSHR for a pump must then be rated by the manufacturer at some stated temperature. For example, the Allis-Chalmers pump on Figure P2.12 is rated at 85°F. For any other temperature, this value must be corrected. For moderate temperature excursions (35°F < T < 160°F), the correction is estimated by NPSHR = $\text{NPSHR}_{85} + \Delta h_{vpT}$, where $\text{NPSHR}_{85}$ is the quoted value and $\Delta h_{vpT}$ is the difference in vapor pressure at 85°F and the value of T being considered ($\Delta h_{vpT} = h_{vp}(T) + h_{vp}(85°)$). This is approximately equal (in water) to:

$$\Delta h_{vp} = 5\left(\frac{T}{100}\right)^2 - 3.6 (\text{ft})$$

The $\text{NPSHR}_{85}$ values for the Allis-Chalmers pump vary, of course, with flow rate and can be approximated by:

$$\text{NPSHR}_{85} = 1.125 + 6.875\left(\frac{Q}{1000}\right)^2$$

with Q in gpm, for the 12.1-inch diameter impeller.
(a) Verify the equation for $\Delta h_{vpT}$.
(b) Verify the equation for $\text{NPSHR}_{85}$.
(c) Develop an equation for $\text{NPSHR}_{85}$ for the 7.0-inch diameter impeller.

3.27. Using the information in Problem 3.26, find the following.
(a) Develop a curve for the maximum allowable flow rate (no cavitation) for the 12.1-inch impeller as a function of temperature, if NPSHA = 34 ft, 24 ft, and 14 ft.

(b) Comment on the allowable suction lift for these conditions.

3.28. For the information in Problem 3.27, discuss the influence of $T < 85°F$ water on the results in (a), and compare to the effects of water at $T > 85°F$.

3.29. For the pump of Problem 3.26, the values of NPSHR at a given flow rate are much higher for the small pump (7-inch diameter impeller) than for the large pump (12.1-inch diameter impeller). Explain this in terms of Thoma correlation for NPSHR. Does the variation for NPSHR for either impeller make sense in terms of Thoma's work? Show all work.

3.30. Rework Problem 2.4 considering compressibility.

3.31. A small double-suction pump has a design point rating of $H = 60$ ft at $Q = 75$ gpm, running at 3600 rpm. It will be installed at some height above an open pond to pump cold water into an irrigation ditch. Atmospheric head is 33 ft, vapor head is 1.4 ft, and the suction line head loss is 4 ft. Estimate the maximum height of the pump above the pond.

3.32. Correct the results of Problem 2.5 for compressibility using both isentropic and polytropic pressure–density laws. Comment on the comparison of these results to each other and to the incompressible calculations. What are the economic implications in using an incompressible result?

3.33. A high-pressure blower supplies 2000 cfm of air at a pressure rise of 42 InWG, drawing the ambient air through a filtering system whose upstream pressure drop is 32 InWG. Assume a polytropic pressure change upstream of the fan inlet with $n = 1.66$ to estimate inlet density from ambient conditions. The ambient air properties are $T = 80°F$ and $p_{amb} = 30$ InHg. Select a single-stage, narrow centrifugal fan for this application, and estimate diameter, speed, efficiency, and sound power level. (Hint: Size a fan using inlet density and correct the required flow and pressure for compressibility.)

3.34. For the pump described in Problem 1.1, develop values of NPSHR at the tabulated flow rates for:
(a) A single-suction pump.
(b) A double-suction pump.

3.35. A pump must operate in the system shown in Figure P3.35 and deliver 1 ft$^3$/s of water. This pump is to be a single-suction type, direct-connected to an a/c induction motor ($N = 600, 900, 1200, 1800,$ or 3600 rpm).
(a) Find the direct-connected speed for which the pump will not cavitate. Use Thoma's criterion to size NPSHR $= 1.1\sigma_v H$ where 1.1 is a "safety factor."
(b) Determine the pump diameter, pump type, and probable efficiency from Cordier.

3.36. One must pump warm gasoline out of an underground tank with a submerged centrifugal pump. The tank is effectively vented to atmospheric pressure (although for environmental correctness, a carbon filter is installed in the vent). If the NPSHR of the pump is 8 m, the vapor pressure of the gasoline is 40.0 kPa, and the upstream friction head losses are 0.8 m, estimate the required submersion depth of the pump to avoid cavitation. The specific gravity of the gasoline is 0.841, and the barometric pressure is 100 kPa.

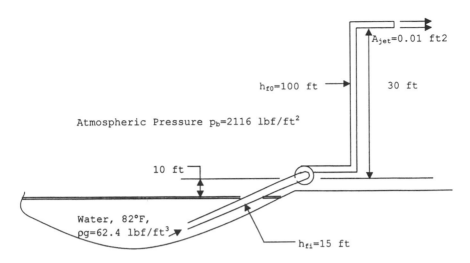

FIGURE P3.35  A pump-system configuration.

3.37. There is a large centrifugal fan, designed to operate at 600 rpm, with an
      impeller diameter of 60 inches. The fan uses 40 hp at 60,000 cfm and 3.0
      InWG static pressure rise. If one installs the fan with a variable-frequency
      motor that can operate at any speed between 600 rpm and 7200 rpm,
      (a) estimate the flow rate, static pressure rise, and power required at 1200
          rpm.
      (b) estimate the change in efficiency at 1200 rpm compared to the value
          at 600 rpm.
      (c) at what speed (in rpm) would the scaled performance need to be
          corrected for compressibility effects?
3.38. A fan is tested for certification of performance with a 1.25-m diameter
      running at 985 rpm. The fan has a BEP performance of 6.0 m³/s at 700
      Pa and a total efficiency of 0.875.
      (a) If the model is scaled to prototype speed and diameter of 600 rpm
          and 2.5 m, what will be the flow rate and pressure at standard density?
      (b) Estimate the prototype efficiency.
3.39. A three-speed fan, when operating at its highest speed of 3550 rpm,
      supplies a flow rate of 5999 cfm and a total pressure rise of 30 lbf/ft² with
      $\rho = 0.00233$ slugs/ft³. Lower speeds of 1750 rpm and 1175 rpm are also
      available.
      (a) Estimate the size of the fan and its efficiency at 3550 rpm.
      (b) Estimate the flow rate and pressure rise at the low speed of 1175 rpm.
      (c) Use the Reynolds Numbers of the fan at 1750 rpm and 1175 rpm to
          estimate the efficiency of the fan at these lower speeds.
3.40. A double-suction pump is designed to provide 500 gpm at 110 ft of head
      in 85°F water. The 9.7-inch impeller operates at 1750 rpm. Estimate the
      NPSHR and maximum suction lift assuming no inlet side friction losses.

# 4  Turbomachinery Noise

## 4.1  INTRODUCTORY REMARKS

In recent years, it has become necessary for designers and users of turbomachinery to understand and predict the acoustic as well as aerodynamic performance of turbomachines. A great deal has been written on the subject of acoustics, as well as acoustics in turbomachinery. This chapter will introduce some general concepts and then examine sound production and predictions in turbomachines. The chapter will focus on the acoustics of fans and blowers, with the understanding that the acoustic phenomena in other types of turbomachines behave in a similar fashion. Interested readers are encouraged to consult the extensive reference list at the end of this book for recommended books and papers on acoustics in turbomachinery.

## 4.2  DEFINITION OF SOUND AND NOISE

In a physical sense, sound is a small amplitude vibration of particles in a gas, solid, or liquid (Broch, 1992). The vibration can be thought of as a transfer of momentum from particle to particle in air. Due to the elastic nature of the bonds between air molecules, sound waves travel at a finite propagation speed. In air at 20°C, the propagation speed, or "speed of sound," is 343 m/s (1126 ft/s). The propagation of the vibration occurs as a density/pressure variation in a longitudinal fashion. The variation is a sinusoidal function, and it is periodic and of constant amplitude. The corresponding "frequency spectrum" associated with the sine waves is often a useful piece of data in acoustic analysis. The spectrum is an illustration of the frequencies and their corresponding magnitude that make up a signal. Spectral analysis is useful in understanding the effect of discrete frequency phenomena as well as the bandwidth of background noise.

One common type of wave formation found in acoustics is the spherical wave. This type of wave can be thought of as a pulsing sphere emanating from a point source and can be used to illustrate the relationship between the noise emitted from a source and the sound pressure at a distance, x, from a point source. Assume a noise of power E is emitted from a point source. The power will radiate in a spherical fashion. The power intensity, I, is given by the following formula:

$$I = \frac{E}{\left(4\pi x^2\right)} = \frac{Power}{Area} \qquad (4.1)$$

At a sufficient distance away from a source, I becomes proportional to the square of the sound pressure p at x. Therefore,

$$p = \frac{k}{x} \qquad (4.2)$$

This relationship is known as the "inverse-distance" law for free sound in an acoustic far field. Acoustic literature often refers to near and far fields (Baranek, 1992).

The nature of a field (i.e., near or far) is important in acoustic measurements. If a source is assumed to be in the far field, it can be treated as a point source. The magnitude of an emission sensed or measured from a point source is solely based on the distance from the source. On the other hand, the magnitude of an emission measured in the near field will be affected by the 3-D location of the transducer. To be certain that a given measurement is in the far field, acoustic measurements should be made 2 to 5 characteristic diameters from the source.

One must keep reflection of waves in mind when working with acoustics. Sound waves will reflect from a surface and affect sound pressure levels. The sound pressure at a point in a field is given not only by the pressure of the original wave but also by an additional effect of the reflected wave.

The physical scale of sound pressures covers a dynamic range of approximately 1 to 1,000,000. Therefore, despite the small magnitude of pressure variations (on the order of Pascals), the relative range of pressure variations is large. It is then convenient to utilize a decibel scale to report or characterize sound pressures. The decibel scale is a relative measure based on a ratio of power. The typical relationship utilized for sound pressure is given by:

$$L_w = 10 \log_{10}\left(\frac{p^2}{p_o{}^2}\right) = 20 \log_{10}\left(\frac{p}{p_o}\right) \tag{4.3}$$

Although this is a ratio of pressures, its use is acceptable due to the relationship between the sound power and the square of the sound pressure (in the far field). $p_o$ is a reference pressure, with a value of $2 \times 10^{-5}$ Pa.

The decibel relationship can also be used with reference to sound power levels. The relationship is given by:

$$L_w = 10 \log_{10}\left(\frac{P}{P_o}\right) \tag{4.4}$$

In this equation, $P_o$ is a reference power level with a value of $10^{-12}$ watts (1 picowatt). The sound power level is, due to its relative nature, essentially the same relationship as the sound pressure level (unless the near field is being considered). Table 4.1 gives examples of the relationship between sound pressures and sound pressure levels of some commonly occurring phenomena.

The reader should try to gain a feel for the decibel scale, noting that 20 decibels account for an order of magnitude difference in pressure. Also note that doubling the sound pressure levels (doubling the sound) results in a change of 6 dB. Therefore, a reduction of 6 dBs would reduce the sound by 1/2.

A simple measurement of sound pressure level at a particular frequency, although useful, is not completely definitive due to the lack of knowledge of the frequency

### TABLE 4.1a
### Typical Noise Levels (Thumann, 1986)

| Noise Type | SPL (Pa) | Noise Level (dBA) |
|---|---|---|
| Hearing threshold | $2 \times 10^{-5}$ | 0 |
| Recording studio | $10^{-4}$ | 20 |
| Normal sleep | $10^{-3}$ | 30–35 |
| Living room | $10^{-3}$ | 40 |
| Speech interference, 4 ft | $10^{-2}$ | 65 |
| Residential limit | $10^{-2}$ | 68 |
| Commercial limit | $10^{-1}$ | 72 |
| Air compressor, 50 ft | $10^{-1}$ | 75–85 |
| OSHA 8-hour limit | 1 | 90 |
| Pneumatic hammer (at Operator) | 3 | 100 |
| Airplane (Boeing 707) | 8 | 112 |
| Concorde SST | 40 | 123 |
| Threshold of pain | 110 | 140 |

### TABLE 4.1b
### Detailed OSHA Limits (Beranek and Ver, 1992)

| Duration per day, hours | Sound Level, dBA |
|---|---|
| 8 | 90 |
| 6 | 92 |
| 4 | 95 |
| 3 | 97 |
| 2 | 100 |
| 1.5 | 102 |
| 1 | 105 |
| 0.5 | 110 |
| 0.25 or less | 115 |

distribution of the entire sound emission (Broch, 1971). To make acoustic measurements more useful, a system of frequency weighting utilizing tone-specific weighting curves has been developed. The curves were developed with the ideal of equal loudness (loudness is defined based on the average human physiological response to an acoustic event) levels at all frequencies by considering the effect of frequency as well as magnitude on the level of loudness of a signal. "A scale" weighting has become a standard in acoustical measurements of sound pressure, with the resulting sound pressures having units of dB(A). Generally, in a sound measurement activity, spectrum analysis techniques are utilized, but the weighting system allows for ready comparison of independent acoustic fields, in a single quantity. The A scale and its associated distribution of reduction values is shown in Figure 4.1, along with several similar scales—the B, C, and D scales.

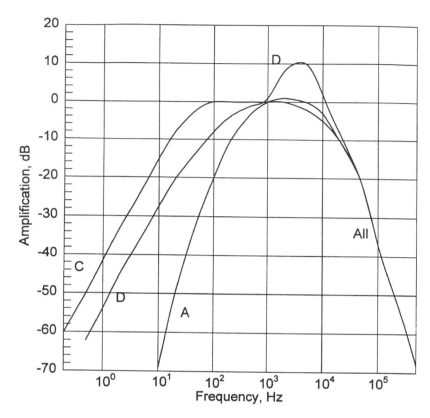

**FIGURE 4.1**   Common "filter" scales for frequency weighting. (Adapted from Broch, 1980.)

## 4.3   FAN NOISE

Axial flow fans are a good example to study with respect to turbomachines due to the wide spectrum of noise associated with fans. There are two general types of noise associated with fans. Broadband and discrete frequency noise are shown in the frequency spectrum illustrated in Figure 4.2.

The broadband is illustrated as the smooth curve, while the discrete frequency noise is illustrated as spikes centered at the blade pass frequency and its harmonics. The contribution of each type of noise is related to the performance characteristics of the fan. Broadband noise arises from sources that are random in nature and can be of two types (Wright, 1976). One source arises from the fluctuations of the aerodynamic forces on the blade, while the second source arises from turbulent flow in the blade wakes. Generally, the force fluctuation source dominates the spectrum.

There are two sources of force fluctuations. The first source arises from the shedding of vorticity at the trailing edge in smooth inflow, which introduces surface pressure fluctuations on the blade. The second source is generated when the blades move through turbulent flow. The turbulence causes random variations in the incidence of flow, relative to the blade, thus creating random pressure loadings and force

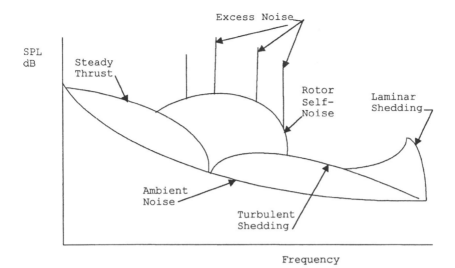

**FIGURE 4.2** Acoustic spectrum for axial flow turbomachines. (Wright, 1976.)

fluctuations. If a body is in non-turbulent flow, the vortex shedding is periodic (the well-known Karman Street). This produces regularly fluctuating lift. As the Reynolds Number is increased, the fluctuations become more random in nature as the flow transitions to turbulence. Also, the effect of turbulent inflow is influenced by the scale of turbulence relative to the chord length.

Since broadband noise is affected by the blade pressure distribution, the incidence angle of the blade will affect the noise levels due to the change in blade load and pressure distribution with the change in incidence angle. In this light, it has been found that fans exhibit minimal noise characteristics when the fan is operated near the design point. An increment of about 5 to 10 dB usually accompanies the fan operating at or near the stall point, or near free delivery. Historically, early investigators tried to model noise solely on the basis of gross performance parameters. Although these methods were a large step forward, they are now being supplanted by schemes that more nearly relate to the aerodynamic and geometric parameters of a machine.

A large number of correlations have been developed to predict noise in various situations. The empirical expressions generally attempt to relate, as simply as possible, the sound output of a rotor to its physical and operating characteristics, with most of the correlations limited to a particular situation.

A general method for the prediction of fan noise (axial and centrifugal) was developed by J.B. Graham beginning in the early 1970s (Graham, 1972, 1991). The method was based solely on performance parameters and was formulated for fans operating at or near design point. Graham tabulated specific sound power levels for various types of fans at octave band center frequencies (see Table 4.2). The specific sound power level was developed to allow for comparison of different types of fans and is defined as the sound power level generated by a fan operating at 1 cfm and 1 InH$_2$O. The table is used with the following correlation:

**TABLE 4.2**
**Tabulated Specific Sound Power Level Data**
**Octave Center Band Frequencies, Hz**

| Fan Type | 63 | 125 | 250 | 500 | 1000 | 2000 | 4000 | 8000 | bpi |
|---|---|---|---|---|---|---|---|---|---|
| Centrifugal airfoil blade | 35 | 35 | 34 | 32 | 31 | 26 | 18 | 10 | 3 |
| Centrifugal backwardly curved blade | 35 | 35 | 34 | 32 | 31 | 26 | 18 | 10 | 3 |
| Centrifugal forward curved blade | 40 | 38 | 38 | 34 | 28 | 24 | 21 | 15 | 2 |
| Centrifugal radial blade | 48 | 45 | 43 | 43 | 38 | 33 | 30 | 29 | 5–8 |
| Tubular centrifugal | 46 | 43 | 43 | 38 | 37 | 32 | 28 | 25 | 4–6 |
| Vane axial | 42 | 39 | 41 | 42 | 40 | 37 | 35 | 25 | 6–8 |
| Tube axial | 44 | 42 | 46 | 44 | 42 | 40 | 37 | 30 | 6–8 |
| Propeller | 51 | 48 | 49 | 47 | 45 | 45 | 43 | 31 | 7–7 |

$$L_w = K_w + 10 \log_{10} Q + 20 \log_{10} \Delta p_s \qquad (4.5)$$

where $L_w$ is the estimated sound power level (dB re $10^{-12}$ watt), $K_w$ is the specific sound power level, Q is the volume flow rate (cfm), and P is the pressure (InWG). In addition, a "blade frequency increment" is added to the octave band into which the blade pass frequency (bpf) falls. This frequency is given by

$$\text{bpf} = \frac{N_B \times \text{rpm}}{60} \qquad (4.6)$$

Therefore, the method is

1. Find the specific $K_w$ values from the table.
2. Calculate $10 \log_{10} Q + 20 \log_{10} \Delta p_s$ and add to the $K_w$ values.
3. Calculate $B_f$ and the bpi and add to the proper bandwidth.

This method is very general, is a good way to estimate noise levels for comparison and selection, and provides an estimate of the spectral distribution.

Although the methods for predicting sound power have historically encompassed the range of theoretical, semi-theoretical, and empirical approaches, most of these methods can be considered too complex for routine analysis (Baade, 1982). In light of Baade's comments, Graham's relatively simple and well-known method for sound power estimation was re-examined. $K_w$ is the vehicle for capturing the intrinsic difference between the various fan types—centrifugal, forward curved, radial bladed, radial tipped, vane axial, tube axial, and propeller. For these groupings of fans, Graham defines expected levels of specific noise $K_w$ in terms of type. Not coincidentally, these types of fans can also be grouped, as noted earlier, in the appropriate "zones" on the $N_s$–$D_s$ Cordier diagram. To use Graham's data directly, it is necessary to assign values of $D_s$ to the fan groupings as delineated by Graham. Using his classifications by type and estimating reasonable specific diameters for them, the

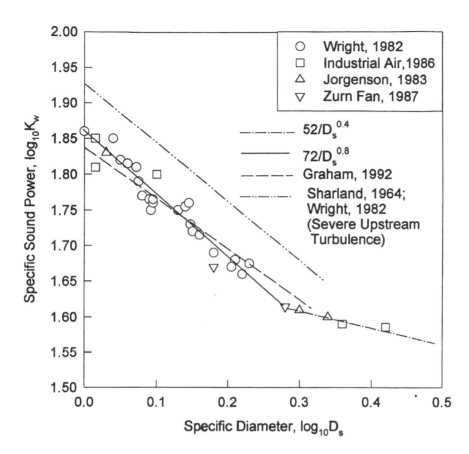

**FIGURE 4.3** $K_w$ results for axial and centrifugal fans.

overall $K_w$ values can be estimated as well. In addition, a large group of axial fans described by Wright (1982), a comparison of centrifugal fans (Beranek, 1992), and additional data from industrial catalogs (Zurn, 1985; Industrial Air, 1986) were analyzed in terms of static pressure, flow rate, and sound power level to generate additional values of $K_w$ across the complete range of $D_s$. The results of these studies are shown in Figure 4.3 as $K_w$ vs. $D_s$. Shown with the data is a linearly regressed curve fit in the form log $K_w = a + b$ log $D_s$. The resulting fit to the data yields

$$K_w = \frac{72}{D_s^{0.8}} \tag{4.7}$$

with a correlation coefficient of −0.90 as reflected in the scatter of the data. The results are reasonably consistent and are within the level of accuracy requirements appropriate to simple selection and preliminary design layout. The data also suggest a leveling off of values near the larger specific diameters. A separate estimate of $K_w$ above $D_s \cong 2.0$ is provided by

$$K_s = \frac{52}{D_s^{0.4}} \tag{4.8}$$

Note that the scatter of the data suggests a probable error in these estimates of perhaps ±4 dB.

Fans with inlet guide vanes (IGVs) and with other forms of extreme inflow disturbances were not included in the database for this correlation. These fans show substantially higher noise levels for the same values of Q and $\Delta p_s$, resulting in levels of $K_w$ far above the rest of the data. If an estimate for fans with IGVs or strong upstream turbulence or disturbances is needed, then the equation for $K_w$ can be modified to

$$K_w = \frac{84}{D_s^{0.8}} \tag{4.9}$$

resulting in an additional 12 dBs (Wright, 1982). This is clearly a punitive result but is relevant to selection or preliminary design when dealing with pre-swirl vanes, severe obstructions, or generally rough inflow.

## 4.4   SOUND POWER AND SOUND PRESSURE

Frequently, the safety and legal constraints applied to noise or sound are written in terms of sound pressure rather than power. For example, the Occupational Safety and Health Administration (OSHA) health and safety standards are given in terms of the sound pressure level that exists at the workstation of an employee. Local regulations may establish maximum values of sound pressure level at the property or boundary line of a plant or other facility. Exposure times per day for a worker are limited to a fixed number of hours per day for a given sound pressure level. Table 4.1 shows these permissible exposure values.

In our development of estimation methods for the noise from fans and blowers, several algorithms for predicting the sound power level, $L_w$, have been discussed. But for purposes of controlling exposure in the workplace, one often needs to be able to estimate sound pressure level ($L_p$) in the vicinity of a fan or blower to meet design or selection criteria. The relation one seeks is based on the diminishing of the signal or wave strength with distance from the noise source.

## 4.5   OUTDOOR PROPAGATION

For a source operating in a "free field," with no reflections from surfaces near the source, the decay rate is given by

$$L_p = L_w + 10 \log_{10}\left(\frac{1}{(4\pi x^2)}\right) + c \tag{4.10}$$

where $c = 0.1$ for x in meters, $c = 10.4$ for x in feet. The distance x is the distance from the source and should be at least three times the characteristic length of the sound source. Although the theory considers the source to exist at a point (a point source), in the case of turbomachinery one can take the impeller diameter, machine length, casing size, or some other physical length of the machine as the characteristic length. Then it is necessary that the point where pressure is to be predicted be at least three or four times that length away from the machine to allow use of the free field equation.

For example, consider a fan whose diameter is 1 m and generates a sound power level of 85 dB. To estimate the free field sound pressure level 5 m away from the fan, one calculates:

$$L_p = 85 + 10 \log_{10}\left[\frac{1}{(4\pi 5^2)}\right] + 0.1 = 60.1 \text{ dB}$$

where dB is the sound pressure level referenced to $p_{ref} = 2 \times 10^{-5}$ Pa. At a distance of 10 m away from the fan, the same calculation will give an $L_p$ value of 54.1 dB. At 20 m away, the sound pressure level would be 48.1 dB, etc. Note that with each doubling of the distance from the source, the sound pressure level is reduced by 6 dB. This is a general result and is a perfect "rule of thumb" for free field noise decay.

Machinery is frequently located in areas where nearby surfaces cause non-negligible reflection of sound waves, so the decay rate of sound pressure in a free field environment must be modified if accurate estimates are to be made. Two factors dominate this effect. First, the presence of reflective walls or a floor near the source can focus the sound waves, yielding an effective increase in pressure. This focusing or concentration is characterized by a Directivity Factor, $\Lambda$, where $\Lambda$ is based on the number of adjacent walls (reflective surfaces).

For example, if the sound source, a fan, is sitting on a hard reflective plane (e.g., a large, smooth concrete pad), then the plane will reflect all of the sound waves that (in a free field) would go down and send them upward. That is, all of the energy will be concentrated into half the original space, thus doubling the sound pressure above the plane. This is expressed as $\Lambda = 2$. A very soft plane, such as grass, sod, or heavy carpet, can absorb much of the downward directed energy rather than reflect it so that the resulting pressure more nearly resembles a free field than a hard single-plane field. (This will be acounted for in a later section.)

If the source is located near two walls or planes, as near a hard floor and a hard, vertical wall, then the pressure waves will not only be reflected up, but they will not be allowed to move both left and right. Instead, the leftward waves will be reflected to the right, causing another doubling of the sound pressure in the field. This is quantified by a Directivity Factor, $\Lambda$, of 4.0. Three walls, as in a corner formed by two vertical, perpendicular walls on a horizontal floor will again double the pressure level for a value of $\Lambda = 8.0$.

These $\Lambda$ values are used to modify the free field decay equation as:

$$L_p = L_w + 10 \log_{10}\left(\frac{\Lambda}{(4\pi x^2)}\right) + c \tag{4.11}$$

If the fan with D = 1 m and $L_w$ = 85 dB were mounted on a large totally reflective, hard plane, the sound pressure 5 m away would be:

$$L_p = 85 + 10 \, \log_{10}\left[\frac{2}{(4\pi 5^2)}\right] + 0.1 = 63.1 \text{ dB}$$

Note that the difference from the free field result is simply $10 \, \log_{10}(2) \cong 3$. For a two-wall situation, the increment is given by $10 \, \log_{10}(4) = 6$, and for the three-wall arrangement, $10 \, \log_{10}(8) = 9$. Clearly, these hard, totally reflective walls have a substantial effect on the local values of sound pressure level and the health and safety risks and annoyance associated with a fan.

If the nearby walls are not really hard (i.e., if they are not totally reflective to sound waves) and some of the pressure energy striking the surfaces is absorbed, then the reflective form of $\Lambda/(4\pi x^2)$ must be modified accordingly. Since a totally absorbing wall or walls would revert the field to a non-reflective environment, the limit of total absorption ($\alpha = 1.0$) would be equivalent to $\Lambda = 1.0$. Similarly, the total reflection value $\alpha = 0.0$ uses the $\Lambda$ values of 2, 4, and 8. When $\alpha$ is "in between," the $\Lambda$ values must be reduced to a modified form. The $\Lambda$ value for a partially absorbing wall can be written as $(2 - \alpha)$. For a two-wall configuration, the factor becomes $\Lambda = (2 - \alpha_1)(2 - \alpha_2)$, where the subscript refers to the properties of each wall. For a three-wall layout, $\Lambda = (2 - \alpha_1)(2 - \alpha_2)(2 - \alpha_3)$ if all three $\alpha$ values are distinct. For walls with the same absorption coefficients, the results simplify accordingly. Examples are given for similar walls in Table 4.3.

If, for example, the three walls have distinct $\alpha$ values of 0.2, 0.4, and 0.6, respectively, then the Directivity Factor would be $\Lambda = (2 - 0.2)(2 - 0.4)(2 - 0.6) = 4.03$. Values of $\alpha$ depend on the wall surface properties and the frequency content of the sound waves striking the surface. Low frequencies may be absorbed to a lesser or greater degree than high frequencies by a given surface. A thorough treatment of absorption or reflection thus requires that one not only know the $L_w$ values but also the details of spectral distribution, as well as the detailed frequency-dependent values of $\alpha$. For a careful treatment of the details, see *Fundamentals of Noise Control* (Thumann and Miller, 1986) or *Noise and Vibration Control Engineering* (Baranek, 1992).

The rougher estimates can be made somewhat cautiously and will be used here to approximate the trends and overall influence of absorption. Thumann and Miller present a range of $\alpha$ values for a variety of surfaces. For example, glazed brick and carpeted flooring are characterized by the values shown here in Table 4.4.

One can use an averaged value or the value near the mid-frequency range, say at 1000 Hz, to characterize the material. Thus, one might assign brick the value $\alpha = 0.04$ or carpet the value $\alpha = 0.37$. Similar non-rigorous values of $\alpha$ for several materials are given in Table 4.5.

As a more complex example, consider the noise level from a vane axial fan in terms of the sound pressure level at a distance of 50 ft from the fan. The fan is resting on a large rough-surfaced concrete pad. Fan diameter is 7 ft, speed is 875

**TABLE 4.3**
**Modified Directivity Factors with Similar Walls**

| No. of Walls | $\alpha = 0$ | $\alpha = 0.5$ | $\alpha = 1.0$ |
|:---:|:---:|:---:|:---:|
| 0 | 1 | 1.0 | 1 |
| 1 | 2 | 1.5 | 1 |
| 2 | 4 | 2.5 | 1 |
| 3 | 8 | 4.5 | 1 |

**TABLE 4.4**
**Sample $\alpha$ Values at Octave Band Center Frequencies**

| Frequency | 125 | 250 | 500 | 1000 | 2000 | 4000 |
|:---|:---:|:---:|:---:|:---:|:---:|:---:|
| Brick, $\alpha$ | 0.03 | 0.03 | 0.03 | 0.04 | 0.05 | 0.07 |
| Carpet, $\alpha$ | 0.02 | 0.06 | 0.14 | 0.37 | 0.60 | 0.65 |

**TABLE 4.5**
**Approximate $\alpha$ Values**

| Material | Overall $\alpha$ Value |
|:---|:---:|
| Brick | 0.04 |
| Smooth concrete | 0.07 |
| Rough concrete | 0.30 |
| Asphalt | 0.03 |
| Wood | 0.10 |
| Heavy carpet | 0.37 |
| Carpet with foam rubber pad | 0.69 |
| Plywood panels | 0.10 |
| Plaster | 0.04 |

rpm, flow rate is 100,000 cfm, pressure is 3.6 InWG (static; 4.0 total), and the fan has 11 blades. Using the simple methods, one can estimate sound power of the source and the decay to sound pressure level 50 ft away. Sound power is given by

$$L_w = \frac{72}{D_s^{0.8}} + 10 \log_{10}(Q) + 20 \log_{10}(\Delta p_s) \qquad (4.12)$$

and $D_s$ is about 1.7. The resulting sound power level for the fan is 108 dB. Using the approximate value of $\alpha = 0.3$ for rough concrete gives $\Lambda_{mod} = 1.7$. Then the $L_p$ calculation yields the approximate value of $L_p = 76$ dB.

Thumann and Miller give the octave bandwidth values for $\alpha$ for rough concrete, and one can estimate the octave bandwidth power levels of the fan using the full

**TABLE 4.6**
**Detailed Bandwidth Calculations**

| Frequency | 63 | 125 | 250 | 500 | 1000 | 2000 | 4000 | 8000 |
|---|---|---|---|---|---|---|---|---|
| $\alpha$ | 0.25 | 0.36 | 0.44 | 0.31 | 0.29 | 0.39 | 0.25 | 0.12 |
| $K_w$ | 42 | 39 | 41 | 42 | 40 | 37 | 35 | 25 |
| $L_w - K_w$ | 61 | 61 | 61 | 61 | 61 | 61 | 61 | 61 |
| bpf | 0 | 6 | 0 | 0 | 0 | 0 | 0 | 0 |
| $\Lambda_{mod}$ | 1.75 | 1.64 | 1.56 | 1.69 | 1.71 | 1.61 | 1.79 | 1.88 |
| $L_{p-oct}$ | 60.3 | 63.0 | 59.3 | 59.3 | 58.4 | 55.1 | 53.6 | 43.8 |

Graham method. Table 4.6 summarizes the details of the calculation. ($L_w - K_w$) is simply $10 \log_{10} Q + 20 \log_{10} \Delta p_s$, and bpf is the blade pass frequency increment in the octave containing the frequency ($N_B$ rpm /60). The octave increments are summed according to $L_p = 10 \log_{10} (\Sigma 10^{Lp-oct/10})$ to yield $L_p = 78$ dB. The approximate result of 76 dB compares favorably with this more detailed, more definitive result.

## 4.6   INDOOR PROPAGATION

When considering the propagation or decay of noise inside a room, the question of the influence of numerous wave reflections and reverberation is raised. One can account for these effects through the use of a reflective "Room Constant" or factor, R. (One could call it a wall factor, but Room Constant is the accepted term.) R is based on the total surface area of the walls, S, and a coefficient of absorption as considered earlier. The same complexities of frequency dependence pertain as before, and the different walls can be made of different materials as well. Keeping things relatively simple for now, and assuming the surfaces to be at least very similar, R is defined as:

$$R = \frac{S\alpha}{(1-\alpha)} \tag{4.13}$$

The pressure level equation is subsequently modified to the form:

$$L_p = L_w + 10 \log_{10}\left[\frac{1}{(4\pi x^2)} + \frac{4}{R}\right] + c \tag{4.14}$$

If, for example, the walls are totally absorptive, R approaches infinity and the "correction" disappears from the equation as expected. Propagation or decay is calculated exactly as in a free field with $\Lambda = 1$. For perfectly reflective walls, the value of $\alpha$ is 0.0 and R becomes 0, implying that the sound pressure will build indefinitely in the room. Fortunately, no materials are totally reflective, so the singular limit is not seen in practice.

Consider the problem of a large room $50 \times 100 \times 300$ ft having a 6-ft diameter fan mounted on one of the $50 \times 100$ ft walls. The fan has a sound power level of $L_w = 105$ dB. It is necessary to estimate the sound pressure at the center of the floor of the room where there is a workstation. The walls and floor are wood and one can use an average value of $\alpha$ for plywood panels of 0.10. Surface area S calculates to $S = 100,000$ ft², so that:

$$R = \frac{S\alpha}{(1-\alpha)} = 11,111 \text{ ft}^2$$

With the fan mounted in an end wall far from the adjacent walls, one approximates, as described earlier, $\Lambda = (2 - \alpha) = 1.9$. Then, calculate $L_p$ as

$$L_p = 105 + 10 \log_{10}\left(\frac{1.9}{\left(4\pi152^2\right)} + \frac{4}{11,111}\right) + 10.4 = 81 \text{ dB}$$

Harder walls made of finished plaster over concrete with $\alpha = 0.04$ yields R = 4167 and $L_p = 85$ dB. For comparison, fully absorptive surfaces would yield $L_p = 61$. A smaller room $30 \times 20 \times 10$ ft with the same fan installed would yield $L_p = 101$ dB with plywood walls and $L_p = 108$ dB with plastered concrete walls. Clearly, the reverberation effects can strongly influence the sound pressure results for a given fan or blower.

In general, if the dimensions of a room and the values for the room surfaces are known, a simple calculation of R allows a convenient calculation of the influence of the installation of a fan or blower. Figure 4.4, taken from Thumann and Miller (1986), allows a quick estimation of the decay of $L_p$ with distance from the source as $(L_p - L_w)$ versus x (in feet) with R as the parameter in ft². As seen in the figure, for small values of R, the room is largely reverberant and not much decay is evident. For example, with R = 50 ft², any position within about 20 ft of the source will see $L_p$ levels greater than $L_w$. Beyond about 20 ft, no further decay occurs. For R = 200 ft², a reduced signal $(L_p - L_w)$ is seen beyond about x = 10 ft. Gradual decay continues until about x = 100 ft, where the decay has reached about 7 dB. These numbers can be used to determine the room properties for a given constraint on $L_p$ if $L_w$ and size are known.

Consider an example where a room has dimensions of $12 \times 36 \times 120$ ft, so that the total surface area is $S = 10,440$ ft². With a sound source in the center of the ceiling (an exhaust fan) of strength $L_w = 86$ dB, the constraint on noise requires that the Lp value near an end wall be less than 60 dB. This constraint is for intelligible normal speech at 3 ft. Since the fan is exposing only the inlet side to the room, one assumes that only half of the total sound power is inserted into the room that $L_w = 83$ dB, and the requirement is for $(L_p - L_w) = -23$ dB. Given that $S = 10,440$ ft² and the distance from the source is about 60 ft, a quick check on the free field noise shows that the maximum attenuation is given by $L_p = 83 + 10 \log_{10} (1/(4\pi\ 3600))$

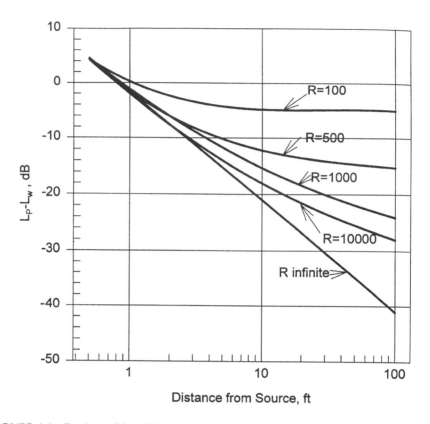

**FIGURE 4.4**  Regions of free field and far field. (Thumann and Miller, 1986.)

+ 10.4 = 46 dB. This value is below the allowable value of 60 dB. For stronger noise sources, greater than 97 dB, the lowest achievable levels—even in an anechoic room ($\alpha = 1.0$)—will be greater than the required 60 dB. The criterion could not be met and the choices would be to select a fan with $L_w$ below 97 dB, or to install a muffler or silencer on the fan as a first step toward meeting the design criterion.

For the 83-dB fan, the ($L_p - L_w$) requirement is −23 dB at x = 60 ft. Figure 4.4 shows that the Room Constant must be about 10,000 ft$^2$ or better. Since $\alpha = (R/S)/(1 + R/S)$, one requires a room whose surface properties yield a value of nearly $\alpha = 0.5$. That probably implies the need for extensive "acoustical treatment" of all room surfaces, but—at cost—the goal can be achieved. As implied, the criterion for the example might best be interpreted as a limitation on the allowable sound power of the fan (or other device), particularly if the room structure is a given. For rough concrete walls ($\alpha = 0.3$), a paneled ceiling ($\alpha = 0.1$), and a carpeted floor ($\alpha = 0.37$), then a weighted absorption value can be estimated according to

$$\alpha = \sum \frac{\alpha_{surf} S_{surf}}{S} \qquad (4.15)$$

where the "surf" subscript implies summing the values for each of the six surfaces of the room. For this example, the calculation is

$$\alpha = \frac{\left(0.3(2 \times 12 \times 36) + (2 \times 12 \times 120) + 0.37(36 \times 120) + 0.1(36 \times 120)\right)}{10,440} = 0.30$$

The Room Constant becomes R = 0.30 × 10,440/0.7 = 4,474. Figure 4.4 indicates the attenuation at 60 ft from the fan will be about 20 dB. The allowable fan source noise in the room is thus limited to about 80 dB. One must select a fan for which $L_w$ = 83 dB or less.

   If dealing with a fan that has a sound power level above 83 dB, then as mentioned before, one has to absorb or attenuate some of the sound power at the source itself—the fan. An example of a fan inlet/outlet silencer for a series of high-pressure blowers is shown in Figure 4.5. Table 4.7 shows the attenuation values for each octave band. Using an estimate of the sound pressure broken down in octave bands allows these silencer values to be subtracted from the fan to arrive at a new estimate of the source sound power level. Table 4.8 summarizes the calculations necessary to estimate the noise reduction achievable using a silencer from Table 4.7 on a pressure blower providing 3450 cfm (inlet density) at 56 InWG pressure rise. One can estimate the specific diameter at about 3, with specific speed of 0.7. The corresponding efficiency is approximately 0.90 based on the Cordier line. Using the high specific diameter formula to predict specific sound power level gives $K_w \cong 52/D_s^{0.4} = 35$ dB. The contribution of flow and pressure rise give another 70 dB, yielding Lw $\cong$ 105 dB. However, to use the silencer attenuation values, the octave band breakdown for the sound power levels is required. These are shown in Table 4.8 along with the Size-10 silencer values. Note that the specific diameter provides an estimate on tip diameter of 26 inches; with d/D = 0.4, the inlet diameter should be about 10 inches or a nominal Size-10. The overall value for the fan sound power level (row 4, Table 4.8) is estimated as $L_w$ = 111 dB. From the net $L_w$ values, the overall sound power level is 103 dB, a reduction of about 8 dB. This reduction could be nearly doubled by adding another Size-10 silencer in series with the first, yielding a source noise of about 95 dB. The process of adding more silencers becomes self-defeating in terms of size and bulk, cost, added flow resistance, and the flow-induced self-noise of the silencers. Figure 4.5 shows the pressure blower and the silencer used in this example.

## 4.7   A NOTE ON PUMP NOISE

The consideration of noise in pumps (see Karasik, 1986) falls roughly into three categories. The first is the airborne or liquid-borne noise as it propagates in the pump environment; it is considered primarily a problem similar to the propagation of fan noise mentioned earlier in the chapter. Given a suitable estimate or manufacturer's rating for a given pump, one can treat airborne noise as fan noise. Potential liquid-generated noise forms include flow separation, turbulence, impeller interaction with the cutwater of the volute, flashing, and cavitation. The broadband noise of a pump

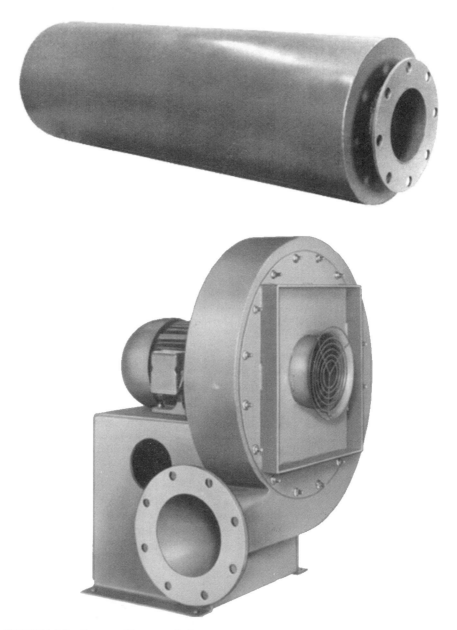

**FIGURE 4.5**   Pressure blower and silencer. (New York Blower.)

generally arises from the turbulent flow over blades and volute surfaces and is worst
at high velocities. Discrete frequency noise is generally associated with the blade
passage frequency, and the cutwater interaction and is worst under conditions of
very high head requirement. Noise associated with flow cavitation in the eye or
impeller will be perceived as intermittent high-intensity noise at relatively high
frequency, typically heard as a snapping or crackling noise. Its appearance in the

**TABLE 4.7**
**Silencer Attenuation Values for Several Sizes (New York Blower)**

| Size | 63 Hz | 125 | 250 | 500 | 1000 | 2000 | 4000 | 8000 |
|------|-------|-----|-----|-----|------|------|------|------|
| 4 | 4 | 18 | 26 | 34 | 37 | 30 | 23 | 21 |
| 6 | 2 | 14 | 23 | 32 | 34 | 29 | 25 | 23 |
| 8 | 1 | 11 | 21 | 30 | 31 | 29 | 26 | 25 |
| 10 | 2 | 14 | 23 | 32 | 31 | 28 | 25 | 24 |
| 12 | 1 | 11 | 24 | 33 | 32 | 28 | 25 | 24 |

**TABLE 4.8**
**Calculations for Silencer with High-Pressure Blower**

| Frequency | 63 | 125 | 250 | 500 | 1000 | 2000 | 4000 | 8000 |
|-----------|-----|-----|-----|-----|------|------|------|------|
| $K_w$ | 35 | 35 | 34 | 32 | 31 | 26 | 18 | 10 |
| $L_w - K_w$ | 70 | 70 | 70 | 70 | 70 | 70 | 70 | 70 |
| bpf | 0 | 0 | 0 | 0 | 3 | 0 | 0 | 0 |
| $L_{w\text{-out}}$ | 105 | 105 | 102 | 102 | 104 | 96 | 88 | 80 |
| I.L. | 2 | 14 | 32 | 32 | 31 | 28 | 25 | 24 |
| $L_w$ Net | 103 | 91 | 70 | 70 | 73 | 68 | 63 | 56 |

spectrum of a pump may be used as a means of monitoring the flow conditions and absence or presence of harmful degradation of the pump impeller and potential loss of performance, or even the pump itself.

A typical pump noise spectrum is shown in Figure 4.6, where the tonal spikes are clearly seen. As in all turbomachines, the sound power will be a strong function of the impeller speed and the correct selection of pump at the proper design performance. Noise mitigation techniques, as discussed by Karasik (1986), include isolation, interdiction of propagation or transmission path, reduction of speed or change of speed to avoid resonance conditions, and avoiding cavitation. This can be achieved by an increase in backpressure or injection of air or another gas (perhaps Argon) into the eye of the pump, if possible. Where the hydrodynamic interaction of the cutwater and the pump blades is a major noise source, the level of the noise generated can frequently be reduced by modifying the shape or location of the cutwater (Karasik, 1986). The distance of the blade tips from the cutwater can be increased, or the cutwater can be made more rounded or shaped to avoid being parallel with the blade trailing edges. Such changes can also reduce the performance level of the pump while alleviating the noise problem.

## 4.8 COMPRESSOR AND TURBINE NOISE

A modest review of gas-turbine engine noise, with emphasis on the high-speed axial compressor component, is available in Richards and Mead (1968) and in more recent papers from the AIAA and ASME journals. The fundamental acoustic properties of

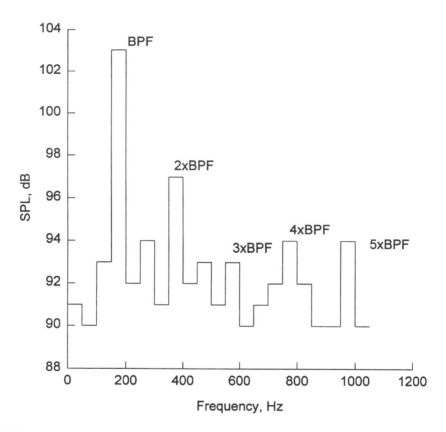

**FIGURE 4.6**   Pump noise spectrum. (Karassic, 1986.)

these machines are characterized by blade pass frequencies and pure tones at much higher frequencies than have been seen in fans and blowers, due to both high rotating speeds and large numbers of rotating blades. The older turbo-jet engines were clearly characterized by strong radiation of compressor noise from the inlet, and very strong turbulent jet noise from the discharge of the engine. A frequency spectrum typical of the older design practice is given in Figure 4.7 (taken from Cumptsy, 1977). Note the very strong magnitudes concentrated in the blade pass and higher harmonic spikes associated with the compressor. Excellent review papers over the past few decades are available (Cumptsy, 1977; Verdon, 1993) and serve as the primary source for this brief section on engine noise.

Considerable work during the 1960s and 1970s went into the reduction and control of these high noise levels as the basic form of the engines continued to evolve. As engines moved to higher, transonic Mach numbers in the compressor rotors, the basic acoustic signatures changed to include the very characteristic multiple pure tone (mpt) noise at frequencies below blade pass or even the rotating speed of the compressor. This behavior, which is related to formation and propagation of shock waves from the compressor rotor, is quite distinctive, and a more contemporary spectrum for such a gas turbine engine is shown in Figure 4.8. This

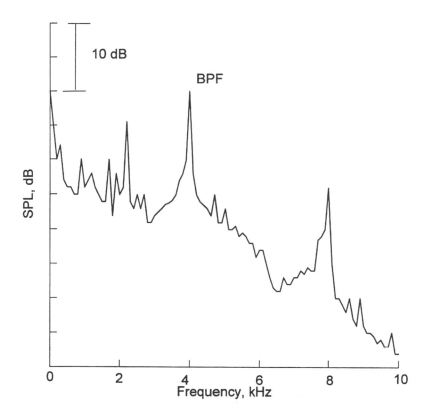

**FIGURE 4.7**   The noise spectrum of a turbo-jet engine. (Cumptsy, 1977.)

mpt component of noise, also known as "buzz-saw" noise, is amenable to duct-lining treatment in the engine intake, and suppression of the component has been achieved (Morfey and Fisher, 1970).

The broadband noise components associated with compressors are more closely related to fan and blower noise as treated earlier in the chapter. Cumptsy shows the clear relation between the diffusion level in a blade row (see Chapter 8) and the magnitude of the broadband noise component, shown here in Figure 4.9 (as developed by Burdsall and Urban, 1973). While the "self-noise" of compressor noise is said to depend on locally unsteady flow on the blade surfaces (Wright, 1976; Cumptsy, 1977), it may also be strongly affected by the ingestion of turbulence into a moving blade row (Sharland, 1964; Hanson, 1974). Such influence is illustrated in Figure 4.10, where an upstream flat plate disturbance was used to introduce turbulence systematically into a compressor blade row (Sharland, 1964). The results show a dramatic increase in the noise level.

The noise contribution associated with the turbine section of an engine has not been investigated or understood as well as the more dominant compressor and jet contributions to the spectra. Noise levels associated with the relatively low blade relative velocities in the turbine are usually masked by strong jet noise (in the older turbo-jet engines) and by aft propagation of fan and compressor noise in the high

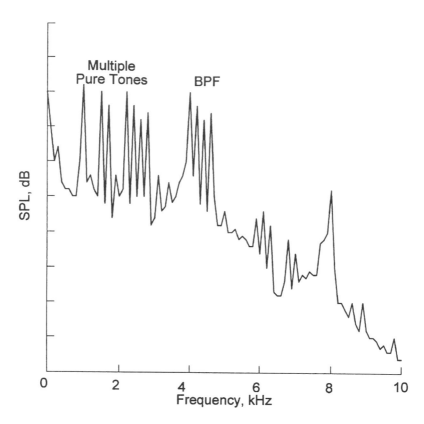

**FIGURE 4.8** Noise spectrum at higher mach number with multiple pure tone noise. (Cumptsy, 1977.)

bypass ratio engines developed more recently. In addition, the downstream turbulence and flow masking have led to a "broadening" of the pure tone spikes of the turbine into less distinguishable "semi-broadband" noise.

Along with the modification to the noise signatures of the engines caused by the high bypass turbo-fan developments, intensive efforts to reduce the high-speed jet noise component had been effective in reducing downstream propagated noise. An example of the performance of a noise suppression nozzle design is shown in Figure 4.11, with reductions of 10 dB or more when compared to a conventional nozzle.

More current trends in the 1980s and 1990s have, of course, developed along the lines of numerical computation and theoretical prediction of aeroacoustic performance of turbomachines. These methods (see Verdon, 1993) seek to accurately cover a wide range of geometries and operating conditions, leading to a broad and very difficult task, demanding great efforts toward minimization of computer storage requirements and highly time-efficient solution algorithms. Verdon characterized this work (in 1993) as being developmental, with actual design

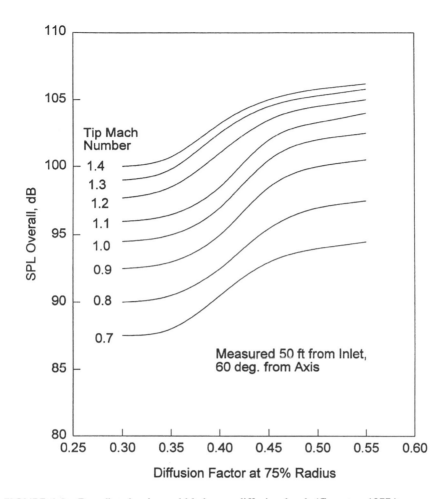

**FIGURE 4.9**   Broadband noise and blade row diffusion level. (Cumptsy, 1977.)

relying on the experimental information and the earlier analytical methods of the 1970s and 1980s.

The fully viscous unsteady approach for numerical solutions of the aerodynamics and aeroacoustics, which attempt to drive an unsteady solution to a converged periodic result, offers the greatest rigor but requires great computational resources as well. The inviscid linear or linearized analyses are still being developed and are capable of close agreement with the more complete approach to such solutions. Of course, the simpler methods remain somewhat more limited in applicability (Verdon, 1993).

There appears to be no real question as to the continued development and application of the modern numerical methods to the problems of turbomachinery acoustics. However, the improved state of experimental measurement will surely lead to an improved and needed experimental database to complement the advances in numerical predictive capabilities.

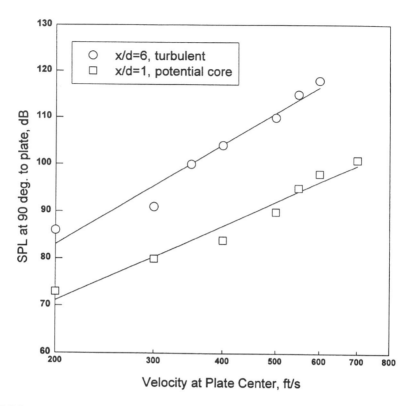

**FIGURE 4.10**   Turbulence ingestion influence on compressor noise. (Sharland, 1964.)

## 4.9   SUMMARY

The topic of noise associated with turbomachines was introduced in fairly basic, conceptual form. The most fundamental concepts of sound wave propagation and the idea of a spectrum or frequency distribution of sound or noise were discussed. These ideas were extended to the free field propagation and diminution of the sound pressure level or intensity of the sound at increasing distances from the noise source. Propagation in the free field was conceptually extended to include the influence of reflective surfaces near the noise source.

The quantitative physical scales of acoustic power and pressure were introduced in terms of reference quantities and decibel levels in power and pressure. Illustrations were provided to introduce decibel levels that are typical of familiar environments in an attempt to develop a "feel" or sense of the loudness of a given noise level.

Fan and blower noise was used as the primary vehicle to discuss and quantify the noise characteristics of turbomachines. The concepts of broadband (wide spectrum) and pure tone (narrow spectrum or spike) noise components were developed in terms of the physical behavior and characteristics of the machines. The Graham method was introduced as a semi-empirical procedure for estimating both the overall sound power level of fans and blowers and means for estimating the spectral

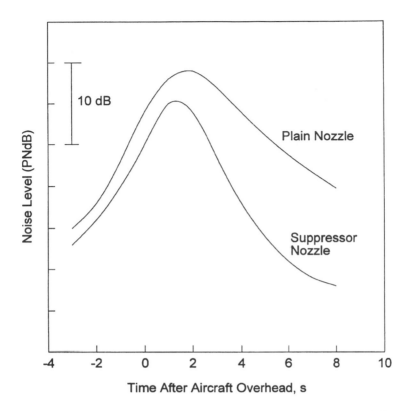

**FIGURE 4.11**   Nozzle noise suppression performance. (Richards and Mead, 1968.)

distribution of the sound power as well. His well-established model was modified to simplify calculation and to unify the concept of specific sound power of a machine as a simple function of specific diameter. Subject to inherent limits in accuracy and some loss in detail, this "modified Graham method" was used in examples to illustrate the process of designing for low noise levels.

Brief sections were provided in the chapter to extend the discussion of turbo-machinery noise to pumps, compressors and turbines. The references provided in the sections may serve as a beginning point for more thorough treatment of these topics as needed by the reader. Finally, the subject of numerical analysis of the fluid mechanics and acoustics in turbomachinery was mentioned briefly and is mentioned again in Chapter 10, along with a few references on the subject.

## 4.10   PROBLEMS

4.1.  Estimate the sound power level for the fans described in Problem 3.5— both the model and the prototype. Develop the octave bandwidth data for both.

4.2.  Estimate $L_w$ for the fan of Problem 3.3 using the modified or simplified Graham method, with $K_w = 52/D_s^{0.4}$. Use the Cordier approach to size this fan and re-estimate the noise using the very simple tip speed model, $L_w = 55 \log_{10} V_t - 24$. Compare with the result using the Graham method.

4.3. From the information given in Problem 2.4, calculate the noise level of the wind tunnel fan ($L_w$). Compare this result to the sound pressure levels shown in Table 4.2. Using appropriate techniques, try to minimize the value of $L_p$ at a distance of 60 ft from the fan inlet (assume a free field).

4.4. Use the results from Problem 2.5 to estimate the noise level of the compressor. Use the simplified Graham method for overall $L_w$.

4.5. From your solution of Problem 2.6, estimate the $L_w$ values at the original speed. Use the tip speed algorithm from Problem 4.2 to approximate the value of $L_p$ at a distance of 30 ft from the fans (assume $\Lambda = 4$). Compare these results to Table 4.2 values and discuss what measures (if any) should be taken for workers operating in the environment of these fans at distances greater than 30 ft.

4.6. A ducted axial fan is tested at 1750 rpm and delivers 6000 cfm of air at 4 InWG (static).
   (a) Calculate the proper size and efficiency for this fan.
   (b) Estimate the power required for the fan motor.
   (c) Calculate the sound power level for this fan using $K_w = 72/D_s^{0.8}$ in the simplified Graham method. Compare this result to an estimate using the onset velocity method, $L_w = 55 \log_{10} W_{1t} - 32$, where $W_{1t} = (V_T^2 + (Q/A)^2)^{1/2}$ is the inlet blade relative velocity at the blade tip. Also compute and compare a sound power level calculation based on the fan shaft power, $P_{sh}$ (in hp), according to $L_w = 20 \log_{10} P_{sh} + 81$, and using a "speed corrected" method based on $L_w = 20 \log_{10} P_{sh} + 81 + 4 \log_{10}(V_t/800)$ ($V_t$ is the tip speed in ft/s). State all assumptions and simplifications.

4.7. A laboratory room $16 \times 10 \times 8$ ft high must be ventilated so that the air in the room is exchanged twice per minute. The resistance to the flow is a short exhaust duct with a set of screened louvers over a vent ($K = 1$). Choose a fan whose level of sound power is less than about 70 dB in order to allow for low speech interference levels within the lab.

4.8. The laboratory room of Problem 4.7 has walls, ceiling, and floor treated with an absorbent padding roughly equivalent to heavy carpeting. Estimate the sound pressure level at the room center ($x = 8$ ft) if the fan is in an end wall. Compare the results to the criterion for speech interference in Table 4.2.

4.9. A large forced draft fan ($D = 10$ ft) supplies cooling air to a steam condenser in an oil refinery. The inlet of the fan is open and faces the boundary of the plant property, where the sound pressure level cannot exceed 55 dB. The fan supplies 100,000 cfm at 8 InWG static pressure rise and is sited against a smooth concrete wall on a smooth concrete pad. The inlet is 250 ft from the boundary. Will the fan inlet need a silencer?

4.10. A large mechanical draft fan supplies cooling air to a primary power distribution transformer in a suburban environment. The fan is a wide, double-width centrifugal type with the inlet of the fan open and facing the boundary of the utility property. At the boundary line, the fan sound pressure level must not exceed 58 dB. The fan supplies 400,000 cfm at 21 InWG static pressure rise and is sited against a smooth concrete wall with deep, soft sod on the ground between the fan and the boundary line which is 65 ft away. Will the fan inlet need a silencer? If so, how much attenuation is required? Could an acoustic barrier or berm be used? (See Baranek or Thumann.) (Hint: Assume $\alpha = 0$ for the wall and $\alpha = 1.0$ for the sod. Also, size the double-width fan as a single-width with half the flow to establish $D_s$. Then double $K_w$, logarithmically of course.)

4.11. There have been many semi-empirical scaling rules proposed for estimation of sound power level for fans and blowers, with many of them based on impeller tip speed. They usually take the form

$$L_w = C_1 \log_{10} V_T + C_2$$

If the speed (rpm) is increased, then the change in the sound power level is given by

$$L_w = L_{w2} - L_{w1} = C_1 \log_{10}\left(\frac{V_{T2}}{V_{T1}}\right)$$

Use the Graham method to show that the $C_1$ value consistent with his correlations is $C_1 = 50$ for N-scaling. That is, restrict the study to variable N and constant D.

4.12. Pursue the same exploration as was proposed in Problem 4.11, but examine the influence of diameter variation with fixed speed (rpm). What value of $C_1$ makes sense in the context of Graham's model? How can you reconcile this result and the result of Problem 4.11 with the typical empirical values of $C_1$ in the range $50 < C_1 < 60$?

4.13. Select a fan that can supply air according to the following specifications:

$$Q = 6 \ \mathrm{m^3/s}$$

$$\Delta p_s = 1 \ \mathrm{kPa}$$

$$\rho = 1.21 \ \mathrm{kg/m^3}$$

with the design constraint that

$$L_w \le 95 \ \mathrm{dB}$$

Choose the smallest fan that can meet the specifications and the constraint on sound power level. Use a Cordier analysis, not a catalog search.

4.14. On large locomotives, axial fans are used to supply cooling air to a bank of heat exchangers. Two of these fans draw ambient air through louvers in the side panels of the locomotive. Each fan draws 5.75 m³/s of air at 28°C and 98.0 kPa and must supply a total pressure ratio of $p_{02}/p_{01}$ = 1.050 ($\Delta p_T$ = 5 kPa).

(a) If locomotive design requirements impose a diameter limit on the fans of D = 0.6 m, calculate the minimum speed required and estimate the sound power level ($L_w$).

(b) If the allowable level of sound pressure level ($L_p$) is 55 dB 15 m to the side of the locomotive, do we need noise attenuation treatment? How much?

(c) Evaluate the need to include compressibility effects on the fan design and selection.

4.15. Calculate the reduction in noise available for the locomotive fans of Problem 4.14 if the constraint on size can be relaxed. Examine the range from 1 m up to 2 m diameter for the stated performance requirements. Develop a curve of $L_w$ and $L_p$ for the stated conditions.

4.16. A small axial flow fan system runs at 1480 rpm to supply a flow rate of 3 m³/s of air with a density of 1.21 kg/m³ and a static pressure rise of 100 mmH₂O.

(a) Determine a proper fan size, efficiency, and power requirement. Adjust the Cordier efficiency estimate for Reynolds Number.

(b) Estimate the value of $L_w$ for the fan using the modified Graham method.

(c) Estimate the sound pressure level, $L_p$, for an open inlet configuration propagating over water. Develop a curve of $L_p$ versus x up to 100 m.

4.17. An axial flow cooling tower fan has a diameter of 32 ft and is a tube axial type. It provides 1,250,000 cfm at a total pressure rise of 1.0 InWG with an air weight density of 0.075 lbf/ft³.

(a) Estimate the sound power level, $L_w$, of the fan using the Modified Graham method and the Tip Speed method.

(b) Calculate the sound pressure level, $L_p$, at a distance of 110 ft from the fan inlet (assume free field propagation).

4.18. A large cooling tower fan has a diameter of 10 m and is a tube axial type. It provides 600 m³/s at a total pressure rise of 500 Pa with an air weight density of 10.0 N/m³.

(a) Estimate the sound power level, $L_w$, of the fan using the Modified Graham method and the Tip Speed method.

(b) Calculate the sound pressure level, $L_p$, at a distance of 50 m from the fan inlet (assume free field propagation).

4.19. Railway locomotives also require air blowers to provide the heat transfer from the coolant "radiators" of the very large diesel engines powering these vehicles. Typical performance numbers for these blowers (there are two for each of the engines) are Q = 3 m³/s and $\Delta p_s$ = 15 kPa with $\rho$ = 1.2 kg/m³.

(a) Estimate the probable unattenuated sound power level of one of these blowers. (Hint: Assume a centrifugal single-width blower of fairly high specific speed, $D_s = 2.4$.)

(b) Estimate the combined sound power level of all four blowers.

(c) Calculate the sound pressure level at 15 m from the side of the locomotive, over soft grassy earth, with only two of these radiator blowers in operation (i.e., assume the two on a given side are the dominant sound power source).

4.20. To reduce the space requirements of the fans within the locomotive, for the engine cooling problem studied in Problem 4.19, one can propose the use of compact three-stage axial fans to replace the blowers used earlier.

(a) Using a specific speed near the high range for axial fans, say $N_s = 3$, size the speed and diameter needed.

(b) For one such stage, estimate the sound power level.

(c) Estimate the combined power level for the multi-stage fan.

(d) Finally, estimate the overall sound pressure level of all four of these multi-stage fans at the 15-m evaluation distance of Problem 4.19, and compare it to the specified maximum level of Problem 4.14 (55dB).

4.21. Cooling tower fans such as the one studied in Problem 4.18 are frequently used in much smaller sizes in adjacent "modules" of several fan-tower units operating in parallel. Typical size will be on the order of 3-m diameter with groups of 5, 10, 20, or more of the smaller fans and tower units. Consider a grouping of 20 fans set in a straight line. Each fan supplies 25 m³/s of air with a static pressure rise of 150 Pa at $\rho = 1.02$ kg/m³.

(a) Estimate the sound power level of one such fan.

(b) Estimate the combined sound power level of all 20 fans.

4.22. For the fans described in Problem 4.21, with a center-to-center spacing of the fans of 4 m:

(a) Estimate the sound pressure level from the line of fans, at a position situated 20 m from the tenth fan, measured perpendicular to the line.

(b) Use line source theory for the arrangement of fans to provide an estimate of the sound pressure of part (a). (Hint: Refer to other references on noise propagation such as Beranek, 1992; Thumann and Miller, 1986.)

(c) From a distance of 10 m from the line, approximately how much error would be incurred in using the line source approximation on the group of 20 fans?

4.23. For fan or blower performance written in m³/s and Pa for volume flow rate and static pressure rise, show that the Graham equation for sound power level can be modified to

$$L_w = K_w + 10 \, \log_{10} Q + 20 \, \log_{10} \Delta p_s - 14.6$$

with $K_w = f(D_s)$ as before.

4.24. An axial cooling tower fan has a diameter of 18 m and is a tube axial fan. It provides a volume flow rate of 4000 m³/s at a total pressure rise of 350 Pa with an air density of 1.10 kg/m³. Estimate the sound power level of the fan using the Graham method and the onset velocity method (see Problem 4.6). Estimate the sound pressure level 120 m from the fan. Assume a ground plane absorption coefficient of 0.25.

# 5 Selection and Preliminary Design

## 5.1 PRELIMINARY REMARKS

In attempts to organize turbomachinery data, Cordier found that machines of different type—axial flow, mixed flow, or centrifugal flow—naturally grouped into different regions on the $N_s$–$D_s$ diagram. On the left side, those machines with low head rise and high flow rate were grouped. These were predominantly axial flow machines of various types at small $D_s$ and large $N_s$. In the middle range of the chart, machines of moderate head and flow with a mixed flow path fell together. Those machines with small flow and large head rise grouped to the right side of the diagram, at low specific speed and high specific diameter. In Figure 5.1, this breakdown of regions of the diagram is refined somewhat beyond the initial observation to depict six regions or subdivisions of the Cordier diagram into distinct types of machines. Bear in mind that the boundaries of each region are indistinct and there is a natural tendency for them to overlap. Nevertheless, this breakdown will help identify the application range, in flow and pressure, for different machines—pumps, fans, blowers, compressors, and turbines.

## 5.2 FLOW REGIONS

In Region A, in a rough range with $6 < N_s < 10$, and $0.95 < D_s < 1.25$, one finds propeller-type machines used for moving huge quantities of fluid with very little change in pressure. Ship and aircraft propulsors, horizontal axis wind turbines, and light ventilation equipment such as floor fans, desk oscillators, or ceiling fans fall into this range. They are usually of the open or unshrouded configuration. Note from the efficiency curve that they also lie in the range of lowest efficiency on the chart, with $0.30 < \eta < 0.60$.

In Region B, the range is roughly $3 < N_s < 6$ and $1.25 < D_s < 1.65$. Here, one considers tube or ducted axial flow equipment such as tube axial fans, vane axials with small hubs, axial flow pumps, or "inducers," and shrouded horizontal axis wind turbines capable of somewhat higher pressure rise (or drop) performance (for a given diameter and rpm). Ventilation applications such as attic fans, window fans, and air-conditioner condenser fans usually use fans in this range. Axial pumps and shrouded propellers (air or water) also fall into this category.

For Region C, the range of $1.8 < N_s < 3$ and $1.65 < D_s < 2.2$ is a region of "full stage" axial equipment. The blade row operates enclosed in a duct or shroud and usually has a set of vanes downstream from the blade row that straighten and recover the swirling flow before the fluid leaves the machine. Typical applications include

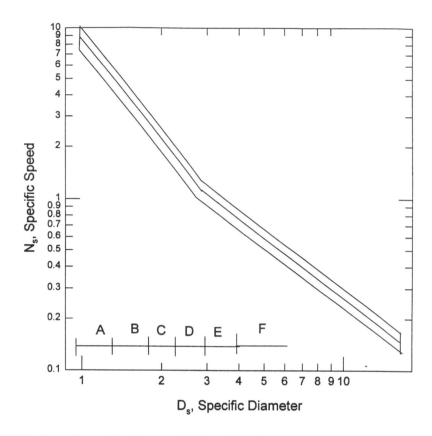

**FIGURE 5.1**   The cordier diagram with the regions for machine types shown.

vane axial fans with fairly large hubs, axial flow pumps, bulb-type hydraulic turbines, and single-stage or multi-stage axial compressors and pumps.

Region D, for $1.0 < N_s < 1.8$ and $2.2 < D_s < 2.80$, includes multi-stage axial flow equipment, mixed flow pumps and blowers, Francis-type mixed-flow hydraulic turbines or possible centrifugal/radial flow pumps or blowers, with unusually wide blading at the exit of the impeller. Applications requiring higher pressure rise with a large flow rate will use this equipment. Heat exchanger applications in HVAC equipment and moderately low-head liquid pumping are typical.

In Region E, with $0.70 < N_s < 1.0$ and $2.80 < D_s < 4$, radial discharge equipment dominates. Centrifugal fans of moderate width and liquid centrifugal pumps lie in this range. Heavy-duty blowers, compressors, and high-head turbines running at fairly high rpm are also typical. Blading will be of the backwardly inclined or backwardly curved type.

Region F is the property of fairly narrow centrifugal or radial flow equipment, with $N_s < 0.70$ and $D_s > 4$. High-pressure blowers, centrifugal compressors, and high-head liquid pumps are typical. For small flow and very high head, one begins to see narrow radially oriented blades and even forward-curved blading.

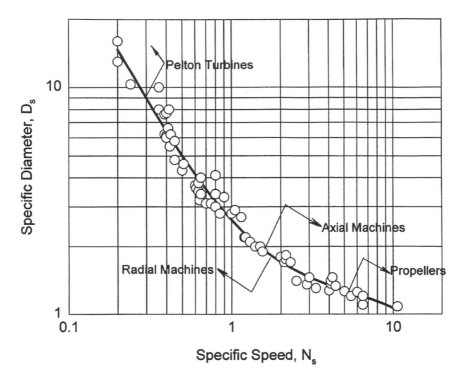

**FIGURE 5.2** The original form of the Cordier line taken from Balje (1981).

To complicate matters, there are some commercially available "hybrid" machines for special-purpose applications. These include very wide centrifugal blowers with forward-curved blading, such as furnace fans or air-conditioner evaporator fans. Radial discharge centrifugal machines, which operate in a cylindrical duct, an essentially axial flow path, are also seen. These fans are often referred to as in-line or plug fans. In addition, a great many centrifugal flow machines are available in double-inlet, double-width, or double-suction configurations and are found to operate well into the nominally axial range of the Cordier diagram. Axial machines with a large ratio of hub to tip diameter (75 to 90%) are capable of crowding into the centrifugal range of the diagram with relatively small flow rate and high pressure. Figure 5.2 shows the original form of the Cordier line (Balje, 1981), and Figures 5.3 through 5.8 illustrate the machine geometry typical of each of the regions.

When it is necessary to meet certain requirements in volume flow rate and pressure rise, one can follow the overall trend prediction of the Cordier diagram to help decide what kind of machine will perform the job. For an initial selection of speed and known fluid density, one can estimate the specific speed. Enter the diagram at Ns, and determine a workable value for the machine size by choosing a value near or on the "Cordier line." This initial procedure will indicate both the size and type of machine one should be considering and the highest value of efficiency one can reasonably expect to achieve.

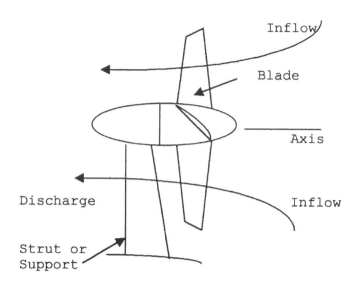

**FIGURE 5.3**   Region A: Open-flow axial fan or pump.

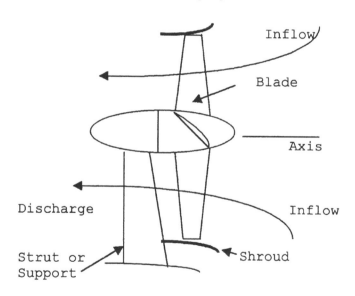

**FIGURE 5.4**   Region B: Tube axial or ducted fan.

For example, if one needs $Q = 5$ m³/s with $\Delta p_T = 1{,}250$ Pa in air with $\rho = 1.20$ kg/m³, then

$$\frac{Q^{1/2}}{\left(\dfrac{\Delta p_T}{\rho}\right)^{3/4}} = 0.0122 \text{ s} \quad \text{and} \quad \frac{Q^{1/2}}{\left(\dfrac{\Delta p_T}{\rho}\right)^{1/4}} = 0.394 \text{ m}$$

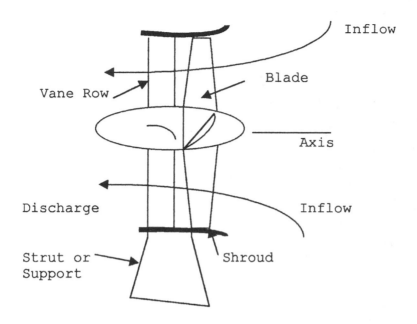

**FIGURE 5.5** Region C: A full-stage or vane axial ducted fan.

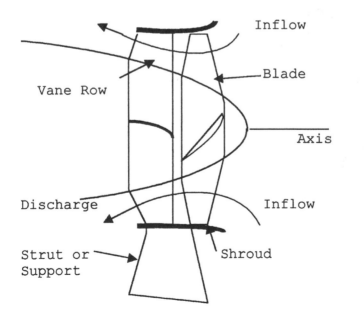

**FIGURE 5.6a** Region D: Mixed flow configurations.

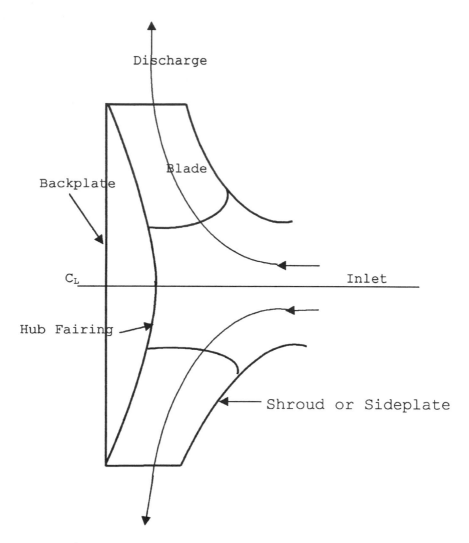

**FIGURE 5.6b**   Another mixed flow configuration.

If $N = 1800$ rpm $= 188.5$ s$^{-1}$, then $N_s = 2.30$. Cordier indicates $D_s \cong 1.91$, for which

$$D = \left[ \frac{Q^{1/2}}{\left( \frac{\Delta p}{\rho} \right)^{1/4}} \right] {}^\circ D_s = 0.753 \text{ m}$$

The probable efficiency is no greater than 0.88 or 0.89.

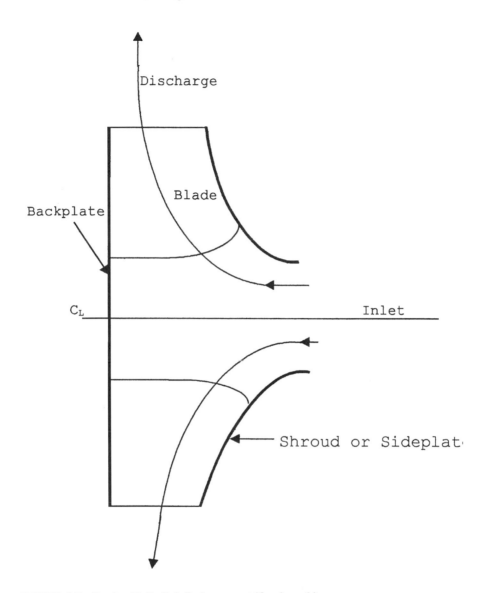

**FIGURE 5.7**   Region E: Radial discharge centrifugal machines.

Further, the fan lies in Region D (to the left edge) and should probably be a mixed flow or a vane axial fan. What if a lower speed is chosen? Let N = 900 rpm. Now $N_s$ = 1.15, with a corresponding $D_s$ of approximately 2.66. This indicates a Region E fan, a radial flow or moderate centrifugal fan with a diameter of about 1.05 m and an efficiency perhaps as high as 0.92 or 0.93. It is more efficient, but the diameter is quite large and will be expensive to manufacture. The point of the example is this: Using the Cordier diagram for guidance, one can very quickly examine some alternatives in machine type, size, and efficiency for a given perfor-

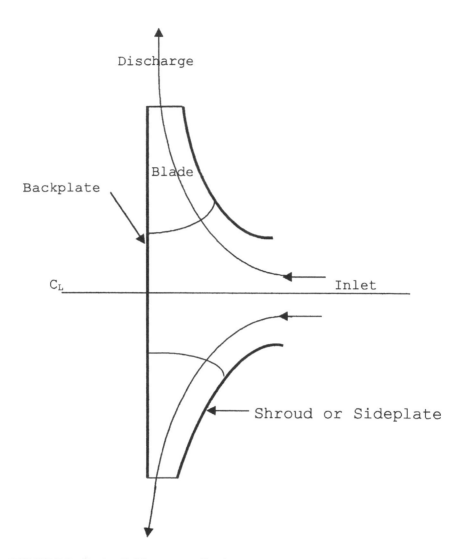

**FIGURE 5.8**  Region F: Narrow centrifugals.

mance requirement and begin to make decisions about selecting or perhaps designing a machine to perform the task.

## 5.3  SELECTION EXAMPLES

This section takes a look at some fairly detailed examples of how to use the Cordier diagram to make an initial estimate of a selection of a turbomachine for a given performance specification. The procedure is as follows: The performance requirements and type of fluid will be known. That is, $\rho$, $\Delta p_T$, and $Q$ are given. Recall that the pressure rise or drop used in the Cordier analysis must be given as total pressure, although system resistance and therefore machine specification are very often stated in terms of static pressure. One must convert the static values to total and if necessary iterate the velocity pressure calculation as machine size changes at constant mass or volume rate. This will be illustrated in the examples. To find out what kind of machine to use, this section will explore combinations of size (D, diameter) and speed (N, rpm or s$^{-1}$) in relation to points on or near the $N_s$–$D_s$ locus of good equipment as indicated by Cordier. Given $\rho$, $\Delta p_T$, and Q, one can either choose D and find N—or choose N and solve for D. This is done according to the definitions of $N_s$ and $D_s$.

When choosing N, use $N_s = NQ^{1/2}/(\Delta p_T/\rho)^{3/4}$ and a choice of N yields $N_s$. Then calculate $D_s$ (from Equation (2.39))

$$D_s \cong 2.84\, N_s^{-0.476}, \quad N_s \geq 1.0 \tag{5.1}$$

and calculate D from Equation (2.37):

$$D_s = \frac{D\left(\dfrac{\Delta p_T}{\rho}\right)^{1/4}}{Q^{1/2}}$$

or

$$D = \frac{D_s Q^{1/2}}{\left(\dfrac{\Delta p_T}{\rho}\right)^{1/4}}$$

If $N_s \leq 1.0$, use

$$D_s = 2.84\, N_s^{-0.888} \tag{5.2}$$

When choosing D, we use $D_s = D(\Delta p_T/\rho)^{1/4}/Q^{1/2}$, and a choice of D yields $D_s$. Then calculate $N_s$ (from Equation (2.39))

$$N_s = 9.0\, D_s^{-2.103}, \quad D_s \leq 2.84$$

and calculate N from the relation

$$N_s = \frac{NQ^{1/2}}{\left(\dfrac{\Delta p_T}{\rho}\right)^{3/4}}$$

If $D_s \geq 2.84$, use $N_s = 3.25\, D_s^{-1.126}$ (Equation (2.37))

Accompanying the determination of D from N or N from D, one can read the efficiency, $\eta_T$, from the top of the Cordier diagram or use the curve-fit equations given in Chapter 2 and repeated here for convenience.

$$\eta_{T-C} = 0.149 + 0.625\,D_s - 0.125\,D_s^2, \qquad \text{for } D_s \leq 2.5$$

$$\eta_{T-C} = 0.864 + 0.0531\,D_s - 0.0106\,D_s^2, \qquad \text{for } 2.5 \leq D_s \leq 5$$

$$\eta_{T-C} = 1.1285 - 0.0529\,D_s, \qquad \text{for } 5 \leq D_s \leq 20$$

Recall that the value from the curve is an upper limit expectation of best efficiency for well-designed, well-constructed machines with smooth finish and tight clearances and that a simple method is provided to de-rate these values for low Reynolds number or for large radial clearance gaps. Note that the values can also be read from the curves with sufficient accuracy for $D_s$, $N_s$, and $\eta_T$. Now, consider some examples.

The first application is to select the basic size and speed for a fan that must move 4.80 m³/s against a resistance of 500 Pa at standard sea level air density ($\rho = 1.21$ kg/m³). For the calculation, it is known that the total pressure rise requirement is about 625 Pa ($1/2\ \rho V^2 = 125$ Pa if $V_d = 21$ m/s). This is a lot of flow and not much pressure rise, so one can start with a fairly high $N_s$ and explore some range around $N_s = 5$.

First calculate:

$$\frac{Q^{1/2}}{\left(\dfrac{\Delta p_T}{\rho}\right)^{3/4}} = 0.0200 \text{ s}$$

and

$$\frac{Q^{1/2}}{\left(\dfrac{\Delta p_T}{\rho}\right)^{1/4}} = 0.455 \text{ m}$$

Then, $N_s = 5$ yields $N = 5/0.0200\text{s} = 250$ s⁻¹ = 2387 rpm

For this value, $D = D_s Q^{1/2}/(\Delta p_T/\rho)^{1/4} = 0.455\,D_s$ where $D_s \cong 2.84\,N_s^{-0.476}$. Here, $D_s = 2.84\,(5)^{-0.476} = 1.32$, so $D = 1.32 \times 0.455$ m $= 0.601$ m. Also, at $D_s = 1.32$, $\eta_T = 0.80$ (at best). Other values of $N_s$ should be checked; some calculations are summarized in Table 5.1.

Some general observations on these results can be made:

1. At the upper range of $N_s$, the efficiency is poor but the fan is small and probably inexpensive. At 4775 rpm ($N_s = 10$), it is very fast and is probably a "screamer." Note that the value of $k_w = 72/D_s^{0.8}$ indicates a very high specific noise level of 75 dB, for a total sound power level of 121 dB.

**TABLE 5.1**
**Cordier Calculations for the Example**

| N$_s$ | D$_s$ | N (rpm) | D (m) | L$_w$ (dB) | η$_T$ | Region | Type |
|---|---|---|---|---|---|---|---|
| 10 | 0.949 | 4775 | 0.436 | 121 | 0.70 | A | Prop |
| 7.5 | 1.088 | 3581 | 0.497 | 113 | 0.75 | B | Tube axial |
| 5 | 1.32 | 2378 | 0.597 | 104 | 0.80 | B | Tube axial |
| 4 | 1.47 | 1910 | 0.668 | 99 | 0.85 | C | Vane axial |
| 3 | 1.68 | 1423 | 0.678 | 94 | 0.89 | C | Mixed |
| 2 | 2.04 | 955 | 0.933 | 88 | 0.91 | D | Radial |
| 1 | 2.84 | 477 | 1.292 | 87 | 0.91 | E | Radial |
| 0.5 | 5.26 | 239 | 2.393 | 84 | 0.82 | F | Narrow centrifugal |
| 0.25 | 9.73 | 119 | 4.425 | 82 | 0.65 | F | Narrow centrifugal |

2.  At the lowest range of Ns, the efficiency is dropping off and the fan is getting very large and expensive, and at 100 to 200 rpm the speed is ridiculously low. It is, of course, very quiet at about 82 dB.
3.  In the middle range, efficiency is good, size and speed are both moderate, sound power level is also moderate, and one may have a good starting point for preliminary design or selection.

One can start the search for a fan with, say, N$_s$ = 2.5 (with D$_s$ = 1.84) such that

$$N = 1200 \text{ rpm}$$

$$D = 0.837 \text{ m}$$

$$\eta_T = 0.90$$

The fan will be vane axial in the middle range of N$_s$.

With D = 0.837 m, the "mean" axial discharge velocity, given by V$_d$ = Q/A$_{fan}$, becomes

$$V_d = \frac{\left(4.80 \text{ m}^3/\text{s}\right)}{\left[\left(\frac{\pi}{4}\right)(0.837 \text{ m})^2\right]} = 8.72 \text{ m/s}$$

which is low (V$_d$ = 8.72 m/s implies a velocity pressure of about 46 Pa). The assumption of 125-Pa velocity pressure was a little conservative and should be corrected in a second iteration.

As an exercise, assume a velocity pressure of 500 Pa, making $\Delta p_T$ = 1 kPa, and rework the example, choosing a suitable initial fan. One should get a high-specific-speed vane axial fan with a diameter of approximately 70 cm running at 1800 rpm (mean discharge velocity pressure is about 450 Pa).

## 5.4    SELECTION FROM VENDOR DATA: FANS

An example of fan performance information in typical catalog vendor form is provided here for a generic vane axial fan. One can use several examples in the process of selecting industrial turbomachinery for a list of specifications in a given flow problem. In general, performance information in the form of dimensionless performance curves can conveniently be used to replace the vendor catalog form.

The selection process will follow the basic Cordier concept to establish a base of candidate size and speed values, which can then be compared to available vendor data. Bear in mind that, although the tables are very limited in the first example, a broad search could involve entire catalogs from many vendors.

For axial fans, Table 5.2 is based on rather generic data for a vane axial fan in the form of a multiple performance rating table or multi-rating table. The fans listed here are belt-driven fans rather than direct-connected or direct-drive fans. Take a moment to examine the structure of Table 5.2. Q in cfm controls the leftmost column entry, with uniquely associated values of velocity pressure ($1/2 \, \rho \, V_d^2$). Increasing values of static pressure rise ($\Delta p_s$ or S.P.) are blocked off to the right of the first two columns. Note that static pressure rise ranges from 0.0 to 1.5 InWG but that the highest pressures are not available for the lower volume flow rates. Examination of corresponding speed values (in rpm) shows the inherent influence of tip speed on available pressure rise. A typical vane axial fan is shown in Figure 5.9.

Consider the following specifications:

$$Q = 3.0 \text{ m}^3/\text{s}; \quad \Delta p_s = 250 \text{ Pa}; \quad \rho = 1.21 \text{ kg/m}^3$$

What about velocity pressure to obtain $\Delta p_T$? One can carry out the initial analysis using D as the independent variable. If it is assumed that d/D is 0.50, one can calculate the annulus area and the velocity pressure so that total pressure is known for each choice of D. Then construct a Cordier analysis table (Table 5.3) using the defining equations for $D_s$ and $N_s$. Since a vane axial fan from Table 5.2 is to be selected, look at the upper range of the Cordier table (Table 5.3). The diameter of such a choice must lie between about 0.6 and 0.8 m, or roughly around 0.7 m. The spread of data underlying the Cordier curve suggests a degree of flexibility in the exact value of $D_s$, within ±5% of 1.64. Thus, one should be able to find a candidate selection with a diameter between 0.66 and 0.74 using the specific speed value of 3.32 or N = 1116 rpm. Since Table 5.2 is in English units, temporarily convert the requirements to Q = 6350 cfm (ft³/min), $\Delta p_s$ = 1 InWG with D ≅ 28 inches at 1116 rpm.

Checking Table 5.2 for the 27-inch size fan, the closest match to the specifications appears to be a fan with 6500 cfm at 1.00 InWG running at 945 rpm requiring 1.23 kW. Another fan is just below this one in the table. It gives 7300 cfm at 1.0 InWG with 1000 rpm requiring 1.64 kW. The next smaller size, the 24-inch fan, yields two choices. At 6400 cfm, a 1.0 InWG static pressure rise at 1200 rpm and 1.5 kW. At 5800 cfm, 1 InWG at 1150 rpm with 1.27 kW. Examination of additional tables for larger and smaller versions of this vane axial fan would provide more choices for fans that can operate close to the required performance point.

## TABLE 5.2
## Multi-Rating Tables for a Vane Axial Fan

**24-inch O.D. Fan:**
**Fan Dia. = 24 inches; Casing Dia. = 24.3 inches**

| Volume (cfm) | V.P. (InWG) | S.P. = 0 (InWG) rpm/hp | 0.25 rpm/hp | 0.50 rpm/hp | 0.75 rpm/hp | 1.00 rpm/hp | 1.25 rpm/hp | 1.50 rpm/hp |
|---|---|---|---|---|---|---|---|---|
| 3850 | 0.09 | 545/0.20 | 663/0.35 | 790/0.56 | | | | |
| 4500 | 0.12 | 640/0.30 | 740/0.50 | 840/0.71 | | | | |
| 5150 | 0.16 | 730/0.44 | 810/0.66 | 900/0.90 | 1000/1.15 | | | |
| 5800 | 0.20 | 820/0.65 | 900/0.90 | 965/1.11 | 1050/1.44 | 1150/1.7 | | |
| 6400 | 0.25 | 900/0.88 | 975/1.12 | 1050/1.40 | 1120/1.75 | 1200/2.00 | | |
| 7000 | 0.30 | 1000/1.16 | 1060/1.45 | 1130/1.70 | 1190/2.00 | 1250/2.40 | 1350/2.80 | 1400/3.10 |

**27-inch O.D. Fan:**
**Fan Dia. = 27 inches; Casing Dia. = 27.3 inches**

| Volume (cfm) | V.P. (InWG) | S.P. = 0 (InWG) rpm/hp | 0.25 rpm/hp | 0.50 rpm/hp | 0.75 rpm/hp | 1.00 rpm/hp | 1.25 rpm/hp | 1.50 rpm/hp |
|---|---|---|---|---|---|---|---|---|
| 4875 | 0.09 | 486/0.24 | 590/0.45 | 700/0.70 | | | | |
| 5676 | 0.12 | 560/0.40 | 650/0.60 | 750/0.90 | | | | |
| 6500 | 0.16 | 650/0.55 | 725/0.80 | 800/1.10 | 885/1.45 | | | |
| 7300 | 0.20 | 730/0.80 | 800/1.10 | 865/1.41 | 950/1.80 | 1000/2.2 | | |
| 8100 | 0.25 | 800/1.11 | 875/1.40 | 950/1.80 | 1000/2.15 | 1000/3.0 | | |
| 9000 | 0.30 | 900/1.50 | 950/1.80 | 1010/2.25 | 1050/2.60 | 1125/3.05 | 1190/3.50 | 1250/4.00 |

Although this is somewhat confusing, the target is surrounded, so to speak, with fan parameters reasonably close to the specifications derived from the analysis. The notable exception is the required power, which is high in every case. As can be shown, both the Reynolds Number and the radial clearances are relatively far from the "excellent" conditions outlined for the $\eta_{T-C}$ estimate. Re = $ND^2/\nu = 0.39 \times 10^7$. From Table 5.2, the 27-inch fan has a blade diameter of 27.0 inches and the casing or inlet diameter is 27.3 inches. Thus, the radial clearance is $\varepsilon = 0.15$ inches and $\delta = \varepsilon/D = 0.0056$ or $\delta_0 = 5.6$. Taken together, these factors reduce the efficiency to 0.62 from a Cordier value of 0.90. The nominal value of required power is about 1.6 kW. That is fairly close to the above values.

The best choice from the tables appears to be the 27-inch fan at 945 rpm and 1.23 kW power required. The 24-inch fan at 6400 cfm and 1.0 InWG could provide the performance, but the power is increased by more than 20%. Also note that the fan tip speed for the 24-inch fan is 125 ft/s, and for the 27-inch fan it is 111 ft/s. Estimating the sound power level on the basis of tip speed, the difference between the two fans can be approximated by 55 $\log_{10}(125/111) = 2.8$ dB. Now, 2.8 dB is a significant difference—but not a commanding one. Estimating the noise for each fan using the Graham method gives 94 dB for the 24-inch fan and 91 dB for the

**FIGURE 5.9**    A vane axial fan that is belt driven with external motor. (New York Blower Company.)

27-inch fan. The choice between the two could be made on the basis of sound power level, but the difference is not very great. Table 5.4 summarizes the parameters of these two fans.

The two fans are very similar and the final choice would depend on the user's preference either for smaller size and perhaps slightly lower initial cost, or a larger fan at increased first cost, but a reduced power requirement running at reduced noise. The preference may well depend on the usage of the fan. If the fan is going to be used more or less continuously over a period of many years, the 27-inch fan with decreased operating cost will be the chosen fan. If the fan is for occasional or seasonal use and noise is not very important, the smaller fan will be the one.

The same kind of Cordier analysis could be used to select a tube axial fan but is deferred to the problem assignments at the end of the chapter.

For centrifugal backwardly inclined fans, the multi-rating tables for the performance of a commercially available centrifugal fan, similar to those for the vane axial fan of Table 5.2, have been reduced to a set of dimensionless performance curves. These curves, given in Figure 5.10, summarize the static pressure rise coefficient, $\psi_s$, and the static efficiency, $\eta_s$, as functions of the flow coefficient, $\phi$. The equivalent total performance is readily generated from the relations given in Problem 2.22

**TABLE 5.3**
**Cordier Analysis**

| D (m) | $\Delta p_T$ (Pa) | $D_s$ | $N_s$ | N (rpm) | $L_w$ (dB) | $\eta_{Tc}$ | P (kW) | Type |
|---|---|---|---|---|---|---|---|---|
| 0.4 | 858 | 1.19 | 6.40 | 4,884 | 101 | 0.77 | 3.31 | Tube |
| 0.6 | 370 | 1.45 | 4.11 | 1,665 | 91 | 0.86 | 1.29 | Tube |
| 0.8 | 288 | 1.82 | 2.52 | 757 | 83 | 0.90 | 0.96 | Vane |
| 1.0 | 266 | 2.23 | 1.70 | 539 | 80 | 0.91 | 0.88 | Vane |
| 1.2 | 258 | 2.65 | 1.31 | 406 | 79 | 0.91 | 0.85 | Mix. |
| 1.4 | 254 | 3.10 | 1.00 | 306 | 78 | 0.90 | 0.85 | Cent. |
| 1.6 | 252 | 3.68 | 0.76 | 231 | 77 | 0.88 | 0.86 | Cent. |

**TABLE 5.4**
**Fan Options**

| Fan | D (m) | N (rpm) | Q (m³/s) | $\Delta p_s$ (Pa) | P (kW) | $L_w$ (dB) |
|---|---|---|---|---|---|---|
| 24-in. | 0.610 | 1200 | 3.03 | 250 | 1.50 | 94 |
| 27-in. | 0.686 | 1000 | 3.07 | 250 | 1.25 | 91 |

**TABLE 5.5**
**Cordier Table for Centrifugal Fan Selection**

| $\phi$ | $\psi_s$ | $\psi_T$ | $D_s$ | $N_s$ | D | N | $\eta_{Tc}$ | P (hp) | $L_w$ (dB) |
|---|---|---|---|---|---|---|---|---|---|
| 0.085 | 0.108 | 0.115 | 2.00 | 1.48 | 2.89 | 1183 | 0.85 | 18.1 | 99 |
| 0.090 | 0.105 | 0.111 | 1.92 | 1.56 | 2.77 | 1305 | 0.86 | 17.9 | 100 |
| 0.095 | 0.099 | 0.106 | 1.85 | 1.68 | 2.61 | 1374 | 0.87 | 17.8 | 101 |
| 0.100 | 0.095 | 0.098 | 1.77 | 1.81 | 2.55 | 1447 | 0.87 | 17.7 | 103 |
| 0.105 | 0.091 | 0.960 | 1.73 | 1.88 | 2.48 | 1497 | 0.86 | 18.0 | 104 |
| 0.110 | 0.086 | 0.094 | 1.67 | 1.95 | 2.41 | 1559 | 0.85 | 18.1 | 105 |
| 0.115 | 0.080 | 0.092 | 1.62 | 2.06 | 2.34 | 1625 | 0.83 | 18.4 | 107 |

**TABLE 5.6**
**Direct-Connected Pump Calculations**

| # Poles | N (rpm) | $N_s$ | $D_s$ | D (in) | $\eta_T$ | $\sigma_v$ | NPSHR (ft) |
|---|---|---|---|---|---|---|---|
| 2 | 3600 | 0.532 | 4.98 | 5.9 | 0.82 | 0.104 | 15.6 |
| 4 | 1800 | 0.266 | 9.20 | 11.3 | 0.68 | 0.041 | 6.2 |
| 6 | 1200 | 0.177 | 13.20 | 15.8 | 0.55 | 0.024 | 3.6 |

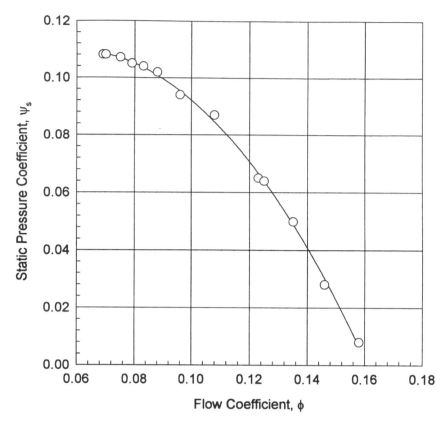

**FIGURE 5.10a**  Non-dimensional performance curves for a SWSI centrifugal fan. Available sizes are 0.46, 0.61, 0.91, 1.02, 1.25, 1.375, 1.50, 1.67, and 1.85 m diameter. Tip speeds are limited to 55 m/s (Class I), 68 m/s (Class II), and 86 m/s (Class III).

$$\Psi_T = \Psi_s + \left(\frac{8}{\pi^2}\right)\phi^2 \tag{5.3}$$

with

$$\eta_T = \left(\frac{\Psi_T}{\Psi_s}\right)\eta_s \tag{5.4}$$

The figure states the available sizes for this fan ranging from 0.46 m to 1.85 m. The limiting tip speeds for structural integrity are stated for three versions of the fan. There is the "light-duty" Class I fan limited to $V_t = ND/2 = 55$ m/s; the "moderate-duty" Class II fan limited to $V_t = 68$ m/s; and the heavy-duty" Class III fan limited to $V_t = 86$ m/s. Clearly, larger sizes or rotating speeds are allowed for the higher classes leading to higher pressure rise and flow rate capabilities. The fans of

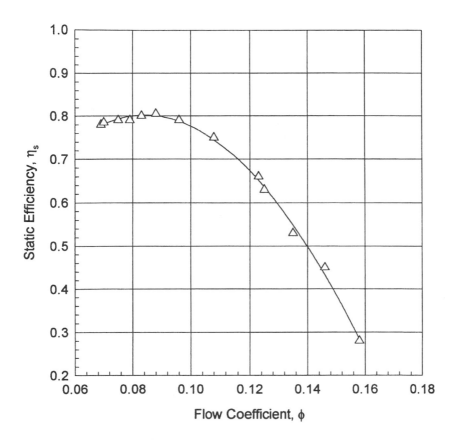

**FIGURE 5.10b** Non-dimensional performance curves for a SWSI centrifugal fan.

Class I, II, and III are increasingly more rugged and more expensive. Given nine distinct sizes, each in three versions and a full $\psi_s, \eta_s$–$\phi$ curve, the combinations are finite but very broad, yet the data is presented in a much more compact fashion than is typified by Table 5.2. Note that the curves of Figure 5.10 are for a single-width, single-inlet fan (SWSI), and that a double-width, double-inlet (DWDI) version is also available with $\phi_{DWDI} = 2\phi_{SWSI}$ (with $\psi$ and $\eta$ unchanged). Figure 5.11 shows a large double-width centrifugal fan impeller on the drive shaft.

As an application example, try to fit a fan to a specification with a noise constraint:

$$Q = 15,000 \text{ cfm}; \quad \Delta p_s = 6.0 \text{ InWG};$$

$$\rho = 0.0023 \text{ slugs/ft}^3; \quad L_w \leq 100 \text{ dB}$$

One can perform the usual analysis using $N_s$ and $D_s$ to see if the performance, size, and speed can be matched to the dimensionless variables. One should restrict the search to the range of better efficiencies of Figure 5.10, say for $0.085 < \phi < 0.115$,

**FIGURE 5.11**   Impeller assembly for a large DWDI centrifugal fans. (Barron Industries.)

so that $0.70 < \eta_s < 0.83$. The BEP for total pressure and efficiency occurs very close
to $\phi = 0.10$ and $\psi_T = 0.106$ with $\eta_T = 0.97$. Thus, $N_s = \phi^{1/2}/\psi_T^{3/4} = 1.70$ and
$D_s = \psi_T^{1/4}/\phi^{1/2} = 1.80$. This point falls at the lower edge of the "scatter band" on the
Cordier line and agrees well with the Cordier efficiency of $\eta_{T\text{-}C} = 0.87$.

One can now center the investigation around $\phi = 0.1$ to try for a match. The
sound power level for these fans is estimated using the modified Graham method.
The first entry at $\phi = 0.085$ comes in below 100 dB with D = 2.90 ft = 0.884 m.
The closest available size for this fan is D = 0.91 m, and the Class I tip-speed limit
of 55 m/s allows a rotational speed of about 1150 rpm. If one uses (in English units)
2.99 ft and 1150 rpm, the flow rate becomes 16,400 cfm at 6.37 InWG static. Both
are a little high, but if one uses 1120 rpm, then one obtains 15,950 cfm and 6.04
InWG static. That is close to the original target (6% high on flow, 1% high on
pressure). Although the selection could be explored in more detail, this one will
work and meet the noise limitation.

## 5.5   SELECTION FROM VENDOR DATA: PUMPS

Single- or end-suction pump data has been supplied as catalog information on Allis-
Chalmers pumps (by ITT Industrial Pump Group, Cincinnati, OH), and a portion of
that catalog is reproduced here. Look at an example and include a constraint on
NPSHR such that Q = 275 gpm and H = 150 ft. The fluid is water (85°F), and
NPSHR must be less than 10 ft. Do a Cordier analysis, but specify direct-driven
pumps and estimate $\sigma_v$ using Thoma's rule to bracket the NPSHR.

Constructing the analysis table with nominal speeds given by

$$N = \frac{(120 \times \text{line frequency})}{(\# \text{ of poles})}$$

04-84

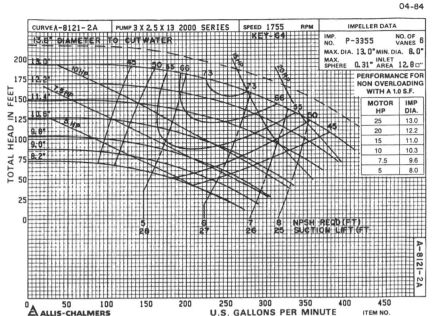

**FIGURE 5.12**   Allis-Chalmers pumps, 1750 rpm. (ITT Industrial Pump Group.)

one can use frequency = 60 Hz (60 cycle AC current) or 50 Hz (for European applications) with # poles = 2, 4, 6, 8, etc. These speeds will then be the synchronous motor speeds at zero load for AC induction motors. Actual running speeds are slightly lower as the motor develops full-load torque (Nasar, 1987). The full-load or design condition for the motor usually requires a "slip" of from 15 rpm to as much as 150 rpm, depending on motor quality and rated power. The smallest values of slip, 15 rpm, are associated with high quality and large rated power (hundreds of horsepower) and the very high values of slip will generally be seen in fractional horsepower, low-cost motors.

One quickly finds, based on the NPSHR constraint, that a 2-pole motor-driven pump will not satisfy the specifications. The 4-pole setup is adequate, as is the 6-pole setup. However, the efficiency is dropping fast and the pump is getting much larger (driving up cost), so 1800 rpm at D = 11 or 12 inches looks like the best possibility. Go to Figure 5.12 to examine the available pumps and begin by looking at 4-pole motor direct-driven pumps. The catalog speed is given as 1750 rpm, implying a 50-rpm slip below synchronous speed to generate the necessary torque. Pick a few pumps:

1. Curve A-8139 shows an 11-inch impeller at 1750 rpm with 275 gpm at 128 ft of head. NPSHR is about 6 ft, and $\eta_T$ is approximately 0.55. Notice the pump is chosen far from the zone of best efficiency and requires a 25-hp motor.

2. Curve A-8131 also shows an 11-inch impeller at 1750 rpm that delivers 275 gpm at 150 ft of head. NPSHR = 8 ft and $\eta_T$ = 0.68. This is better, but it is still to the left of the zone of best efficiencies. It is closer on head and still has a 25-hp motor.

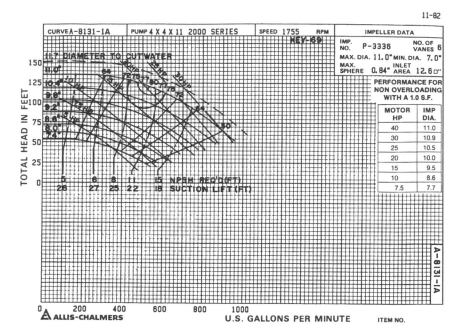

**FIGURE 5.12 (continued)**    Allis-Chalmers pumps, 1750 rpm. (ITT Industrial Pump Group.)

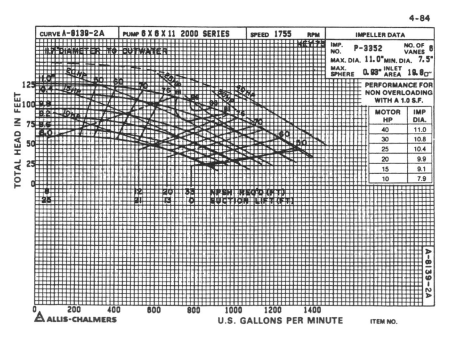

**FIGURE 5.12 (continued)**    Allis-Chalmers pumps, 1750 rpm. (ITT Industrial Pump Group.)

3. Curve A-8121 shows a somewhat larger pump with D = 12.2 inches at N = 1750 rpm, which delivers 275 gpm at 148 ft of head. NPSHR = 6 ft and $\eta_T = 0.73$. This is the best selection candidate thus far. Some efficiency is gained, but it is at the expense of a moderate increase in size and a small increase in NPSHR (which is still less than 10 ft). More importantly, there has been a drop-down to a motor requirement of 20 hp, which will reduce the cost substantially. This appears to be a good choice based on review of very limited catalog data.

## 5.6 SELECTION FROM VENDOR DATA: VARIABLE PITCH FANS

There are many applications where mass flow or volume flow rate must vary over fairly wide ranges or must be very precisely controlled. For such requirements, the use of variable pitch blading in a vane axial fan can provide the variability and control required over very wide ranges of performance. An example of these fans, the Variax Series of fans from Howden Variax, range in size from 794 to 1884 mm in diameter at 1470 rpm (50 Hz line frequency), with some smaller sizes available at 2950 rpm. Flow ranges from about 4 to 110 m³/s, at total pressure rises from 250 to 3000 Pa. These ranges are illustrated in Figure 5.13, and an example performance map is shown in Figure 5.14. Variable pitch fans are constructed with blades mounted in bearing assemblies that allow the blades to rotate around their radial axes so that the blade or cascade pitch angle can be continuously varied to change the flow and pressure rise while the fan is in operation. The bearing assemblies, linking mechanisms, and feedback control systems may add somewhat to the complexity and cost of a given machine, but the resulting configuration is capable of adjustment and control to meet very exacting requirements. The alternative system that can compete with the flexibility of the variable pitch fan is one with continuously variable speed, requiring perhaps a steam turbine drive system or continuously variable line frequency controllers to vary the fan motor speed.

The particular fan illustrated in Figures 5.15 and 5.16 has a diameter of 1000 mm or 1.0 m. The fan hub diameter is 630 mm, there are 10 blades, and the fan runs at 1470 rpm. The pressure rise vs. flow rate graph gives total pressure in Pa and flow rate in m³/s for a machine without the optional downstream diffuser (catalog information is also available for fans with diffusers). Also given is a graph of shaft power required, in kilowatts, and efficiency contours superposed on the pressure-flow curves, with total efficiencies shown. The distinguishing property of these graphs is the multiple performance curves given, one for each of a set of blade pitch angle settings ($\alpha_g$, the geometric pitch). The angle is generally measured from the plane of rotation to the blade chord line for the section located at the hub or base of the blade. $\alpha_g$ ranges for this fan from 10° to 55°, and flow rates can vary from about 0.7 to 28 m³/s, with total pressure rises up to about 1800 Pa. The zone of best efficiency performance (replacing the BEP) lies in an oval-shaped region within the 80% total efficiency contour, roughly between 1000 and 1500 Pa and 7 to 11 m³/s.

**TELLUS fans without diffuser**

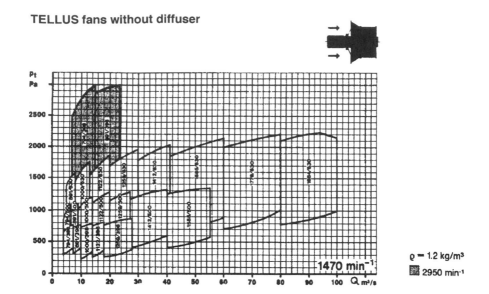

**FIGURE 5.13**   Performance range of variable pitch axial fans. (Howden Variax.)

In selecting a variable pitch fan, the performance point at which the system operates most of the time should be near the basic best efficiency point of the machine and certainly must lie within the best efficiency contour. If possible, the identifiable alternative points of operation should fall within this zone as well. In no case should an operating fan be allowed to lie outside of the dashed curve define the locus of stall entry, above which the various performance curves are not defined. As discussed previously, operation of an axial fan within the defined stall regions will yield rough, inefficient, and very noisy operation. Structural damage and eventual failure is possible if the fans are used in the stalled mode of operation. In addition, any choice of operation within the unstalled region of the fan map should provide an adequate margin of operation away from the stall boundary. This margin is usually defined as being at a pressure rise no greater than 90% of the stall pressure at that flow rate. For example, if the ASV-1000/630-10 fan of Figure 5.14 operates at 15m³/s, the allowable pressure rise is about 1620 Pa (0.9 × 1800 Pa), so the blade pitch would be about 49°.

Consider a simple selection problem for which the ASV-1000/630-10 fan can be used. Define the normal operating point (the NOP) as Q = 9 m³/s at 1100 Pa, a test block or maximum required point of operation (TBOP) as 13 m³/s at 1500 Pa, and a first alternate operating point (AOP1) as 6 m3/s at 1000 Pa. Other requirements on alternate points of operation are of course possible. For these three points, choose the ASV-1000/630-10 fan and obtain

NOP: 9m³/s at 1100 Pa, with $\alpha_g = 32°$ and $\eta_T = 0.81$, 73% of stall
TBOP: 13 m³/s at 1500 Pa, with $\alpha_g = 45°$ and $\eta_T = 0.78$, 85% of stall
AOP1: 6 m³/s at 1000 Pa, with $\alpha_g = 24°$ and $\eta_T = 0.76$, 77% of stall

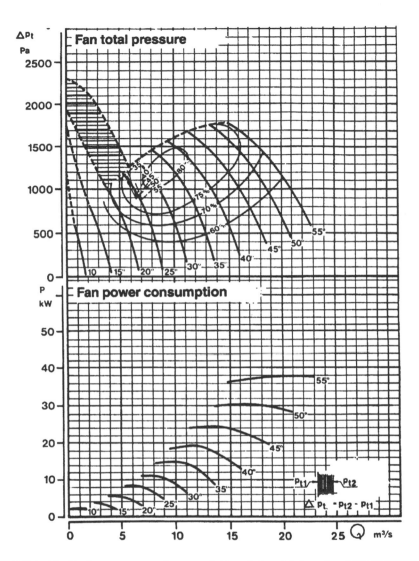

**FIGURE 5.14** Performance map for a variable pitch axial fan, the ASV-1000/630-10. (Howden Variax A/S.)

These three points may be identified as points in Figure 5.14. Clearly, these points work with the chosen fan and exhibit good efficiencies and stall margins. In general, one does not have such a simple starting point but simply is presented with the operating point values.

A real example might consist of

NOP: 50 m³/s at 1200 Pa
TBOP: 60 m³/s at 2000 Pa
AOP1: 20 m³/s at 1100 Pa

**FIGURE 5.15** A variable-pitch axial fan with diffuser. (Howden Variax A/S.)

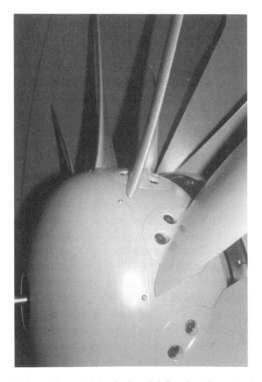

**FIGURE 5.16** Detail view of a variable-pitch axial fan showing circular blade base mounts for the controllable blades. (Howden Variax A/S.)

The Quick Selection Chart for a NOP of 50 m³/s at 1200 Pa indicates that the ASV-1585/630 would be adequate as a starting selection for this fan.

## 5.7 SUMMARY

The chapter began with a discussion of the breakdown of the Cordier diagram into zones of performance capability. These zones or regions were identified with particular kinds of turbomachines, ranging from a region on the left side of the diagram (low values of specific diameter) to a region at the right edge (high values of specific diameter). The leftmost region was associated with lightly loaded axial flow machines, and the rightmost region with heavily loaded radial flow machines. The intervening zones were subsequently assigned to the more heavily loaded axial flow equipment, mixed flow machines, and moderately loaded centrifugals. The hybrid or crossover turbomachines were discussed and the practice of multi-staging both axial flow and centrifugal impellers was introduced.

The concept of these classification zones was then extended to develop a logical treatment for determining the appropriate machine configuration for a given set of performance specifications. The process was quantified in a set of useful but approximate algorithms written in terms of the variables of the Cordier diagram, the specific speed, specific diameter, and the total efficiency. Thus, the process is defined as the relationship of speed (N), diameter (D), and machine type to flow rate (Q), pressure rise ($\Delta p_T$), and fluid density ($\rho$). The following illustrations of the technique were used to show the procedure as well as to emphasize the approximate nature of the algorithms and the need for flexibility in the selection and determination of efficiency.

Several rather extensive selection exercises, using manufacturers performance and geometric information, illustrated the procedure with fans, blowers, and pumps. The procedure was shown to produce an optimal choice of equipment with a variety of constraints.

## 5.8 PROBLEMS

5.1. A ventilating fan is required to provide 15,000 cfm of standard density air with a total pressure rise of 2 InWG. Use a Cordier analysis to select the cheapest fan possible with the following constraints:
(1) Use a total efficiency 10 points lower than Cordier.
(2) Motor cost is estimated by dollars/horsepower by number of poles ($/hp = 13.33 + 0.3 (\# \text{ poles} - 4)^2$.
(3) Use direct-connected motors of 2, 4, 6, 8, and 10 poles.
If the fan is to be operated for 200 hours per year, find the 5-year cost for buying and using the fan. Assume the initial cost is twice the motor cost, and use $0.08/kWH.

5.2. Use a Cordier analysis to evaluate candidate fan configurations for a flow rate of 10,000 ft³/min with a total pressure rise of 5 InWG (standard air). Assume a belt-driven configuration and cover a wide range of equipment. Choose a best candidate based on total efficiency. If discharge velocity pressure is not recovered (base $V_d$ on impeller outlet area), choose a best fan based on static efficiency.

5.3. A chemical process fan must deliver 5000 cfm of air ($\rho = 0.0023$ slugs/ft³) at 32 InWG total pressure rise. Select the proper single-stage centrifugal fan for this application, and compute the diameter, speed, efficiency, and power. Also select a simple two-stage centrifugal fan arrangement (in series arrangement) and compute the diameter, speed, efficiency, and power. Compare the two selections as quantitatively as possible, and discuss the advantages and disadvantages of both configurations (Hint: Neglect compressibility effects.)

5.4. Rework Problem 5.3 using a multi-stage axial configuration. Justify your number of stages.

5.5. Re-constrain the analysis of Problem 5.4 to include a minimization of the overall fan noise.

5.6. Use Table 5.2 with a Cordier analysis to select a vane axial fan with requirements of $Q = 12,000$ cfm and $\Delta p_s = 0.50$ InWG at standard density. Assume initially that the velocity pressure is 0.75 InWG and iterate the solution once only.

5.7. To select a tube axial fan that delivers 2 m³/s at a static pressure rise of 70 Pa, do a Cordier analysis aimed at Figure P5.7. (Hint: Analyze based on available diameters and speed limitation. Select based on best fit to Figure P5.7 and minimum power required.)

5.8. Following a Cordier analysis, use Figure P5.7 to choose a tube axial fan suitable for $Q = 200,000$ cfm at $\Delta p_s = 3.75$ InWG, with a low noise as a selection criterion.

5.9. Use Table 5.2 to try to find a fan to provide $Q = 6250$ cfm at $\Delta p_T = 0.75$ InWG, in standard air.

5.10. We need to pump 900 gpm of 85°F water at 100 ft of head from an open pond requiring 8 ft of lift. Select a pump that can perform to this specification with an adequate cavitation margin. Try for the best efficiency attainable.

5.11. Select a fan to deliver 15,000 cfm of air (68°F at 1 atmosphere) that provides a pressure rise of 4.0 InWG static. To keep costs down, try to avoid a need for more than 12 hp, and try to find a comparatively quiet fan.

5.12. A pump is required to supply 50 gpm of 85°F water at 50 feet of head. NPSHA is only 10 ft. Look for the smallest pump available.

5.13. Find a fan that can supply 100,000 cfm of 134°F air with $P_b = 28.7$ InHG at a pressure rise of 1.0 InWG. Use minimum noise as a criterion for the best choice.

5.14. Identify a minimum sound power level blower that can deliver 25 m³/s of standard density air ($\rho = 1.2$ kg/m³) at 500 Pa static pressure rise.

5.15. A small parts assembly line involves the use of solvents requiring a high level of ventilation. To achieve removal of these toxic gasses, a mean velocity of 8 ft/s must be maintained across the cross-section of the room as sketched (Figure P5.15). The toxicants are subsequently removed in a wet scrubber or filter, which causes a pressure drop of 30 lbf/ft². Using a

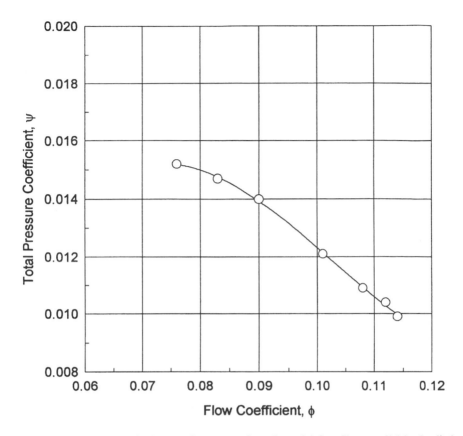

**FIGURE P5.7a** Dimensionless performance of a tube axial fan. Sizes available (in-dia.): 12, 15, 18, 21, 24, 27, 32, 36, 42, 48, 54, 60, 72, 84, 96. Maximum tip speed, $V_T \leq 590$ ft/s.

   Cordier analysis, select the design parameters for a fan that will provide the required flow and pressure rise, while operating at or near 90% total efficiency and that minimizes the diameter of the fan. Give dimensions and speed of the fan, and describe the type of fan chosen.

5.16. For the performance requirements stated in Problem 5.15, constrain your selection of a fan to less than $L_w = 100$ dB. Relax the size and efficiency requirements in favor of this noise requirement.

5.17. A water pump must supply 160 gpm at 50 ft of head, and the pump is provided with 4.0 ft of NPSHA. For a single-suction pump direct-connected to its motor, determine the smallest pump one can use (defining D, N, and $\eta$) without cavitation. (Hint: The direct connected speeds are at or near synchronous for 2-pole, 4-pole, ..., up to 12-pole motors on 60 Hz current.)

5.18. A small axial flow fan has its performance at BEP defined by $Q = 2000$ cfm, $\Delta p_s = 0.10$ InWG, $\rho = 0.00233$ slugs/ft$^3$ with a diameter of 20 inches.
   (a) Estimate the specific speed and required rpm.
   (b) What type of fan should one use?

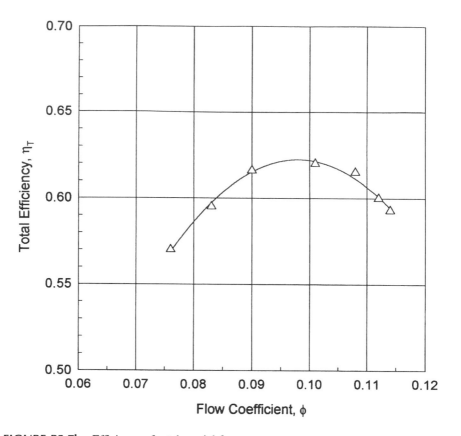

**FIGURE P5.7b**   Efficiency of a tube axial fan.

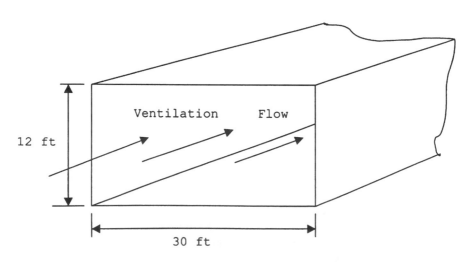

**FIGURE P5.15**   Room schematic.

(c) How much motor (shaft) horsepower is needed for the fan?

(d) Estimate the sound power level in dB.

5.19. An aircraft cabin must be equipped with a recirculation fan to move the air through the heaters, coolers, filters, and other conditioning equipment. A typical recirculation fan for a large aircraft must handle about 0.6 m³/s of air at 1.10 kg/m³ (at the fan inlet) with a total pressure rise requirement of 5 kPa. The electrical system aboard the aircraft has an alternating current supply at a line frequency of 400 Hz, and the fan must be configured as a direct-connected impeller for reliability.

Size the fan impeller to achieve the best possible static efficiency.

5.20. If one constrains the fan of Problem 5.16 to generate a sound power level of less than 85 dB, how are design size and efficiency affected?

5.21. For the recirculation fan of Problem 5.19, if one must keep the sound power level below 90 dB, what are the design options?

5.22. The performance requirements for an air mover are stated as flow rate, $Q = 60$ m³/min; total pressure rise, $\Delta p_T = 500$ Pa. Ambient air density, $\rho = 1.175$ kg/m³. The machine is constrained not to exceed 90 dB sound power level.

(a) Determine a suitable speed and diameter for the fan and estimate the total efficiency.

(b) Correct the total efficiency for the effects of Reynolds Number, a running clearance of 2.5 × Ideal, and the influence of a belt drive system (assume a drive efficiency of 95%).

5.23. Use the dimensionless performance curves of this chapter to select three candidates for the requirements of Problem 5.22 that closely match the results of your preliminary design decision in that problem.

5.24. To evacuate a stone quarry (to preclude the pleasure of unauthorized swimming), one needs a pump that can move at least 0.5 m³/s against a head resistance of 80 m. The pump must be floated on a barge with its inlet 1.125 m above the surface of the 12°C water. The atmospheric pressure during pumping may fall as low as 98 kPa. Select a non-cavitating pump for this application and match your modeling of size and speed to the best available pump from Figure P5.24.

5.25. Repeat the fan selection exercise of Problem 5.7 using the flow rate of 2 m³/s but changing the static pressure rise to 140 Pa and then 35 Pa. Compare the results to each other and the results of Problem 5.7. Use the dimensionless curves of Figure P5.7 for the fans.

5.26. A water pump, which is required to provide 21 m³/hr with 16 m of head, must operate with 1.25 m of NPSHA. For a single-suction machine, direct-connected to its motor, determine the smallest pump one can use, defining the D, N, and η. What would the minimum size be for an NPSHA of 0.5 m?

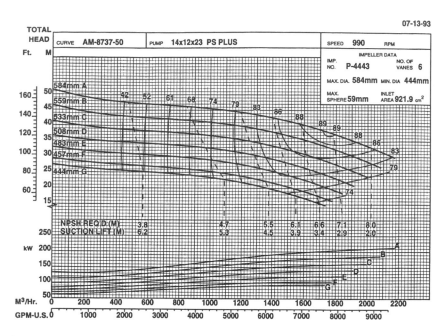

**FIGURE P5.24a**  Performance curves for an ITT A-C pump. (ITT Industrial Pump Group, Cincinnati, OH.)

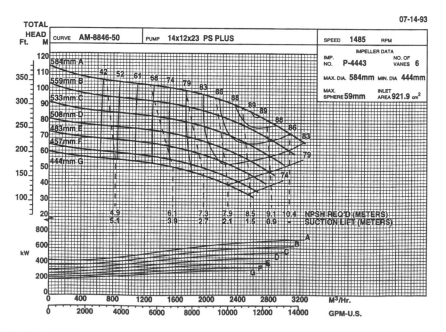

**FIGURE P5.24b**  Performance curves for an ITT A-C pump. (ITT Industrial Pump Group, Cincinnati, OH.)

# 6 Energy Transfer and Diffusion In Turbomachines

## 6.1 FUNDAMENTAL CONCEPT OF ENERGY TRANSFER

Transfer of energy to or from the fluid passing through the moving blade rows of a turbomachine can be accounted for through application of the Reynolds Transport Theorem to the conservation of angular momentum. Recall from earlier study of fluid mechanics that the equation for an arbitrary extrinsic property of a system, B, can be analyzed through the relation

$$\frac{dB}{dt}\bigg|_{sys} = \frac{\partial}{\partial t}\left(\iiint_{cV}\rho\beta dV\right) + \iint_{cS}\beta\rho(\mathbf{V}\bullet\mathbf{n})dA \tag{6.1}$$

where $\beta = dB/dm$, cV is the control volume, and cS is the control surface. Applying the general conservation law for control volumes to the angular momentum $H_o$ where

$$\mathbf{H}_o = \int(\mathbf{r}\times\mathbf{V})dm \tag{6.2}$$

one obtains

$$\beta = \frac{d}{dm}\int(\mathbf{r}\times\mathbf{V})dm = (\mathbf{r}\times\mathbf{V}) \tag{6.3}$$

and

$$\frac{d\mathbf{H}_o}{dt} = \sum\mathbf{M}_{axis} = \sum(\mathbf{r}\times\mathbf{F}) \tag{6.4}$$

Using the Reynolds Transport Theorem, one can write this as

$$\sum\mathbf{M}_{axis} = \sum(\mathbf{r}\times\mathbf{F}) = \frac{\partial}{\partial t}\left(\iiint_{cV}\rho(\mathbf{r}\times\mathbf{V})dV\right) + \iint_{cS}\rho(\mathbf{r}\times\mathbf{V})(\mathbf{V}\bullet\mathbf{n})dA \tag{6.5}$$

or

$$\sum\mathbf{M}_{axis}\frac{\partial}{\partial t}\left(\iiint_{cV}\rho(\mathbf{r}\times\mathbf{V})dV\right) + \sum(\mathbf{r}\times\mathbf{V})\dot{m}'_{out} - \sum(\mathbf{r}\times\mathbf{V})\dot{m}'_{in} \tag{6.6}$$

155

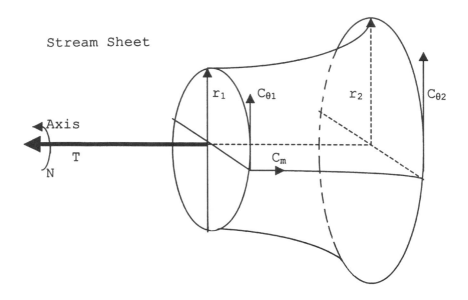

**FIGURE 6.1**   Control volume for angular momentum analysis.

Here, $m'$ is the mass flow rate. For steady flow, $\partial/\partial t = 0$, and one can drop the first term. Then, defining $\Sigma M_{axis} = T_o$, the torque, one can rewrite the equation as

$$T_o = \sum (\mathbf{r} \times \mathbf{V}) m'_{out} - \sum (\mathbf{r} \times \mathbf{V}) m'_{in} \qquad (6.7)$$

As shown in Figure 6.1 for the inlet side,

$$(\mathbf{r} \times \mathbf{V})_{in} = r_1 C_{\theta 1} \qquad (6.8)$$

and

$$(\mathbf{r} \times \mathbf{V})_{out} = r_2 C_{\theta 2} \qquad (6.9)$$

Further, one-dimensional mass conservation states that $m'_{in} = m'_{out}$ for steady flow. Thus, $T_o$ becomes

$$T_o = m'(r_2 C_{\theta 2} - r_1 C_{\theta 1}) \qquad (6.10)$$

or in terms of the power (P)

$$P = T_o N = m'(N r_2 C_{\theta 2} - N r_1 C_{\theta 1}) \qquad (6.11)$$

Conventionally (Shepherd, 1964), these are written separately for turbines and pumping machines (compressors, fans, blowers) as

$$\frac{P}{m'} = U_2 C_{\theta 2} - U_1 C_{\theta 1} : \qquad \text{pump} \qquad (6.12)$$

$$\frac{P}{m'} = U_1 C_{\theta 1} - U_2 C_{\theta 2} : \qquad \text{turbine} \qquad (6.13)$$

where $U = Nr$. This is a general relation for incompressible or compressible flow, ideal or frictionless fluid. However, it is restricted to steady and adiabatic flow with no extraneous, external torques associated with bearing drag, seal drag, or disk shear. For these conditions, $h_{02} - h_{01} = U_2 C_{\theta 2} - U_1 C_{\theta 1}$ is another form for writing the result. The more general formulation using

$$\iint_{cS} \rho (\mathbf{r} \times \mathbf{V})(\mathbf{V} \bullet \mathbf{n}) dA \qquad (6.14)$$

allows an analysis across inlets and outlets with varying properties to be carried out by integration or summation.

Consider a simple planar flow example as displayed in Figure 6.2. The flow is a model of what occurs in a single stage of a turbine. For the example, let $m' = 20$ kg/s, $U_1 = U_2 = U = 1047$ m/s. The flow enters axially in the turbine nozzle section and exits axially from the turbine rotor section. The nozzle approach flow is turned by the nozzle vanes through 45° to an absolute velocity of $C_1 = 500$ m/s. The flow then approaches the moving blade row for which $U = 1047$ m/s. One calculates the power from

$$P = m'(U_1 C_{\theta 1} - U_2 C_{\theta 2})$$

If one assumes an axial discharge and the known blade-row inflow,

$$C_{\theta 1} = C_1 \sin 45° = 0.707(500 \ \text{m/s}) = 352.5 \ \text{m/s}$$

$$C_{\theta 2} = 0$$

So,

$$P = m'(U_1 C_{\theta 1} - 0) = 20 \ \text{kg/s}\left[(1047 \ \text{m/s})(353.5 \ \text{m/s}) - 0\right]$$

$$= 7,402,290 \ \text{kgm}^2/\text{s}^3 = 7.40229 \times 10^6 \ \text{Nm/s} = 7.40 \ \text{MW}$$

The Euler equation allows one to relate performance, in this case power, directly to the pattern of flow in the machine. Consider a pumping flow example (see Figure 6.3)

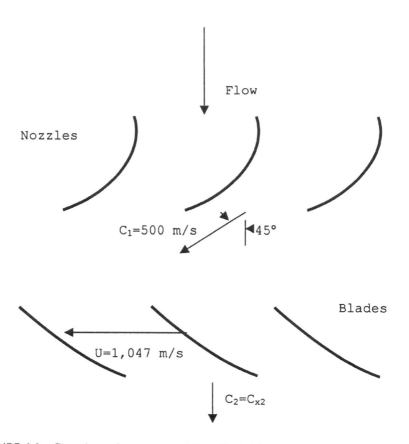

**FIGURE 6.2**   Cascade or planar representation of a turbine stage.

where m′ = 12 slugs/s, the flow approaches the blade row axially at C = 100 ft/s, and the blades are moving at U = $U_1$ = $U_2$ = 300 ft/s. The flow exits the blade row at a 15° angle to the axial (x) direction, $C_{x2}$ = 100 ft/s, and the blades have turned the flow through 15° of "fluid turning."

What is the power input to the fluid? Use Euler's equation:

$$P = m'\left(U_2 C_{\theta 2} - U_1 C_{\theta 1}\right); \quad C_{\theta 1} = 0$$

and

$$C_{\theta 2} = C_{x2} \tan 15° = 26.8 \text{ ft/s}$$

Thus,

$$P = 12 \text{ slugs/s}\left(300 \text{ ft/s}\right)\left(26.8 \text{ ft/s}\right) = 96,480 \text{ lbf ft/s}$$

$$P = 175.4 \text{ hp}$$

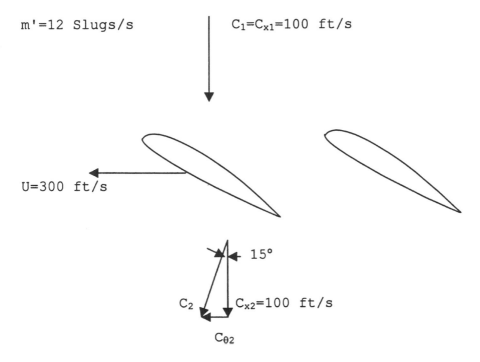

**FIGURE 6.3**  Cascade for a pumping machine.

The amount of shaft power (which must account for the viscous flow losses) that must be input depends on the efficiency of the machine. Typically,

$$P_s = \frac{P}{\eta_T} \text{ for a "pump" (input)} \tag{6.15}$$

$$P_s = P \cdot \eta_T \qquad \text{for a turbine (output)} \tag{6.16}$$

One can also expand the $P$ value in terms of the performance variables as

$$P = m'gH = \rho QgH = \Delta p_T Q \tag{6.17}$$

so that the Euler equations can also be written as

$$\frac{P}{m'} = gH = \frac{\Delta p_T}{\rho} = \left(U_2 C_{\theta 2} - U_1 C_{\theta 1}\right) \tag{6.18}$$

for pumps, and similarly for turbines.

Next, take a look at some more involved examples for typical flow through turbomachines and analyze what happens to the velocity vectors as flow moves

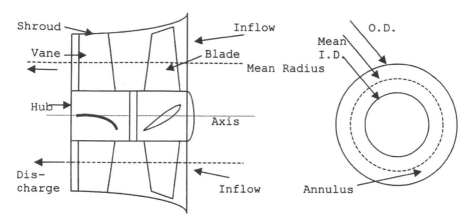

**FIGURE 6.4**  Mean radius stream flow surface in an axial fan.

through the machine. These vectors can be related to the performance variables and
also to the shape of the blade and vane elements.

## 6.2  EXAMPLE OF ENERGY TRANSFER IN AN AXIAL FAN

Consider a fan with an outer diameter (o.d.) of $D = 1$ m, and an inner or hub diameter
$d = 0.67$ m. We are going to analyze blading, vanes, work input, and pressure rise
along the "mean radius" $(r_m)$, as shown in Figure 6.4, where $r_m = (D + d)/4$. The
fan operates at 900 rpm, producing $\Delta p_T = 1250$ Pa, and $Q = 2.0$ m³/s with an air
density of $\rho = 1.2$ kg/m³. The mean radius is $r_m = 0.417$ m and $N = 94.3$ s⁻¹. Calculate:

$$U_m = U_1 = U_2 = \left(94.3 \text{ s}^{-1}\right)(0.417 \text{ m}) = 39.3 \text{ m/s}$$

Axial velocity, $C_{x1}$, is given by the flow over the annulus area:

$$C_{x1} = \frac{Q}{A_{ann}} = \frac{Q}{\left(\frac{\pi}{4}\right)\left(D^2 - d^2\right)} = 4.62 \text{ m/s}$$

One can combine these two to look at the blade relative velocity, $W_1$, at the leading
edge of the blade as shown in Figure 6.5. This yields

$$W_1 = \left[U_m^{\ 2} + C_{x1}^{\ 2}\right]^{1/2} = \left[(39.3)^2 + (4.62)^2\right]^{1/2} = 39.6 \text{ m/s} \qquad \textbf{(6.19)}$$

with

$$\beta_1 = \tan^{-1}\left(\frac{U_m}{C_{x1}}\right) = \tan^{-1}\left(\frac{39.3}{4.62}\right) = 83.3° \qquad \textbf{(6.20)}$$

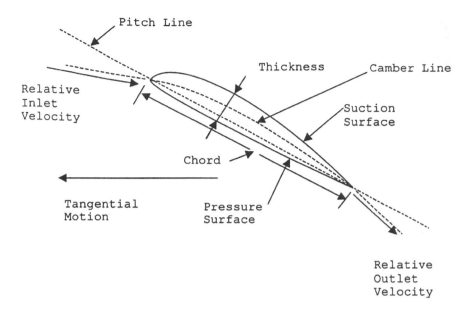

**FIGURE 6.5**   Sketch of an airfoil blade with vectors, angles, and nomenclature.

measured from the direction of the axis.

The blade chord should be aligned at its leading edge with the relative velocity $W_1$, for widely spaced blades, such that the flow turns upward slightly and moves smoothly over the blade with minimum initial disturbance. For closely spaced blades, the proper alignment moves to become tangent to the camber line at the leading edge. This alignment of the velocity vector and blade is illustrated in Figure 6.5, along with some basic airfoil nomenclature. Again, if the blades are closely spaced and thin, the proper alignment of blade and inlet vector is along the camber line at the nose of the blade (leading edge). For more widely spaced blades, this alignment moves toward the chord line, a straight line joining the nose and tail (trailing edge). A later chapter takes a look at this alignment question in more detail. For now, approximate the "cascades" of blades as being fairly closely spaced with thin blades or vanes.

Now use the Euler equation to calculate the outlet absolute flow by doing two things:

1. Calculate $C_{x2}$ by imposing conservation of mass for incompressible flow and writing

$$C_{x2} = \frac{Q}{A_{ann}} = C_{x1} \tag{6.21}$$

2. Use the Euler equation, assuming that the fan efficiency $\eta_T = 1.0$ with $C_{\theta 1} = 0$.

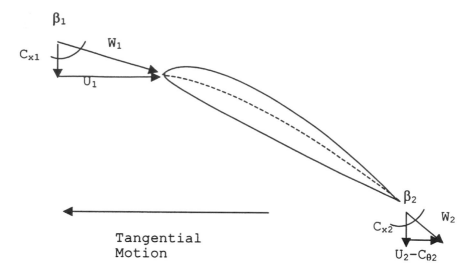

**FIGURE 6.6**   Outlet and inlet velocity vectors and flow angles for an axial flow blade.

Then,

$$\Delta p_T = \eta_T \rho U_m C_{\theta 2}$$

$$C_{x2} = 4.62 \ \text{m/s} \tag{6.22}$$

$$C_{\theta 2} = \frac{\Delta p_T}{(\rho U_m \eta_T)} = 21.2 \ \text{m/s}$$

The above assumptions are the "simple stage" assumptions requiring uniform incompressible axial velocity components everywhere in the fan "stage" and requiring that $C_{\theta 1} = 0$. Now use $C_{\theta 2}$, $U_m$, and $C_{x2}$ to look at the absolute and relative velocities at the exit of the blade row (at the blade trailing edge). This is illustrated in Figure 6.6. $W_2$ can then be written as

$$W_2 = \left[ (U_m - C_{\theta 2})^2 + (C_{x2}^2) \right]^{1/2} = 18.6 \ \text{m/s} \tag{6.23}$$

and the flow angle is

$$\beta_2 = \tan^{-1} \left[ \frac{(U_m - C_{\theta 2})}{C_{x2}} \right] = 75.7° \tag{6.24}$$

Now, we can relate the blade relative inlet flow vector to the outlet vector using

$$W_1 = 39.6 \text{ m/s}; \quad \beta_1 = 83.3°$$

$$W_2 = 18.6 \text{ m/s}; \quad \beta_2 = 75.7°$$

These are conventionally combined into:

$$\frac{W_2}{W_1} = \frac{18.6}{39.6} = 0.470$$

(6.25)

$$\theta_{fl} = \beta_1 - \beta_2 = 83.3 - 75.7 = 7.6°$$

$W_2/W_1$ is called the de Haller ratio and $\theta_{fl}$ is the angle of deflection caused by the blade acting on the fluid. $\theta_{fl}$ is also called the flow turning angle. Both provide a measure of how hard the blade is working. The larger the flow turning angle, the harder the blade or vane is working, and, conversely, the smaller the de Haller ratio (Wilson, 1984; Bathie, 1996), the harder the work is.

One can also examine the vane row using the absolute velocity vector coming out of the blade row as the vane row inlet velocity (see Figure 6.7). From before, $C_{\theta 2} = 21.2$ m/s and $C_{x2} = 4.62$ m/s. Combining, one obtains:

$$C_2 = \left(C_{x2}^{2} + C_{\theta 2}^{2}\right)^{1/2} = 21.7 \text{ m/s}$$

(6.26)

at an approach angle of

$$\beta_3 = \tan^{-1}\left(\frac{21.2}{4.62}\right) = 77.7°$$

(6.27)

Since the purpose of a vane row, at least for a single-stage fan, is to recover or straighten the flow to a pure axial discharge, the flow must be turned through 77.7° by the vanes so that $\beta_4 = 0°$. That is,

$$\frac{W_4}{W_3} = \frac{C_4}{C_3} = \frac{4.62}{21.7} = 0.213$$

(6.28)

and

$$\theta_{fl} = 77.7°$$

## 6.3  DIFFUSION CONSIDERATIONS

In the previous example, the vane appears to be working considerably harder than the blade. To quantify what is happening, take a look at the concept of velocity

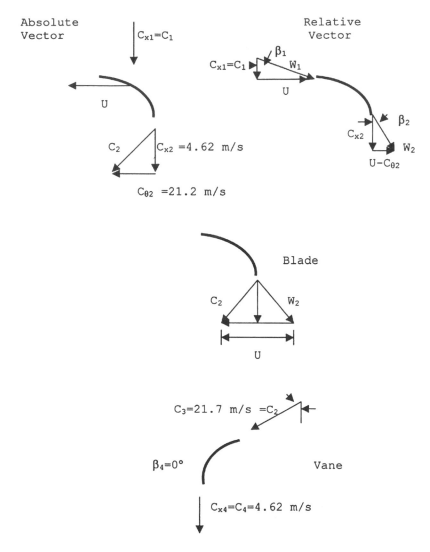

**FIGURE 6.7** Blade and vane geometry with velocity components.

deceleration along a solid boundary, or diffusion along a wall. Most books on fluid mechanics present a brief treatment of diffusion flow (e.g., Fox and McDonald, 1995) and illustrate that only a limited amount of deceleration or diffusion can occur. Beyond this limit, the flow undergoes high momentum loss associated with boundary layer separation. The flow streamlines no longer follow along the direction of solid surfaces constraining the flowfield, and the flow will generally become unsteady and unstable in a stationary or moving flow passage, for example, in the blade or vane row of a turbomachine. This breakdown is referred to as "stall," and it is a flow condition that should be carefully avoided in design, selection, and operation of a machine.

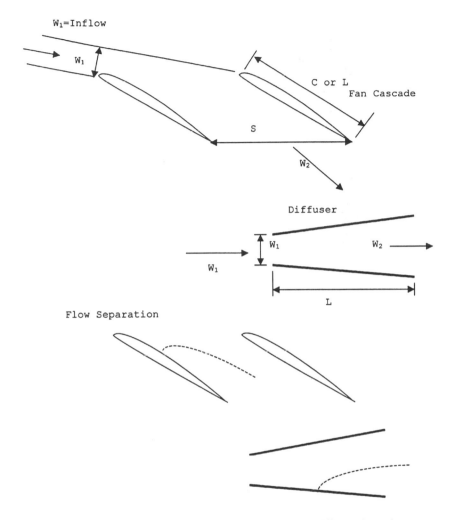

**FIGURE 6.8**   A blade-to-blade channel compared to a planar diffuser channel.

A blade is compared to a two-dimensional internal flow analog in Figure 6.8. When excessive deceleration in either device is attempted, the results are similar. For a diffuser, a pressure recovery coefficient is developed using Bernoulli's equation, such that for inlet and exit station 1 and 2, respectively,

$$C_p = \frac{(p_2 - p_1)}{\left(\left(\frac{\rho}{2}\right)V_1^2\right)} = 1 - \left(\frac{V_2}{V_1}\right)^2 \tag{6.29}$$

That is, the static pressure difference is normalized by the inlet velocity pressure or dynamic pressure to form a dimensionless coefficient.

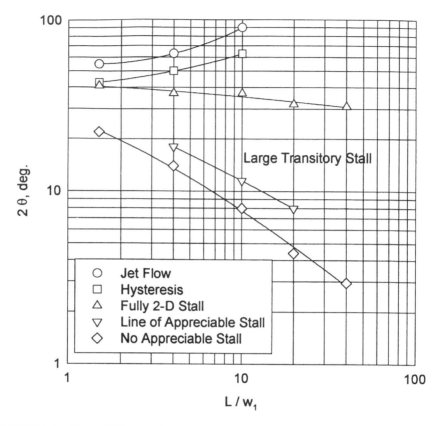

**FIGURE 6.9**  Planar diffuser performance map with blade-to-blade channel analogy. (Reneau et al., 1967.)

For our blades or vanes, this is equivalent to

$$C_p = 1 - \left(\frac{W_2}{W_1}\right)^2 \qquad (6.30)$$

If $W_2$ becomes small compared to $W_1$, $C_p$ approaches 1. Before that happens, as illustrated by experimental data for planar diffusers, the diffuser or blade stalls.

Two figures on planar diffusion stall are presented in Figure 6.9. To illustrate this limit, use the stall limit for planar diffusers as a working analog for a blade row by looking at very small values of $L/w_1$, say $1 < L/w_1 < 2$. Based on conservation of mass, one can relate the diffuser angle and length ratio to inlet and outlet velocities, at least ideally. That is,

$$\frac{A_1}{A_2} = \frac{w_1}{w_2} = \frac{W_2}{W_1} \qquad (6.31)$$

which is the velocity ratio we have been considering as the de Haller ratio. From Figure 6.9, at $L/w_1 = 1$, $2\theta = 24°$ at the line of appreciable stall. This line serves as a conservative lower limit on $W_2/W_1$ for stable, unstalled flow. For $\theta_{stall} = 12°$ and $L/w_1 = 1$, $W_2/W_1 = 0.70$. One infers that, with blades arranged so that "$L/w_1$" is about 1.0 (see Figure 6.8), one should use a limit on the velocity ratio of approximately 0.70. The limit stated by de Haller is a little more conservative; he suggested that $(W_2/W_1)_{min}$ should be at least 0.72 for good, clean, stall-free flow over the blading. Note that the "$L/w_1$" value used here is referred to as the blade row solidity, usually denoted by $\sigma$. Chapter 8 investigates the influence of this parameter extensively.

Using the diffuser analogy, one sees that, for $L/w_1 = 1.5$, $\theta_{stall} = 10°$, so $(W_2/W_1)_{min}$ can be as low as 0.65. The implication is that increasing blade row solidity allows an increase in blade loading or diffusion. The limits on $W_2/W_1$ are equivalent to pressure coefficients of $C_p = 0.51$ for $W_2/W_1 = 0.70$, and $C_p = 0.57$ for $W_2/W_1 = 0.65$.

## 6.4   DIFFUSION IN AXIAL FLOW

In axial flow machines, a practical limit on solidity in terms of construction weight and surface area for skin frictional losses is about 2 or a little more (Johnson and Bullock, 1965). This implies (from the diffuser analogy) an overall lower limit on $W_2/W_1$ of about 0.61. For now, use the conservative level chosen by de Haller of $W_2/W_1 > 0.72$ as a criterion for judging acceptable blade loading.

In the last example on energy transfer, $W_2/W_1$ was 0.470 for the blade and was 0.213 for the vane. Thus, one concludes that the blade is much too heavily loaded, and the blade row is probably stalled, since the de Haller ratio is far less than 0.72. The vane row is in even more distress at a de Haller ratio of only 0.213, and one must conclude that the cascades for the fan are probably poorly designed and operating with a low efficiency. One can examine this contention by comparing the fan to the Cordier diagram. For the given data,

$$N_s = \frac{Q^{1/2}}{\left(\left(\frac{\Delta p_T}{\rho}\right)^{3/4}\right)} N = 0.72$$

and

$$D_s = \frac{\left(\left(\frac{\Delta p_T}{\rho}\right)^{1/4}\right)}{Q^{1/2}} D = 4.02$$

This places the fan near the Cordier line, but it lies in Region E. This is the upper range for radial discharge machines, whereas the axial fan as a full-stage machine should be in Region C. There appears to be a simple explanation for the blade overloading problem. The design is trying to achieve unrealistically high pressure rise for an axial fan, without sufficient tip speed to generate the pressure rise.

If (perhaps for cost reasons) one insists on using an axial fan, then it will be necessary to look for one with a different size and speed to do this job, or design one's own. From the Cordier diagram, one can choose $N_s$ or $D_s$ values from Region C and find out what the axial fan should look like. Choose $D_s = 1.75$ from the middle of Region C, so that $N_s = 9D_s^{-2.103} = 2.77$. Inverting the definition of $N_s$ yields

$$N = N_s \left( \frac{\left( \frac{\Delta p_T}{\rho} \right)^{3/4}}{Q^{1/2}} \right) = 3433 \text{ rpm}$$

and

$$D = D_s \left( \frac{Q^{1/2}}{\left( \frac{\Delta p_T}{\rho} \right)^{3/4}} \right) = 0.435 \text{ m}$$

That is, choose a much smaller fan running at a much higher speed (smaller $D_s$, larger $N_s$). Now rework the mean-radius blade load calculations. Using $d = 0.67D$ as before, calculate the new $r_m$ as $r_m = (0.435 + 0.291)/4 = 0.182$ m. With $N = 3433$ rpm $= 359$ s$^{-1}$, $U_m = 359$ s$^{-1} \times 0.182$ m$) = 65.4$ m/s.

From d and D, one obtains $A_{ann} = 0.0819$ m$^2$, so

$$C_{x1} = \frac{Q}{A_{ann}} = \frac{2}{0.0819} = 24.4 \text{ m/s}$$

Then $W_1 = 69.8$ m/s and $\beta_1 = 70.7°$. Using the Euler equation,

$$C_{\theta 2} = \frac{\Delta p_T}{(\rho \eta_T U_m)} = 15.9 \text{ m/s}$$

So,

$$W_2 = \left( (65.4 - 15.9)^2 + (24.2)^2 \right)^{1/2} = 55.1 \text{ m/s}; \quad \beta_2 = 63.9°$$

For the blade row,

$$\frac{W_2}{W_1} = \frac{55.1}{69.8} = 0.79$$

compared to 0.470, and

$$\theta_{fl} = 6.75°$$

compared to 7.6°. This is much better and indicates acceptable blade loading.

For the vane, $W_3 = C_2 = [(15.9)^2 + (24.4)^2]^{1/2} = 29.1$ m/s and $\beta_3 = 33.1°$. Then, $W_4 = C_{x4} = C_{x1} = 24.4$ m/s, yielding $W_4/W_3 = 24.4/29.1 = 0.84$ versus 0.213. Finally, $\theta_{fl} = 33.1$ (vs. 77.7°). These values for the revised speed and size for the fan are a little conservative, since 0.72 is a slightly conservative number in itself. However, one can conclude that the fan stage properly chosen by the Cordier criterion is a reasonable choice. If an axial fan were chosen for this duty, one would look for a fan with a diameter somewhat less than 0.5 m but running at very high speed. Such a choice would yield a sound pressure level, from Chapter 4, of about 96 dB, which is rather loud for this level of performance, due to the relatively high tip speed.

For this example, the hub diameter was chosen rather arbitrarily to be 2/3 of the tip diameter. Had another value for d/D been chosen, a different value for $C_{x1}$ and $C_{\theta2}$ would have been obtained. The entire vector field would have changed along with the de Haller ratio $W_2/W_1$. For example, if the hub-tip diameter ratio d/D = 0.5, there will be a decrease in $C_{x1}$ and an increase in $C_{\theta2}$, so that the de Haller ratio decreases from 0.790 to 0.727. Conversely, if the relative hub size increases to d/D = 0.75, the de Haller ratio increases to 0.820. This is a general result that can be used to establish a constraint on d/D for a given specification of performance.

A general study of the dependence of the level of diffusion on the value of d/D can be carried out over a range of specific speed or diameter appropriate to axial fans to establish such a constraint. Work with the de Haller ratio calculated at the hub station of the blade in order to examine the most strenuous diffusion conditions on the blade or vane row. At the hub station, the value of U is the smallest so that the value of $C_{\theta2}$ required to achieve the specified value of head or pressure rise will be the largest value needed. (The underlying assumption here is that the work done or the total pressure rise achieved is the same at each radial position in the fan annulus. This is not an absolute restriction on the design of axial blade rows, but it is an approximate constraint, since large variations in the work distribution across the blade will lead to serious difficulties in design. This distribution question will be considered at length in Chapter 9, for three-dimensional flow. For now, a somewhat conservative constraint on minimum hub size can be established using this assumption.)

The fan in the previous example, with d/D = 0.67, would have a value of U at the hub of $U_h = Nr_h = 52.0$ m/s. The Euler equation with $\Delta p_T = 1250$ Pa with 1.2 kg/m³ density would require

$$C_{\theta 2} = \frac{1250}{(1.2 \times 52.3 \times 1.0)} = 19.9 \text{ m/s}$$

Thus, the de Haller ratio becomes $W_2/W_1 = 40.6/57.7 = 0.703$, significantly smaller than the mean radial station value of 0.790. The result is again a general one, but it can be mollified by allowing a less conservative velocity ratio criterion (as discussed earlier). One can allow the hub station value of $W_2/W_1$ to decrease to 0.60 (on the presumption of an increase in blade row solidity), but continue to use the 0.70 or 0.72 value as a guideline for mean station analysis. For the reduced hub case considered earlier with $d/D = 0.50$, the hub station de Haller ratio would be reduced to $W_2/W_1 = 21.7/42.9 = 0.506$. Thus, a hub size that seemed to be acceptable based on the mean station calculations now appears to be too small, even allowing a diffusion level of 0.60. Clearly, the acceptable value of hub size for this fan is between 0.50 and 0.67 times the tip diameter. An interpolation suggests $d/D = 0.58$, which yields $W_2/W_1 = 0.607$. Then the smallest hub size for this fan is about 0.25 m. This corresponding level of required performance can be represented by the specific speed or diameter ($D_s = 1.75$ for this case), or $N_s = 2.77$).

If one carries out a systematic study (Wright, 1996) to find the smallest hub–tip ratio capable of generating a de Haller ratio of 0.6 for a large number of $N_s$ values (from 1 to 8), the results can be plotted as shown in Figure 6.10. Along with the diffusion-based values, a curve from Fan Engineering (Jorgenson, 1983) is shown as well. This curve is somewhat more conservative. The diffusion-based data points were heuristically fitted to a curve given by

$$\frac{d}{D} \cong \left(\frac{1}{2}\right)\left\{1 - \left(\frac{2}{\pi}\right)\tan^{-1}\left[\left(\frac{2}{\pi}\right)(N_s - 3.8)\right]\right\} \qquad (6.32)$$

Also shown in the figure are results adapted from Balje's work (Balje, 1968). Balje's estimates are based on constraining $C_2^2 \leq gH/\eta_T$ and with $D_s$ and $N_s$ matched from the Cordier line. Nevertheless, these results are in reasonable agreement.

## 6.5   DIFFUSION IN RADIAL FLOW

The diffusion analogy for the blade or vane row can be readily applied to the radial flow cascade of a centrifugal fan or pump impeller. As Chapter 9 points out, in a more thorough discussion of radial flow cascades, the analogy is of somewhat limited value except for blades that are backwardly inclined for excellent high efficiency, moderate pressure rise impeller design. In many cases, particularly for relatively highly loaded cascades, the flow is significantly separated from the blade trailing faces (suction surfaces). The flow itself is smooth and stable, forming the classic jet-wake or low-speed core flow pattern without the unsteady pulsating flow seen in axial flow machines (Moore, 1984; Balje, 1968). The influence on overall machine performance is acceptable in light of the large amount of head rise obtained in a single stage.

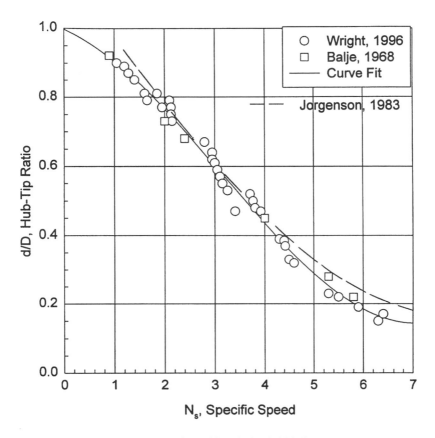

**FIGURE 6.10**  Hub size constraints for uniformly loaded blades.

Figure 6.11 shows a radial discharge cascade as a cutaway view of the blade row of a centrifugal impeller, in this case a centrifugal fan with backwardly inclined blading. The relationship shown between $W_1$ and $W_2$ to inlet and outlet blade geometry suggests an inherently high rate of diffusion when the geometry is viewed in two dimensions. However, in the design process, the rate of increase of available flow area for the relative velocity can be controlled through variation of the axial depth of the channel (into the page). Specifically, one can control the magnitude of $C_{r1}$ (at $r_1$) relative to $C_{r2}$ (at $r_2$) through choice of the channel widths $w_1$ and $w_2$. In terms of head rise as governed by the Euler equation, the relative velocity through the blade channel must also be decreased. The equations for $W_1$ and $W_2$ are (assuming $C_{\theta1} = 0$)

$$W_1 = \left(U_1^{\,2} + C_{r1}^{\,2}\right)^{1/2} \quad \text{and} \quad W_2 = \left[\left(U_2 - C_{\theta2}\right) + C_{r2}^{\,2}\right]^{1/2} \qquad (6.33)$$

where the radial components and the peripheral speeds may all vary and usually do. The flow areas, normal to the radial velocities, can be defined by

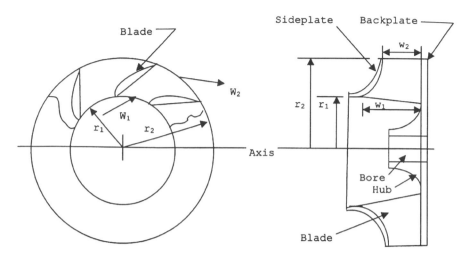

**FIGURE 6.11**   Radial flow cascade layout for a centrifugal fan.

$$A_1 = \frac{2\pi r_1 w_1}{N_B} \qquad \text{(at the inlet)} \qquad (6.34)$$

and

$$A_2 = \frac{2\pi r_2 w_2}{N_B} \qquad \text{(at the outlet)} \qquad (6.35)$$

Conservation of mass requires that

$$\rho_1 C_{r1} A_1 = \rho_2 C_{r2} A_2 \qquad (6.36)$$

Assuming incompressible flow and noting that $r_1 = d/2$ and $r_2 = D/2$, one can write

$$\frac{C_{r2}}{C_{r1}} = \left(\frac{d}{D}\right)\left(\frac{w_1}{w_2}\right) \qquad (6.37)$$

This velocity ratio is a critical parameter in determining the diffusion ratio through the radial flow cascade so that, equivalently, d/D and $w_1/w_2$ become critical design parameters for the centrifugal blade row. d/D is analogous to the hub–tip ratio that plays such a strong role in axial flow machine design, and $w_1/w_2$ provides an additional variable to maintain good flow conditions.

Looking back to the equations for $W_1$ and $W_2$, one can see that in order to control the negative influence of $C_{\theta2}$, required to generate $\Delta p_T$, one can manipulate $C_{r1}$ and $C_{r2}$ by varying the d/D and $w_1/w_2$ geometry. However, the best approach to ensure a good starting point seems to be to establish a minimum value of $W_1$ so that when $W_2$ is forced to be small through $C_{\theta 2}$, there is a decent chance of achieving an acceptable de Haller ratio. The minimization also serves to reduce inlet Mach number in compressors to avoid premature choke or shock wave formation, and it will reduce cavitation sensitivity at the leading edges of the blades in liquid pumps.

One can rewrite $W_1$ as

$$W_1 = \left( \left( \frac{Q}{A_0} \right)^2 + \frac{N^2 d^2}{4} \right)^{1/2} \tag{6.38}$$

Here, $A_0$ is the area of the inlet throat, $\pi d^2/4$, and the assumption or approximation is made that $C_{r1} \cong C_{x0}$ (or that $A_0 \cong A_1$). An equivalent minimization for this discussion is to minimize $(W_1/U_2)^2$ by setting $d[(W_1/U_2)^2]/d(d/D)$ to zero. The argument becomes

$$\left( \frac{W_1}{U_2} \right)^2 = \frac{\phi_{BP}^2}{\left( \frac{d}{D} \right)^4} + \left( \frac{d}{D} \right)^2 \tag{6.39}$$

so the optimum value of $d/D$ is

$$\frac{d}{D} = \phi_{BP}^{1/3} 2^{1/6} = \left( \frac{2^{7/6}}{\pi^{1/3}} \right) \phi^{1/3} \tag{6.40}$$

where $\phi_{BP}$ is a flow coefficient based on tip speed and disk or backplate area of the impeller according to

$$\phi_{BP} = \frac{Q}{\left( \left( \frac{\pi}{4} \right) D^2 U_2 \right)}; \quad \phi = \frac{Q}{(ND^3)} \tag{6.41}$$

One can adopt this result, constraining $d/D \leq 1$

$$\frac{d}{D} \cong 1.53 \, \phi^{1/3} \tag{6.42}$$

This algorithm (Wright, 1996) supplies for centrifugal impellers the equivalent to the hub–tip ratio guideline for axial flow machines. It allows an intelligent starting point for laying out the centrifugal cascade and leaves the $w_1/w_2$ ratio as an additional parameter for fine-tuning the diffusion level.

As an example, consider a centrifugal fan required to supply 4 m³/s of air with $\rho = 1.2$ kg/m³. The required pressure is $\Delta p_T = 3$ kPa. If a large specific diameter in

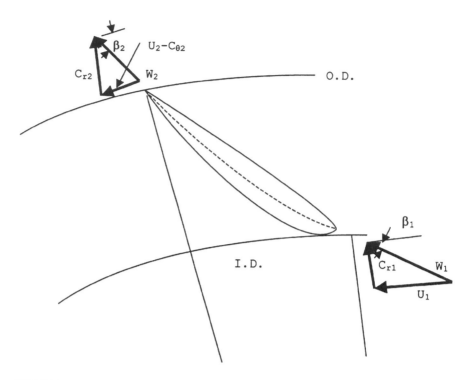

**FIGURE 6.12**   Relative velocity components and flow angles for a centrifugal cascade.

region E of the Cordier diagram (Figure 5.1) is chosen, one obtains $D_s = 3.4$. The corresponding $N_s$ is about 0.91 and $\eta_T \cong 0.90$. With the given data, the fan diameter should be 0.962 m and the rotating speed about 1150 rpm. Using the throat size algorithm gives $d = 0.492$ m. The areas through the cascade are assumed to be equal $(A_0 = A_1 = A_2)$ so that $w_1 = d/4 = 0.123$ m and $w_2 = (d/D)w_1 = 0.063$ m. Then, $C_{r1} = C_{r2} = 21.0$ m/s. Also, $U_1 = 29.6$ m/s, $U2 = 57.9$ m/s, and $C_{\theta 2} = 48.0$ m/s. $W_1 = 26.3$ m/s, $W_2 = 23.0$ m/s, and the de Haller ratio is 0.65. Reducing $w_2$ slightly to 0.057 m raises the de Haller ratio to about 0.704. This is a very reasonable-looking impeller layout, and 0.65 or 0.71 are probably fairly conservative values. Further details of the blade layout will be largely deferred to later chapters, but an approximate sketch, based on alignment of $W_1$ and $W_2$ with the blade mean camber line, is given in Figure 6.12. The flow angles are given by

$$\beta_1 = \tan^{-1}\left(\frac{C_{r1}}{U_1}\right) \quad \text{and} \quad \beta_2 = \tan^{-1}\left(\frac{C_{r2}}{(U_2 - C_{\theta 2})}\right) \quad \text{(6.43)}$$

Note that the convention for defining these angles is the opposite of that used in the axial flow cascades, but the distinction is traditional in the literature.

## 6.6 SUMMARY

This chapter introduced the fundamental concept of energy transfer through the mechanism of angular momentum. The basic Reynolds Transport Theorem was employed to develop the Euler equations for turbomachines in the form of the product of peripheral speed and induced swirl velocity. The resulting relations are general, subject to the steady, adiabatic flow restrictions and the exclusion of extraneous torques. Simple planar examples were used to illustrate the flow behavior related to the formal equations.

A more extensive example of energy transfer in axial flow was used to relate the energy transfer process to the geometric features and mechanical speed of the axial flow machines. Through the example, the concept of blade and vane loading in a machine was developed in terms of the blade or vane relative velocities and the fluid turning angles.

The concept of loading was extended to the introduction of blade or vane row diffusion. The analogy of channel flow to flow in a planar diffuser was used to establish the concepts of flow separation and stall as limitations on achievable performance within a machine.

Axial flow examples were extended in a discussion of the influence of blade row solidity on the allowable levels of diffusion. The example being studied was continued by using the guidance of the Cordier algorithms to develop an acceptable flow path for an axial fan to meet both the performance requirements and the de Haller diffusion constraint. The diffusion concept was also used to establish an algorithm relating the ratio of hub diameter to fan diameter required for acceptable diffusion levels, and it was written in terms of the fan specific speed or specific diameter.

Radial flow blade rows were also examined to determine acceptable geometric layouts based on required performance and acceptable diffusion levels. It was noted in doing so that, for radial flow machines of very high specific diameter, the velocity ratios or diffusion levels are not as critical to good design as they are for axial flow machines. However, the constraint remains an important one for radial flow paths of moderate specific diameter with backwardly inclined blading.

The ratio of impeller inlet diameter to exit diameter, similar to the hub–tip ratio for axial machines, was examined in terms of minimization of the relative velocity at the impeller inlet. An algorithm was established to relate this geometric property of the impeller to the flow coefficient, and hence to the specific speed and specific diameter of the radial flow machine.

## 6.7 PROBLEMS

6.1. A centrifugal pump is configured such that $d = 30$ cm, $D = 60$ cm, $w_1 = 8$ cm, $w_2 = 5$ cm, $\beta_1 = 20°$, and $\beta_2 = 10°$. Pumping water (20°C) at $N = 1000$ rpm, determine the ideal values of flow rate, head rise, and required power.

6.2. A pump delivers 2700 gpm at 1500 rpm. The outlet geometry is given as $w_2 = 1.5$ inches, $r_2 = 8$ inches, and $\beta_2 = 30°$. Estimate the head rise and ideal power required.

6.3. A centrifugal fan with a total efficiency of 85% running at 1800 rpm has $w_1 = w_2 = 6$ cm, $d = 12$ cm, $D = 40$ cm, $\beta_1 = 40°$, and $\beta_2 = 20°$. Estimate the flow rate, total pressure rise, and input horsepower ($\rho = 1.2$ kg/m³).

6.4. An axial water pump has a total pressure rise of 20 psig, a volume flow rate of 6800 gpm, and tip and hub diameters of 16 and 12 inches, respectively. The speed is 1200 rpm.
   (a) Estimate the total efficiency and required power input.
   (b) Construct a mean station velocity diagram to achieve the pressure rise, including the influence of efficiency.
   (c) Lay out the blade shape at the mean station, and at the hub and tip stations. Assume zero incidence and deviation.
   (d) Calculate the de Haller ratio at each station.

6.5. A tube axial fan provides a static pressure rise of 1250 Pa with a volume flow rate of 120 m³/s and an air density of 1.201 kg/m³. The fan has a diameter of 3.0 m with a 1.5-m hub.
   (a) Develop a vector set (velocity diagram) at the inlet and outlet of the fan at the hub station of the blade, assuming ideal loss-free flow.
   (b) Estimate the total efficiency and power for the fan and correct the outlet velocity vectors for the implied losses.

6.6. Develop an appropriate vane row configuration for the tube axial fan of Problem 6.5, given that the absolute velocity leaving the blade row of that fan was $C_{x2} = 23.7$ m/s and $C_{\theta2} = 16.3$ m/s (at the mean radius). Estimate the increase in static efficiency and static pressure rise based on recovery of the velocity pressure associated with the swirl velocity at the hub.

6.7. Air flow ($\rho = 0.00233$ slugs/ft³) enters the blade row of an axial fan through a set of inlet guide vanes (IGVs) and exits the blade row with pure axial velocity. (See Figure P6.7.) Calculate the ideal power and total pressure. The exit angle from the IGVs is $-32°$, and the mass flow rate through the fan is 10.0 slugs/s.

6.8. Rework Problem 6.7 if the exit angle from the IGV row is:
   (a) 0°; (b) $-10°$; (c) $-20°$; (d) $+20°$.

6.9. For the wind turbine studied in Problem 3.19, one wants to force the turbine to have a pure axial flow discharge. Develop the required velocity diagrams for both rotor and vanes at the 70% radial station. Do the vanes make much difference? Are they really justified? What is gained, both qualitatively and quantitatively?

6.10. An axial flow fan must produce 450 Pa total pressure and provide a flow rate of 5.25 m³/s. Choose a reasonable specific speed and diameter for a vane axial fan layout. Vary the hub–tip diameter ratio, d/D, while checking the de Haller ratio at the hub station of the blade. Using 0.6 as the lowest allowable value of $W_2/W_1$, determine the smallest hub–tip ratio one can use. Compare this value to the information provide in Figure 6.10.

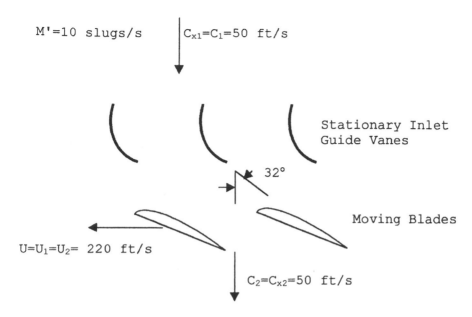

M'=10 slugs/s                        $C_{x1}=C_1=50$ ft/s

32°

Stationary Inlet
Guide Vanes

Moving Blades

$U=U_1=U_2=$ 220 ft/s

$C_2=C_{x2}=50$ ft/s

**FIGURE P6.7**   A blade cascade with IGVs.

6.11. A large axial-flow pump must supply 0.45 m³/s water flow rate with a pressure rise of 140 kPa. The hub and tip diameters for the impeller are D = 0.4 m and d = 0.3 m, respectively, and the running speed is 1200 rpm.
 (a) Use the Cordier diagram to estimate the pump's total efficiency, correcting the Cordier number for the influence of Reynolds number.
 (b) Lay out an axial flow velocity diagram for the mean radial station of the blade for this pump to achieve the required pressure rise, allowing for the influence of efficiency on the vectors.
 (c) Approximate the shape of the mean station of this blade based on the vector diagram of part (b).
 (d) Is the de Haller ratio for this station acceptable?
6.12. Develop vector diagrams, blade shapes, and de Haller ratios for the hub and tip stations of the pump examined in Problem 6.11.
6.13. Repeat the development outlined in Problem 6.5, using d/D = 0.75 and the "optimal" value from the curve fit in Figure 6.10. Compare these results to those of Problem 6.5 and comment on the most significant differences.
6.14. A centrifugal fan provides 0.4 m³/s of air ($\rho$ = 1.21 kg/m³) with a pressure rise of $\Delta p_T$ = 1.75 kPa at 1150 rpm. Use D = 0.75 m, d = 0.35 m, and the Cordier value for total efficiency to develop the velocity vectors at the blade leading and trailing edges. Assume uniform radial velocity through the blade row.
6.15. Re-examine the velocity vectors and approximate blade layout of the fan impeller of Problem 6.14 using a d/D ratio based on the minimum $W_1$ analysis for centrifugal impellers (d/D = $2^{1/6}\phi_{bp}^{1/3}$).

6.16. For the fan of Problem 5.22, let 500 kPa be the static pressure requirement and
   (a) Estimate the change in power, total and static efficiency, and the sound power level, $L_w$.
   (b) Choose a suitable hub–tip diameter ratio for the fan and repeat the calculations of (a) and (b).
   (c) Calculate the de Haller ratios for the hub, mean, and tip stations for both the blades and the vanes.

6.17. A lightly loaded fan in a tube-axial configuration must deliver 1500 ft³/s of 0.0020 slugs/ft³ air with a static pressure rise of 5 lbf/ft². The fan has a diameter of 5 ft. Determine a reasonable speed for the fan and select a suitable hub–tip ratio.
   (a) Develop a vector diagram at the inlet and outlet positions on the fan blade at the 70% radial station of the blade, at the blade hub and at the tip station.
   (b) Estimate the total efficiency and power and correct the outlet side vectors to account for the total pressure losses implied by the total efficiency.

6.18. Further examine the velocity vectors and impeller shape of the fan of Problem 6.14 by considering blade widths that yield values of $C_{r2} = kU_2$, with k = 0.2, 0.3, and 0.4. These increasing k values increase the blade angles and reduce the performance sensitivity to precise accuracy requirements in blade manufacturing tolerances. Develop inlet and outlet vectors and approximate blade shapes for each case.

6.19. In Problem 6.18, narrowing the impeller to keep the blade angles fairly large results in significant acceleration of the flow from the eye to the blade leading edge. Does this matter greatly? Can one control this behavior by modifying the throat size? If so, does one gain or lose in terms of efficiency or cavitation margin?

6.20. Show that the flow and pressure rise coefficients can be written respectively as $\phi = 1/(N_s D_s^3)$ and $\psi = 1/(N_s^2 D_s^2)$.

6.21. A 0.4-m diameter tube axial fan has a performance target of 0.8 m³/s and 225 Pa static pressure rise with $\rho = 1.0$ kg/m³. Determine a good hub size and a reasonable speed for the fan. Develop velocity vector diagrams for the inlet and outlet station at 2r/D = 0.75 (the 75% blade radial station). Estimate static and total efficiencies and modify the outlet vector diagram to account for total pressure losses in the blade row (based on $\eta_{T-C}$).

# 7 Velocity Diagrams and Flowpath Layout

## 7.1 AXIAL FLOW PUMPING

Chapter 6 examined the relationship between the energy exchange and the flow patterns within the blade and vane rows of turbomachines. Blade relative (W) and absolute (C) velocity vectors were employed to analyze blade layout and loading. In order to formalize that process, one can now introduce some new non-dimensional variables and a way to concisely present velocity vector information. Thus, one can define:

1. Work Coefficient: $\Psi = gH/U^2 = (\Delta p_T/\rho)/U^2$; where U is the local value of peripheral speed previously defined. From the Euler equation $\Delta p_T/\rho = gH = \Delta C_\theta U = C_{\theta 2} U$ if $C_{\theta 1} = 0$, so

$$\Psi = \frac{\Delta C_\theta}{U} \qquad (7.1)$$

2. Axial Velocity Ratio, of flow coefficient

$$\Phi = \frac{Q}{(UA)} = \frac{C_x}{U} \qquad \text{and} \qquad \beta = \tan^{-1}\left(\frac{1}{\Phi}\right) \qquad (7.2)$$

   which is the value used in finding the flow angles.
3. Reaction (R'), a new variable defined as the ratio of the static pressure rise to the total pressure rise: $R' = \Delta(gH_s)/\Delta(gH_o)$; where $gH_s$ is the static value, and $gH_o = \Delta gH_s + (C_2{}^2 - C_1{}^2)/2$. Note that $\Delta gH_o = gH = U_2 C_{\theta 2} - U_1 C_{\theta 1}$, so

$$R' = 1 - \frac{\left(C_2{}^2 - C_1{}^2\right)}{\left[2\left(U_2 C_{\theta 2} - U_1 C_{\theta 1}\right)\right]} \qquad (7.3)$$

For the "simple-stage" fan:

$$U_1 = U_2 \qquad \text{and} \qquad C_{x1} = C_{x2} \qquad (7.4)$$

$$C_1{}^2 = C_{x1}{}^2 + C_{\theta 1}{}^2 \qquad \text{and} \qquad C_2{}^2 = C_{x2}{}^2 + C_{\theta 2}{}^2 = C_{x1}{}^2 + C_{\theta 2}{}^2 \qquad (7.5)$$

and

$$U_2 C_{\theta 2} - U_1 C_{\theta 1} = U\left(C_{\theta 2} - C_{\theta 1}\right) \tag{7.6}$$

This simple stage assumption reduces R′ to

$$R' = 1 - \frac{\left(C_{\theta 1} + C_{\theta 2}\right)}{(2U)} \tag{7.7}$$

Note that if $C_{\theta 1} = 0$

$$R' = 1 - \frac{\Psi}{2} \tag{7.8}$$

For a stage, one can sketch an inlet vector and an outlet vector to form the "flow diagram," as shown in Figure 7.1. Look at the vectors for an example. $U = U_m = 149$ ft/s; $C_{x1} = C_{x2} = 45.1$ ft/s; $C_{\theta 1} = 0$; $C_{\theta 2} = 45$ ft/s. We calculate $W_1 = 155.7$ ft/s and $W_2 = 113.4$ ft/s, with $C_1 = C_{x1}$ and $C_2 = W_3 = 63.6$ ft/s. These can all be combined, along with the various angles, into the stage velocity diagram in Figure 7.2. For convenience, the diagram can be normalized by U and the definitions just established, as shown in Figure 7.3, can be incorporated into the diagrams. It is a handy device because it completely describes what is entering and leaving a blade row and the entrance conditions for the vane row. Note that

1. The distance between the peaks is always the work coefficient.
2. The height is always the axial velocity ratio or flow coefficient.
3. The reaction is visually measured from the right edge to the point midway between the peaks.
4. All relative and absolute flow angles are contained in the diagram, including the flow turning angle.
5. The de Haller ratio is essentially available to scale as the ratio of relative vectors for the blade and absolute vectors for the vane.

One can generalize the velocity diagram for other cases that are not quite so simple. For example, a rotor in axial flow with non-axial entry conditions typifies a pump or fan with pre-swirl velocity created by a set of inlet guide vanes (IGVs), as shown in Figure 7.4. The IGVs turn the flow, in this case, toward the moving blades, giving a velocity vector layout as illustrated in Figure 7.5. This leads to a velocity diagram that looks like Figure 7.6, with the non-dimensionalized form shown in Figure 7.7.

This particular layout or velocity diagram was chosen such that the magnitudes of $C_1$ and $C_2$ are equal, $|C_1| = |C_2|$, and represents $R' = 1$, or a "100% reaction" diagram. Note that $|C_{\theta 1}| = |C_{\theta 2}|$, although they have opposite algebraic signs. The function of the inlet guide vane shown here is to increase the blade loading, as is seen from a comparison of $W_1$ and $W_2$. Another layout, shown in Figure 7.8, is used to decrease blade loading and provide a mechanism for controlling rotor blade

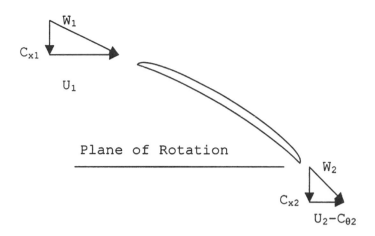

**FIGURE 7.1**  Vector diagram for axial flow blading.

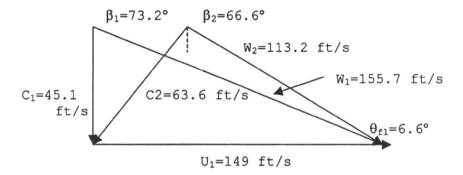

**FIGURE 7.2**  Dimensional stage velocity diagram.

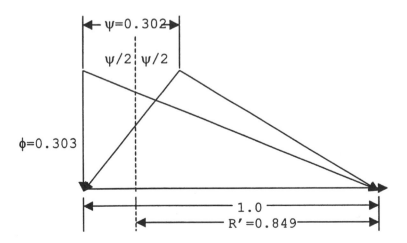

**FIGURE 7.3**  Dimensionless velocity diagram.

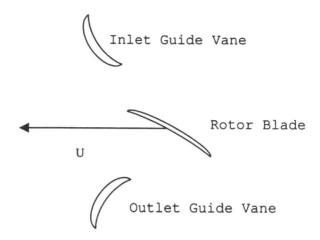

**FIGURE 7.4**  Velocity diagram with non-axial inlet flow.

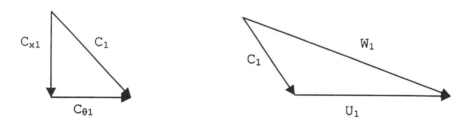

**FIGURE 7.5**  Velocity vectors for negative inlet swirl.

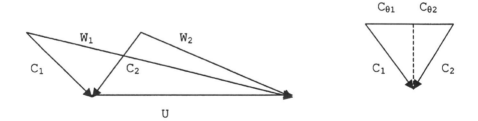

**FIGURE 7.6**  Velocity diagram for negative inlet swirl.

reaction and vane row diffusion. The IGVs are set to turn the incoming flow away from the moving blade such that $C_{\theta 1}$ and $C_{\theta 2}$ are in the same direction.

Recall that $[1 - (C_{\theta 1} + C_{\theta 2})]/(2U) = R'$, so that in a non-dimensional diagram, the $R'$ is still located at the mean value of $C_\theta$, $C_{\theta m}$. This kind of pre-swirl serves to reduce the "degree of reaction," while the previous example shows an increase in $R'$ for the same value of $\Psi$ (as sketched). Note, however, that if one considers $C_2$ as the approach vector to the outlet guide vanes (OGVs), the turning requirement imposed on the vane becomes much larger, which will lead to a greatly decreased,

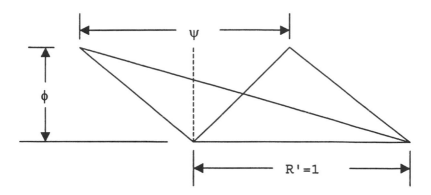

**FIGURE 7.7** Dimensionless velocity diagram for negative swirl.

probably excessive, de Haller ratio. In a sense, one is shifting loading from the blade to the vane. This can be useful up to a point but, in general, not very much is gained. The real reason for using "positive" pre-swirl can be seen using an alternate vane row that recovers only part of the swirl. This makes sense only for a multi-stage machine for which the outlet vector from the vane row is exactly the same as the inlet vector to the blade row. In this configuration, there are any number of "repeating stages" that see the same inlet-outlet conditions on blade and vane rows. For the last stage, some sort of adjustment must be made. If recovering all of the swirl places too much loading on the vane, the use of an extra vane row to capture the residual swirl is a possible solution, as shown in Figure 7.8b.

Another approach to multi-stage (or single-stage) layout is to use the pre-swirl of an IGV to turn the flow into the rotor and add just enough opposite swirl in the blade row to yield an axial discharge downstream of the blade. Such a configuration is shown in Figure 7.9, along with its velocity diagram. $C_2$ is simply equal to $C_x$ and, for a given $W_2/W_1$, is no worse than the value expected for a typical rotor-OGV stage. The penalty arising from having a higher W1 (higher losses) is offset by having acceleration instead of diffusion through the vane row (lower losses). A secondary penalty will be associated with having the rotor pass through a cyclic series of vane wakes from the upstream IGV elements. This "slapping" of the blades through the vane wakes will yield an increase in the noise level generated by the fan stage (see Chapter 4). In a multi-stage configuration, the extra noise effect may be negligible.

## 7.2 AXIAL TURBINE APPLICATIONS

For analysis of velocity vectors in axial turbines, it is conventional to reverse the diagram, as seen in Figure 7.10. One notices that $C_{\theta 1}$ is a finite number, while $C_{\theta 2}$ is zero. Using Euler's equation for turbines, one obtains

$$\frac{P}{m'} = gH = U_1 C_{\theta 1} - U_2 C_{\theta 2} = U_1 C_{\theta 1} \qquad (7.9)$$

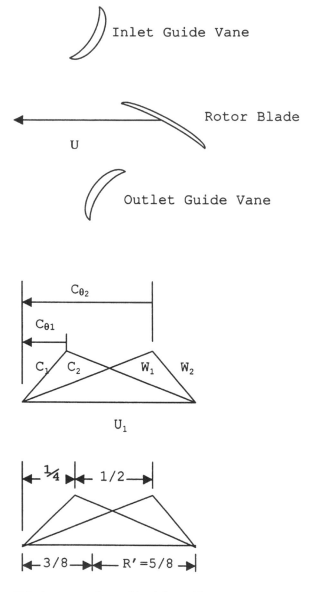

**FIGURE 7.8a**   Velocity vectors for positive inlet swirl.

and the turbine blade essentially recovers the swirl induced by the nozzles and discharges the flow with a pure axial outlet vector. Note that the diagram is symmetrical and represents a "50% reaction turbine," $R' = 0.50$. Note also that $W_2/W_1 > 1.0$, as it generally will be in a turbine cascade. If one "normalizes" the diagram as done with pumps, the peaks will be separated by the distance $\Psi$ and the height will be $\Phi$; the geometric angles are also described as before. The fact that

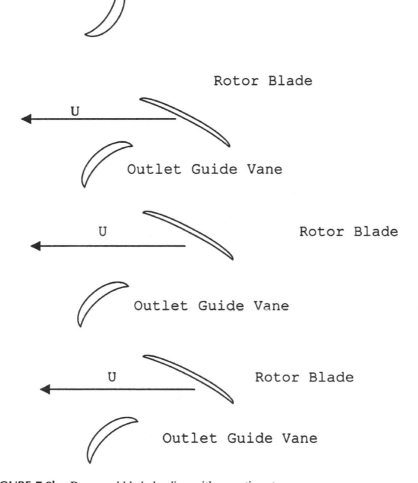

**FIGURE 7.8b**   Decreased blade loading with repeating stages.

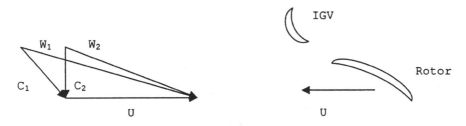

**FIGURE 7.9**   Inlet swirl without outlet guide vanes.

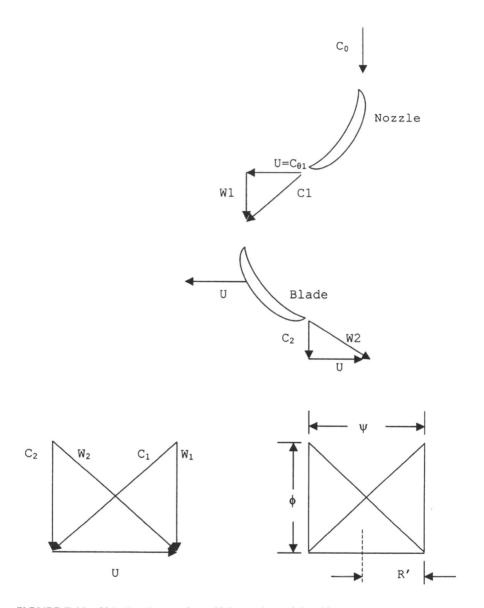

**FIGURE 7.10**   Velocity diagram for a 50% reaction axial turbine.

$W_2/W_1 > 1$ is simply a reminder that the turbine blade is a device that accelerates the flow (blade relative) and, as such, is not subject to the diffusion limitations present when dealing with pumping machines.

A more probable turbine layout is sketched in Figure 7.11. In this layout, the nozzle row accelerates the flow very sharply and relative velocities on the blade row are very high. The diagram is still symmetrical, with $R' = 0.50$ and $\Phi$ about the same, but note that $\Psi$ is much larger. Also notice that the discharge is no longer

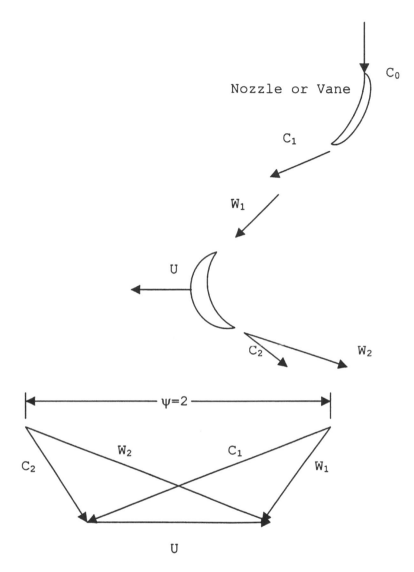

**FIGURE 7.11** Symmetrical turbine diagram with high velocities.

purely axial but has a swirl component opposite to the swirl entering the blade row. This contributes part of the value of $\psi$ as seen from the Euler equation (here, $C_{\theta 2} < 0$). In a 50% reaction axial turbine, the machine will frequently be a multi-stage configuration with similar velocity diagrams for each succeeding stage (neglecting changes in density, which may be accommodated by increasing the annulus area of the next stage by the ratio of incoming $\rho$ divided by outgoing $\rho$).

Not all turbines are 50% reaction designs, but they can be characterized by nonsymmetrical vector diagrams and blade-vane geometries. A common example is represented by the "zero reaction" turbine, also known as the "impulse" turbine.

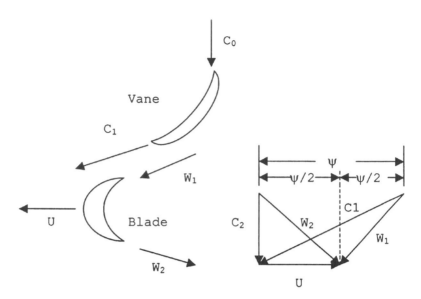

**FIGURE 7.12** Impulse turbine diagram.

Here, symmetry occurs between the relative vectors $W_1$ and $W_2$. They have equal magnitude and symmetrical direction, as shown in Figure 7.12. Vanes and blades are sketched on the left for this vector diagram. Notice that the blade has geometric symmetry about its 50% chord. The impulse blade or impulse "bucket" shape immediately identifies what kind of diagram one is dealing with and that the reaction is zero. Here, there is no geometric symmetry between the blades and vanes. The diagram, as shown in Figure 7.12, has $\Psi \cong 2.0$. Higher or lower values are, of course, possible with corresponding changes in turbine geometry.

Consider an example with $\Psi = 0.50$ and $\Phi = 0.50$ using a 50% reaction layout. The diagram is sketched in Figure 7.13, with the angles shown. We could do an initial blade-vane layout using, for example, circular arcs, to conform to the required vectors. Blades angles are obtained from $W_1$ and $W_2$ as shown in Figure 7.13. Bear in mind that the blade-vane layout given here is somewhat approximate and would have to be corrected for the effects of incomplete or imperfect guidance of the flow through a two-dimensional passage of finite geometric aspect ratio.

Consider the example of a hydraulic turbine-generator set with the performance specification given by

$$H = 20 \text{ m}; \quad Q = 115 \text{ m}^3/\text{s}; \quad \# \text{ of poles} = 72 \ (N = 100 \text{ rpm})$$

We calculate $N_s = N(Q^{1/2})/(gH)^{3/4} = 2.14$ and $D_s = 2.84 \ N_s^{-0.486} = 1.96 = D(gH)^{1/4}/Q^{1/2}$. Extract D as

$$D = \frac{1.96(115)^{1/2}}{(20 \times 9.81)^{1/4}} = 5.62 \text{ m}$$

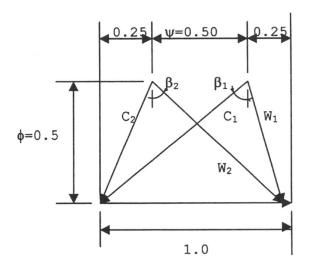

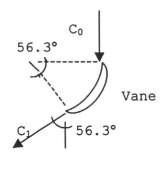

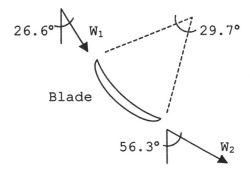

**FIGURE 7.13** Velocity diagram with blade and vane shapes for the example turbine study.

Look at the mean station with $d/D = 1/3$ or $d = 1.873$ m. Then, $r_m = 1.873$ m. $N = 10.47$ s$^{-1}$, so $U_m = r_m N = 19.6$ m/s. $C_x = Q/A_{ann} = 5.22$ m/s and $\Phi = C_x/U = 0.266$

What is $\Psi$? Assume an axial discharge turbine such that $C_{\theta 2} = 0$ and $C_{\theta 1}$ consumes the head. The Euler equation gives

$$gH = C_{\theta 1}U_1 - C_{\theta 2}U_2 = U_1 C_{\theta 1} = 1962 \ \text{m}^2/\text{s}^2$$

Therefore,

$$C_{\theta 1} = \frac{1962\left(\dfrac{\text{m}^2}{\text{s}^2}\right)}{U\left(\dfrac{\text{m}}{\text{s}}\right)} = \frac{1962}{19.62} = 10 \ \text{m/s}$$

And, $\Psi = C_\theta/U = 0.510$ and the velocity diagram and blade layout look like Figure 7.14. The blade is sketched with only the "mean-line" or camber lines shown for clarity. The arrangement shown here is schematically correct, but it more nearly resembles an arrangement typical of steam or gas turbine nozzle and blade rows, where elements are arranged in an axial cascade configuration. In axial flow hydraulic turbines, a more usual arrangement consists of the axial turbine with a radial inflow arrangement of turning vanes or "wicket gates" upstream of the axial blade row as sketched in Figure 7.15. The incoming stream flows past the adjustable wickets, which impart a swirl motion or tangential velocity $C_{\theta o}$. Following the streamline down through the blade row, one notes that, by conservation of angular momentum, the swirl velocity seen by the blade row at the radial mean station is given by

$$r_w C_{\theta o} = r_m C_{\theta 1} \qquad \text{or} \qquad C_{\theta 1} = \left(\frac{r_w}{r_m}\right) C_{\theta o} \qquad\qquad \textbf{(7.10)}$$

where $C_{\theta o}$ is controlled by the wicket setting $\alpha_w$. That is, $C_{\theta o} = C_{ro} \tan \alpha_w$, where $C_{ro} = Q/A_o$, with $A_o$ given as the wicket cylindrical area, $A_o = 2\pi r_w h_w$. In a full-blown setup, both wickets and turbine blades could have adjustable angles, allowing for a broad range of power extraction levels, depending on available flow rate and power demand from the utility network.

For the previous example, the required wicket gate angle $\alpha_w$ can be estimated from

$$\alpha_w = \tan^{-1}\left(\frac{C_{\theta 1}}{\left(\dfrac{Q}{(2\pi r_w h_w)}\right)}\right) \qquad\qquad \textbf{(7.11)}$$

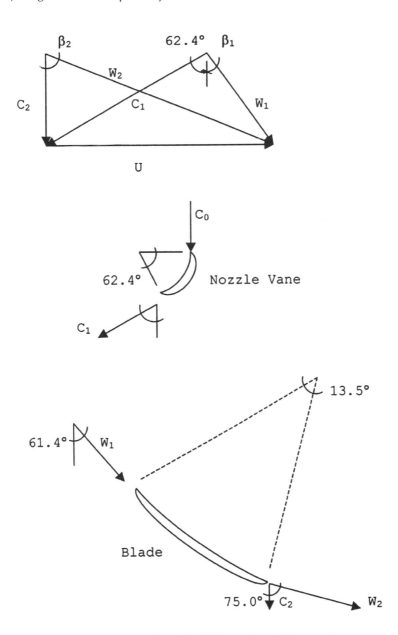

**FIGURE 7.14** Layout for hydraulic turbine blades and axial-flow nozzle vanes.

For example, if $h_w = r_w = 1.873$ m, then $Q = 115$ m³/s and $C_{\theta 1} = 10$ m/s. Thus, $\alpha_w = 22.6°$. Setting at this angle would serve the same function as the axial nozzle vanes in the axial inflow with 62.4° of camber or fluid turning. $h_w$ and $r_w$ can be used as design variables to minimize either the frictional losses on the wicket gates or to control the sensitivity of power extraction to wicket gate angle, and excessive incidence.

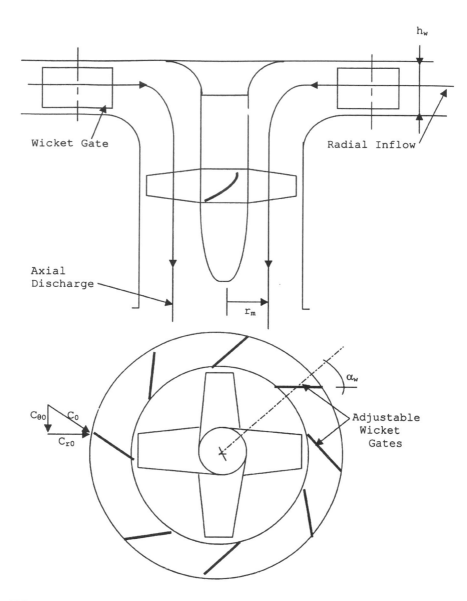

**FIGURE 7.15**   Wicket gate concept for axial flow hydraulic turbine.

## 7.3   CENTRIFUGAL AND RADIAL FLOW

To consider velocity diagrams in mixed flow or radial flow regimes in a general way, the concept of a "meridional" flow path or stream surface through the impeller must be introduced. The meridional path through a machine is the stream surface along which the mass flow occurs. In an axial flow machine, the surface is cylindrical, with subsequent stream surfaces described by concentric cylinders whose axis is the

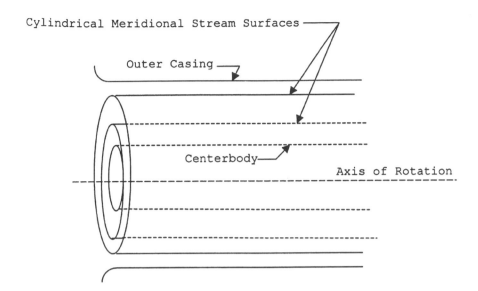

**FIGURE 7.16** Meridional stream surface with velocities.

axis of rotation of the impeller. The flow over these surfaces can be circumferentially averaged, and the vector for the axial machine will consist of averaged values of axial and tangential flow velocity components in the absolute frame of reference. The cylindrical path is shown in Figure 7.16.

For the pure radial flow situation shown in Figure 7.17, the meridional path, at least through the blade row, can be approximated by flow along planes perpendicular to the axis (or parallel to the backplate of the machine). If the flow path is a mixed-flow type, the stream surface will be partially axial and partially radial. A simple example can be considered for an impeller whose outer casing and center body consist of sections of simple cones concentric about the rotational axis of the impeller, as seen in Figure 7.18. The most general case must be described by concentric surfaces of revolution along curved paths that move from axial to approximately radial, as seen in Figure 7.19.

For both the conical and general shapes, the meridional velocity consists of both radial and axial components. For the extreme cases of axial and radial flow, the meridional velocity is simply the axial and radial velocity component, respectively. In any of these cases, the tangential component of absolute velocity can be added to form the overall absolute velocity and to describe a mean stream path that sweeps along and winds around the meridional surface, as illustrated in Figures 7.20 and 7.21. It is on these increasingly complex surfaces that one must describe the absolute and blade relative velocity vectors to be able to construct the velocity diagrams.

Blade relative vectors can be generated by superposing a rotation of the stream surface about its axis of revolution (the rotational axis of the machine). The general case is sketched in Figure 7.22 in relative motion. The cross-sections of several blades, as cut by the surface of revolution, are shown along with the rotational component $U = Nr$, and the resultant relative vector. Blade "inlet" and "outlet"

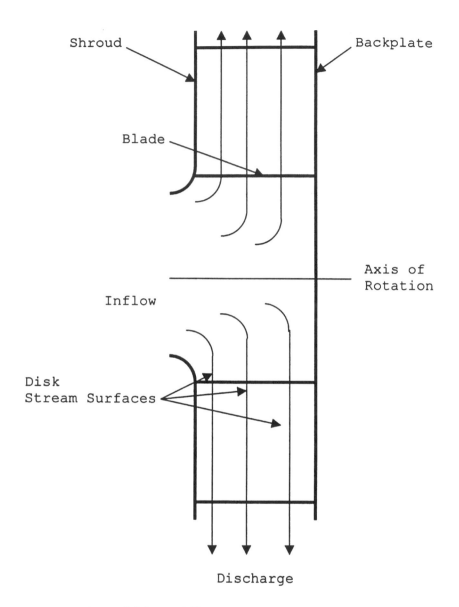

**FIGURE 7.17**   Pure radial flow path illustration.

vectors are shown to illustrate the idea of proper layout in a fairly complex situation. The blade is shown with $C_{\theta 1} = 0$ and $C_{\theta 2} > 0$ for simplicity.

In practice, a machine with pure radial flow is the easiest one to examine, so begin by looking at the case of a centrifugal blower having an essentially radial path along the entire blade. This is shown in Figure 7.23. The radial flow surface is sketched next to a cutaway of the impeller showing the backwardly inclined airfoil

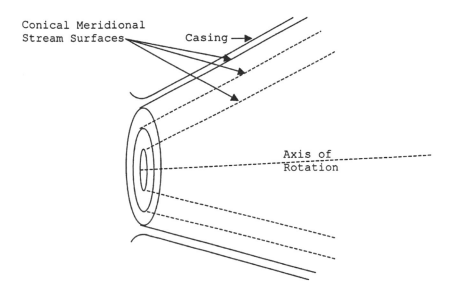

**FIGURE 7.18** Mixed flow machine with conical stream path.

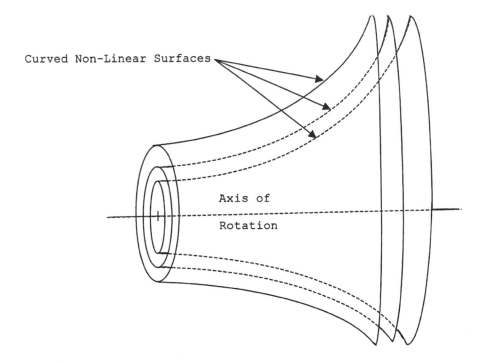

**FIGURE 7.19** A generalized stream surface and meridional flow path.

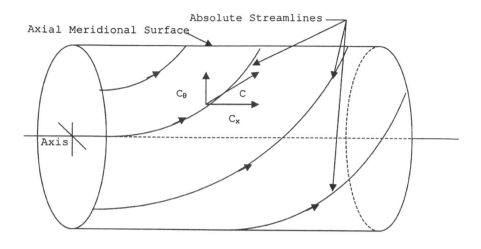

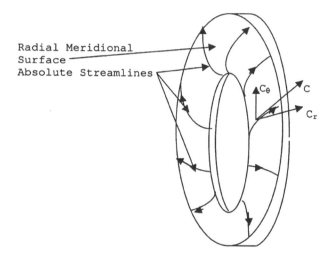

**FIGURE 7.20** Tangential velocity component related to meridional surface.

blading typical of blower practice. Figure 7.24 provides a blowup of a single blade and draws the absolute and relative velocity vectors at the blade row extreme and outlet. The inlet vector is relatively simple, consisting of $C_{r1} = Q/(2\pi r_1 w_1)$, and $U_1 = Nr_1$. The resultant $W_1 = (U_1^2 + C_{r1}^2)^{1/2}$ lines up with the blade leading edge camber line as shown. At the outlet, $W_2$ aligns with the trailing edge mean camber line and must be equal to the vector sum of $U_2$ and $C_2$, as shown. The magnitude of $W_2$ is controlled by the size of $C_{r2}$, and the blade orientation as specified by $\beta_2$. As can be seen from Figure 7.25, the size of $C_2$ and its direction are in turn governed by the vector addition of $W_2$ and $U_2 = Nr_2$, thus fixing the size of $C_{\theta 2}$. One can calculate $C_{\theta 2}$ from all of this, since

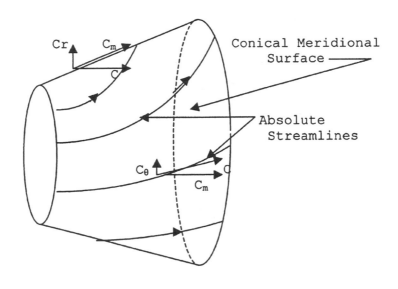

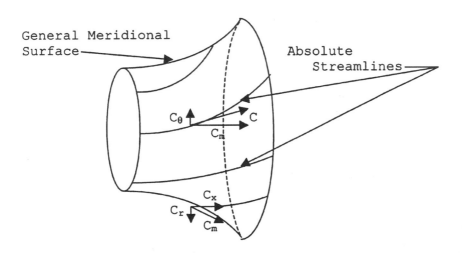

**FIGURE 7.21**  Velocity components on meridional stream surface.

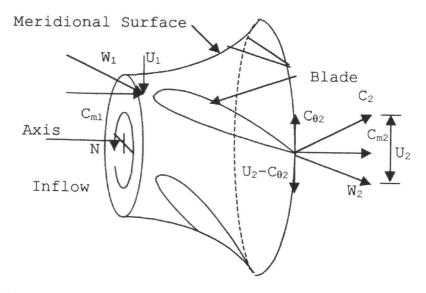

**FIGURE 7.22**   General velocity description in three-dimensional flow.

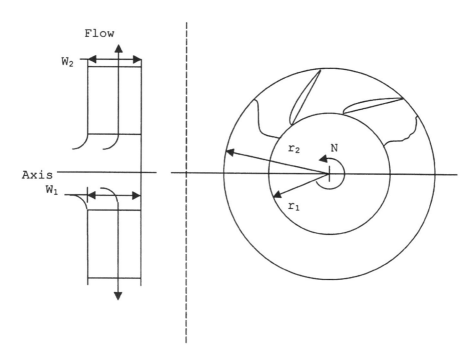

**FIGURE 7.23**   Flowfield for a radial flow layout and cascade.

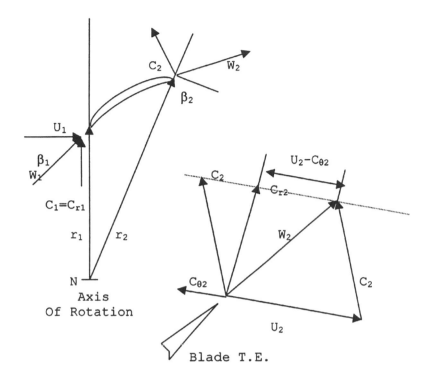

**FIGURE 7.24**   Velocity vectors in a radial cascade.

$$\tan\beta_2 = \frac{C_{r2}}{\left(U_2 - C_{\theta2}\right)} \qquad \text{or} \qquad C_{\theta2} = U_2 - C_{r2}\cot\beta_2 \qquad (7.12)$$

Because of the change in $C_r$ through the blade row, it is not conventional to combine the inlet and outlet vectors into a velocity diagram, although this can be done. As shown in Figure 7.26, if one fixes $C_{r2}$ and $U_2$, then increasing $\beta_2$ acts to increase $C_{\theta2}$ and the head rise (gH or $\Psi$) according to $\Psi = C_{\theta2}/U_2$ (with $C_{\theta1} = 0$). Analytically, one writes

$$\Psi = \frac{\left(U_2 - C_{r2}\cot\beta_2\right)}{U_2} \qquad (7.13)$$

or

$$\Psi = 1.0 - \left(\frac{C_{r2}}{U_2}\right)\cot\beta_2 = 1.0 - \Phi\cot\beta_2 \qquad (7.14)$$

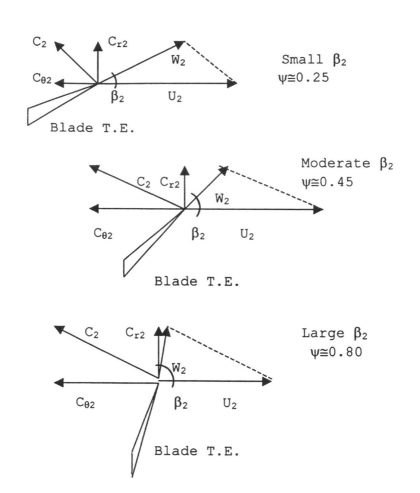

**FIGURE 7.25**   Outlet vectors from a radial cascade.

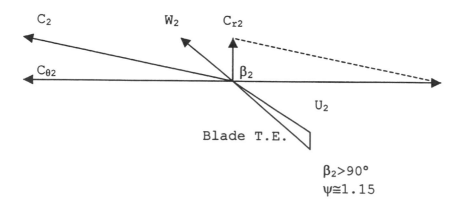

**FIGURE 7.26**   Forward-curved blading layout.

**TABLE 7.1**
**Relationship between**
$\beta_2$ **and** $W_2/W_1$

| $\beta_2$ | $W_2/W_1$ |
|-----------|-----------|
| 30 | 1.0 |
| 40 | 0.78 |
| 50 | 0.653 |
| 60 | 0.577 |

where $\Phi_2$ now needs the subscript to be general. As $\beta_2$ approaches 90°, $\cot\beta_2$ goes to zero so that $\Psi$ approaches 1.0, the ideal head rise or wake coefficient for a "radial-tip" fan or blower.

Higher values of $\psi$ are of course possible, as illustrated by the "forward-curved" blade with $\beta_2 > 90°$, as shown in Figure 7.26. Scaling from the sketch, the configuration has a value of $\Psi \cong 1.14$. Note from the equation for $C_{\theta 2}$ that $C_{\theta 2}$, and thus gH or $\Psi$, goes to zero at a finite value $\beta_2$, normally $\beta_2 = \tan^{-1}\Phi_2$. That is, for a fixed combination of $C_{r2}$ and $U_2$ ($\Phi_2$), the head rise goes to zero at the same value of $\beta_2$. This is equivalent to saying that, for a fixed $\beta_2$ and $U_2$, there is a value of $C_{r2}$ for which the head rise vanishes—the free delivery point in terms of $\Delta p_T$.

Inlet and outlet vector diagrams can be used to examine blade surface diffusion, as for the axial blade rows. It is necessary to express $W_1$ and $W_2$ in terms of geometry, flow, and pressure rise and examine $W_2/W_1$ in terms of de Haller's limiting value of 0.72. $W_1$ can be written as $W_1 = (Cr_1^2 + U_1^2)^{1/2}$, where $C_{r1} = Q/(2\pi r_1 w_1)$; $U_1 = Nr_1$. $W_2$ is written as $[C_{r2}^2 + (U_2 - C_{\theta 2})^2]^{1/2}$ where $C_{r2} = Q/(2\pi r_2 w_2)$ and $U_2 - C_{\theta 2} = C_{r2} \cot\beta_2$. Thus, with some rearrangement, one can write

$$\frac{W_2}{W_1} = \left(\frac{w_1 r_1}{w_2 r_2}\right)\left(\frac{\sin\beta_1}{\sin\beta_2}\right) \tag{7.15}$$

An example of application is to consider the constant width impeller with $w_2 = w_1$. If one fixes $r_1/r_2$ and $\beta_1$ (i.e., fix Q), then $W_2/W_1 = \text{constant}/(\sin\beta_2)$. Thus, driving $\beta_2$ upwards causes a nearly linear decrease in $W_2/W_1$; that is, trying to increase the head rise leads directly to a decrease in $W_2/W_1$ and eventual problems in efficiency and flow separation on the blading.

Another example that reflects a realistic design approach is seen for the case where $w_1 r_1 = w_2 r_2$, which tapers the blade row width such that $C_{r1} = C_{r2}$ and $C_r$ is constant throughout the impeller flow channel. An example of this layout is shown in Figure 7.27.

For such a case, $W_2/W_1 = (\sin\beta_1)/(\sin\beta_2)$. A typical value for a high specific speed backwardly inclined fan might be $\beta_1 = 30°$, with $\Phi_1 = C_{r1}/U_1 = \tan\beta_1 = 0.50$. So, $W_2/W_1 = 0.5/\sin\beta_2$. A conservative upper limit on pressure rise for this case is seen (using $r_1/r_2 = 0.65$ and hence $\Phi_2 = 0.33$), $\Psi = 0.66$. As shown in Table 7.1, for the constant width machine, any value of $\beta_2$ above about 42° will lead to viscous

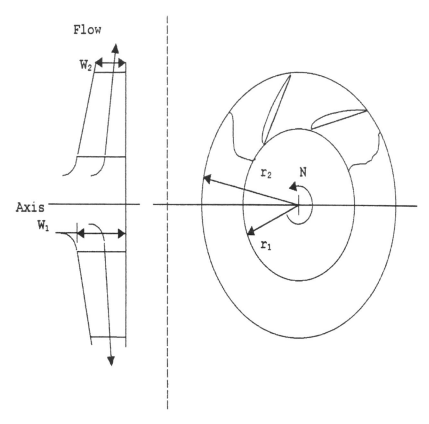

**FIGURE 7.27**  Flow field for a tapered radial flow layout and cascade.

flow problems on the blading. Such a result is consistent with the low values of efficiency commonly seen in impellers with "radial sideplates" or shrouds. Note that the zero head rise value of $\beta_2$, with $\beta_1 = 30°$, is given by $\beta_2 = \tan^{-1}\Phi_2$, where

$$\frac{\Phi_2}{\Phi_1} = \frac{\left(\dfrac{Q}{2\pi r_2 w_2 N r_2}\right)}{\left(\dfrac{Q}{2\pi r_1 w_1 N r_1}\right)} = \frac{r_1}{r_2} \qquad (7.16)$$

Thus, for $\beta_1 = 30°$, $r_1/r_2 = 0.7$; $\beta_2 = \tan^{-1}(0.70 \tan 30°) = 16°$. Clearly, the ratios $w_2/w_1$ and $r_2/r_1$ are vital parameters in controlling the level of diffusion in the impeller for a specified level of flow and head rise. For the case where $r_1 w_1 = r_2 w_2$ or $C_{r1} = C_{r2}$, $\Phi_2/\Phi_1 = (r_1/r_2)$, and the zero value of $\beta_2 = 22°$. The "acceptable" diffusion value of $\beta_2 = 44°$ yields $\Psi = 0.50$ if $r_1/r_2 = 0.70$. This is a reduction of 12%. As was the case for the axial cascade layout, it is possible (up to a point) to reduce $W_2/W_1$ below the level of 0.72 through increases in solidity and careful, detailed design considerations.

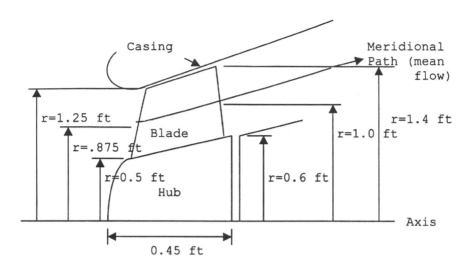

**FIGURE 7.28**   Impeller layout for a mixed flow machine.

In particular, for all but the lowest specific speed centrifugals, the flow behavior is dominated by a stable jet-wake pattern as discussed in Chapter 6. In such cases, diffusion is not a dominant factor in design (Johnson and Moore, 1980). However, one can still use de Haller's limit as a conservative level in examining basic blade layouts in centrifugals with highly backwardly inclined blading.

## 7.4   MIXED FLOW EXAMPLE

Suppose that one wants a fan to provide 15,000 cfm of air at $\rho = 0.00233$ slugs/ft$^3$ with a total pressure rise of 5 InWG. Wanting to save cost with a 4-pole motor, one can specify 1750 rpm as the running speed ($N = 103$ s$^{-1}$). Calculate $N_s = 2.33$ from this information and the corresponding $D_s$ value from the Cordier line is 1.90. Thus, one is in the center of the mixed-flow region (Region D), so try to lay out a sort of conical flow path to suit the need. The given data reveals that the diameter is 2.8 ft, and one can sketch a flow path as shown in Figure 7.28. The mean radius meridional streamline has a slope of about 25.0°, so this is the path along which the blade will be laid out. At the inlet, $r_{m1} = 0.875$ ft, so $U_1 = 160$ ft/s. The inlet area is $A_1 = 4.12$ ft$^2$, such that $C_{x1} = Q/A_1 = 56.9$ ft/s and $C_{m1} = C_{x1}/\cos 25° = 62.7$ ft/s. Then calculate $W_1$ and $\beta_1$ as $W_1 = 172$ ft/s and $\beta_1 = 68.6°$. At the outlet, $r_{m2} = 1.0$ ft, so $U_2 = 183$ ft/s. $A_2 = 5.03$ ft$^2$ and $C_{x2} = 49.7$ ft/s, yielding $C_{m2} = 51.6$ ft/s. $C_{\theta2}$ is found from $gH/U_2 = (\Delta p_T/\rho)/U_2 = 61.0$ ft/s. $W_2 = [C_{m2}^2 + (U_2 - C_{\theta2})^2]^{1/2} = 132.5$ ft/s $\beta_2 = \tan^{-1}[(U_2 - C_{\theta2})/C_{m2}] = 67.1°$. Thus, $\theta_{fl} = 1.5°$ and $W_2/W_1 = 0.77$. These are very reasonable values for the blade row, although $W_2/W_1$ might be a little high. Note that the angles and the blade itself must be laid out along the meridional conical path, not on a cylinder.

If one tries to lay out a simpler axial flow path using $D = 2.8$ ft, $N = 1750$ rpm, and $d = 1.4$ ft, one obtains (try these calculations) $W_2/W_1 = 0.67$ and $\theta_{fl} = 8.1°$,

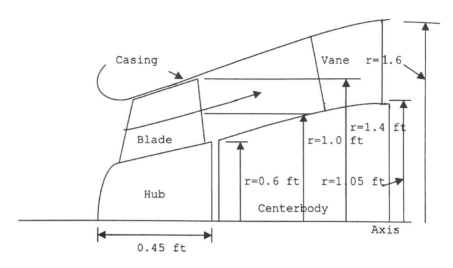

**FIGURE 7.29**  Mixed flow discharge geometry.

which is a blade working significantly harder. Part of the problem can be overcome by trying a larger hub diameter, $d = 1.6$ ft. The resulting numbers are $W_2/W_1 = 0.701$ and $\theta_{fl} = 7.4°$, which becomes more acceptable. Note that $C_x = 60.2$ ft/s while $C_{x2} = 50$ ft/s for the mixed flow configuration. Axial outlet velocity pressures are 0.81 InWG for the mixed flow and 0.56 InWG for the axial flow. This may become a significant advantage for the mixed flow fan in terms of static efficiency. In addition, if we use a simple vane row for the axial fan with $C_x = 60.0$ ft/s and $C_{\theta 2} = 66.7$ ft/s, we obtain $C_4/C_3 = 0.576$, which is unacceptably low. For the mixed flow fan, one can readily control the vane row diffusion by constraining the conical flow path to an exit condition with $D = 3.2$ ft, $d = 2.1$ ft, and turning to an axial discharge direction (see Figure 7.29). This yields $C_x = 55$ ft/s and $C_{\theta 3} = (r_{m2}/r_{m3})C_{\theta 2} = 46$ ft/s. As a result, $C_4/C_3 = 0.767$, which is a conservative level without driving $C_x$ to values that may be too high. The apparent compromise is the creation of a larger machine.

## 7.5  PUMP LAYOUT EXAMPLE

Consider a pump that delivers 450 gpm at 100 ft of head and do a rough layout of the impeller. Changing to basic units, one obtains (for water) $N_s = 0.002341\ N$; $D_s = 7.530D$. If a 4-pole direct-drive motor running at 1750 rpm is used, one obtains $N_s = 0.428$, for which $D_s = 6.029$ from the Cordier diagram. From this, $D = 0.8$ ft $= 9.6$ inches. Now go ahead and try to lay out the impeller, but we need some preliminary dimensions first. Conceptually, the pump is going to look like the one in Figure 7.30, but one needs some idea of what $w_2$ and $D$ should be. Looking at drawings of pump impellers in catalogs, $w_2/D$ is typically a fairly small number, say 0.03 to 0.1, depending on the specific speed. Also, typical blade outlet angles range from about $\beta_2 = 20°$ to approximately 35°, where $\beta_2$ is measured from a line tangent to the impeller. So, one can do some rough sizing of the relative velocity vector at the impeller discharge to size $w_2$ or $\beta_2$.

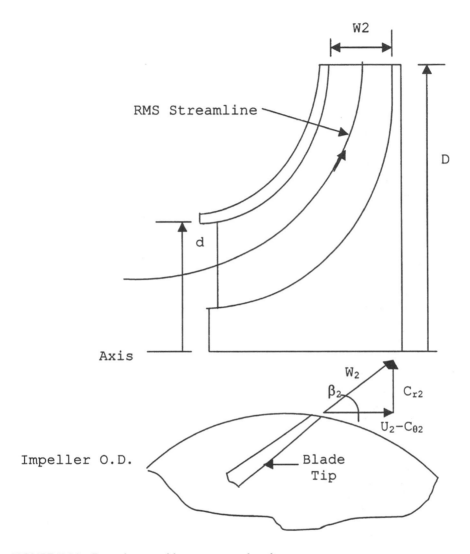

**FIGURE 7.30** Pump layout with geometry and outlet vector.

Since $gH = 3217$ ft$^2$/s$^2$ and $N = 183$ s$^{-1}$ with $D = 0.8$ ft, one can calculate $C_{\theta 2} = gH/U_2$. Better yet, calculate $C_{\theta 2} = gH/(\eta_T U_2)$, where $\eta_T = 0.8$ from Cordier, so the NET head rise, after losses, will be 100 ft. This gives $C_{\theta 2} = 55$ ft/s, and $U_2 - C_{\theta 2} = 73.2 - 55 = 18.2$ ft/s. This value can be used to lay out $W_2$ if one assumes a reasonable $\beta_2$. Choosing $\beta_2 = (20 + 35)/2 = 27.5°$ calculate $C_{r2} = (U_2 - C_{\theta 2})$ $\tan\beta_2 = 9.5$ ft/s $= Q/(\pi D w_2)$. Therefore, $w_2 = Q/(\pi D C_{r2}) = 0.042$ ft $= 0.503$ inches; with $\beta_2 = 27.5°$. Note that $w_2/D = 0.052$, which looks reasonable.

Now, what about d or d/D? Typical pumps seem to range from $0.25 < d/D < 0.50$, so one might choose d/D = 0.353. To be more systematic, use the algorithm presented in Chapter 6.

$$d = 2^{1/6} \Phi^{1/3}{}_{BP} D = 2^{1/6} (0.02718)^{1/3} (0.8 \text{ ft})$$

$$d = 0.270 \text{ ft} = 3.24 \text{ inches}$$

With provisional values for $w_2$ and d chosen, finish laying out the inlet and outlet vectors and the blade angles. As observed, one is using a radial discharge, $C_{r2}$, at the outlet and an axial influx flow, $C_{x1}$, at the inlet. The inlet vector and angle will resemble the technique used for axial flow machines. At the shroud, one can calculate $C_{x1} = Q/((\pi/4)d^2) = 11.2$ ft/s and $U_1 = Nd/2 = 30.9$ ft/s. $W_1 = 32.8$ ft/s. From the outlet vector layout, $C_{r2} = 9.5$ ft/s and $U_2 - C_{\theta2} = 18.2$ ft/s, so $W_2 = 20.5$ ft/s. The de Haller ratio along the shroud streamline is $W_2/W_1 = 0.63$. Along a mean stream-line, $W_1 = 24$ ft/s, yielding $W_2/W_1 = 0.85$. The diffusion picture is not very clean-cut, but here the value is acceptably high. The root mean square value of the inner diameter was used to establish a value to weight the inflow mean position, since very little activity is associated with the region in which the radius approaches zero. Thus, the "mean" value of $W_2/W_1 = 0.85$ provides a measure of diffusion in this case. The inlet setting angle for the axial "inducer" section of the blade becomes $\beta_1 = \tan^{-1}(U_1/C_{x1})$, measured relative to the axial direction and is equal to $62.6°$ at the 70% radial station. Since an axial entry geometry coupled with a radial exit geometry is being used, there must be a compound twist or curvature of the blade shape to connect the ends in a smooth manner.

## 7.6   RADIAL FLOW TURBOCHARGER COMPONENTS

As an example, consider the blade layouts for a radial inflow exhaust gas turbine coupled to a radial discharge compressor supplying compressed air to the induction system of an internal combustion engine. (A thorough treatment of these systems is available in internal combustion engine textbooks (Obert, 1950)). The radial inflow turbine configuration can be chosen based on the exhaust energy available. Then choose a compressor that is matched in power requirement to the actual shaft power output of the turbine. In addition, the two components are constrained to run at the same speed on the same shaft and with mass flows related to the engine air flow and fuel flow requirements. An exception to this last requirement exists when a "waste gate" or bypass exhaust flow path allows partial exhaust gas supply to the turbine.

Given the following turbine inflow exhaust flow properties:

$$T_{01} = 850 \text{K}$$

$$p_{01} = 152 \text{ kPa}$$

$$\rho_{01} = 0.625 \text{ kg/m}^3$$

$$m' = 0.200 \text{ kg/m}^3$$

$$p_b = 101.3 \text{ kPa}$$

one can select appropriate values for size and speed through $N_s$ and $D_s$ using inlet values. The only constraint placed on the turbine is the outlet static pressure, which can be set equal to the barometric pressure (neglecting any downstream losses associated with mufflers, tailpipes, or other unknown components). The turbine variables are written as

$$\Delta p_T = p_{01} - p_{02} = 152 \text{ kPa} - 101.3 \text{ kPa} = 52.7 \text{ kPa}$$

$$gH = \frac{\Delta p_T}{\rho} = 81,100 \text{ m}^2/\text{s}^2$$

Selecting the $N_s$ value of 0.3 as a narrow radial inflow turbine, one can calculate the turbine as

$$N = \frac{N_s(gH)^{3/4}}{Q^{1/2}} = \frac{0.3 \times 81,100^{3/4}}{0.320^{1/2}} \text{s}^{-1}$$

$$N = 24,365 \text{ rpm}$$

The Cordier value of $D_s$ corresponding to $N_s = 0.3$ is 7.0 and the total efficiency is estimated at $\eta_T = 0.75$. The diameter of the turbine is $D = 0.236$ m. That seems a little large but can readily be reduced by choosing a higher specific speed. The tip speed of the turbine becomes $U_1 = 300$ m/s, so one can estimate the required inlet swirl velocity as

$$C_{\theta 1} = \frac{gH}{(U_1 \eta_T)}$$

having assumed that the single-stage turbine will have no outlet swirl ($C_{\theta 2} = 0.0$). Then, $C_{\theta 1} = 360$ m/s in the direction of the motion of the turbine blades.

Now, to begin a layout of the turbine one needs to establish values for d/D, $w_1$, $w_2$, $\beta_1$, and $\beta_2$. Begin with the throat diameter ratio of 0.36 (approximately equal to 1.53 $\phi^{1/3}$ so that d = 0.086 m. Next, make the simplifying, conservative assumption that the blade entry and outlet areas are the same as the throat areas that $w_2 = d/4$ (at r = d/2) and $w_1 = w_2(d/D)$ (at r − D/2). These yield $w_1 = 0.0078$ m and $w_2 = 0.0215$ m. These values permit calculation of the radial velocities at the inlet and discharge of the turbine as

$$C_{r1} = \frac{m'}{(\rho_{01}\pi Dw_1)} \qquad \text{(inlet)}$$

$$C_{r2} = \frac{m'}{(\rho_{02}\pi Dw_2)} \qquad \text{(outlet)}$$

The inlet density was given in the specifications, but one must compute the outlet density from the turbine pressure ratio: $p_{01}/p_{02}$ = 152 kPa/101.3 kPa. The correct relation is given by the polytropic process equation (Chapter 1) as $\rho_{01}/\rho_{02} = (p_{01}/p_{02})^{1/n}$ where n is the polytropic exponent given by

$$\frac{n}{(n-1)} \cong \frac{\eta_T \gamma}{(\gamma-1)}$$

If $\gamma = 1.4$ (for diatomic gasses at moderate temperatures) and $\eta_T = 0.75$, then n = 1.61. With these results, one can calculate the $\rho_{01}/\rho_{02}$ = 1.287 so that $\rho_{02}$ = 0.486 kg/m³. Then the radial velocities become $C_{r1}$ = 53.3 m/s (in) and $C_{r2}$ = 70.8 m/s (out). With these results, the inlet and outlet velocity values and flow angles can be calculated as well.

$$W_1 = \left[ C_{r1}^2 + (U_1 - C_{\theta 1})^2 \right]^{1/2} = 81.6 \ m/s$$

$$W_2 = (C_{r2}^2 + U_2^2)^{1/2} = 130.1 \ m/s$$

Flow angles are given by

$$\beta_1 = \tan^{-1} \left[ \frac{C_{r1}}{(U_1 - C_{\theta 1})} \right] = 42.7°$$

$$\beta_2 = \tan^{-1} \left( \frac{C_{r2}}{U_2} \right) = 33.0°$$

Figure 7.31 shows the geometry developed for the turbine rotor.

Finally, the actual power available from the turbine shaft is given by

$$P_{sh} = \eta_T m'gH = 0.75 \times 0.20 \times 81100 \ W = 12.165 \ kW$$

This value is, of course, the input power to the compressor side of the turbocharger. The mass flow rate to the compressor must be less than that of the turbine by the mass flow rate of the fuel for the IC engine. Assuming that the air-to-fuel mass ratio for the engine is 15 to 1, the compressor mass flow rate will be 0.188 kg/s. Having already fixed the shaft speed at 24,365 rpm—the same as the turbine—it remains to determine the diameter of the compressor impeller. Since the power that can be imparted to the air is

$$P = \eta_T P_{sh} = m'gH \tag{7.17}$$

Turbine Inlet Vector

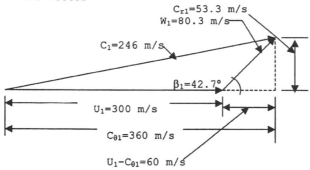

Turbine Outlet Vector

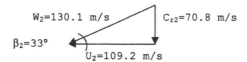

Cascade

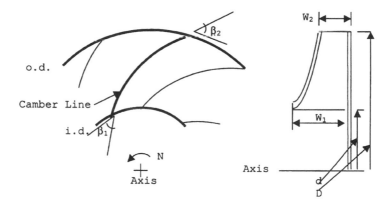

**FIGURE 7.31**   Geometry developed for the turbine rotor.

then the head rise becomes

$$gH = \frac{\eta_T P_{sh}}{m'} \tag{7.18}$$

However, one needs the Cordier value of the compressor specific speed to estimate efficiency and the efficiency to estimate gH and hence $N_s$. If it is assumed that the value of compressor efficiency is the same as the turbine efficiency, one can start an initial estimate and try to iterate to a converged value for $\eta_T$. Then, $gH \cong (0.75 \times 12{,}165)/0.188 = 48{,}531 \ m^2/s^2$. Using a standard sea level air inlet

**TABLE 7.2**
**Iterative Calculation for the**
**Outlet Density**

| # | $\rho_{02}$ (kg/m³) | $\rho_{02}/\rho_{01}$ | $\Delta p_T$ (Pa) |
|---|---|---|---|
| 1 | 1.21 | 1.543 | 54,810 |
| 2 | 1.569 | 1.704 | 75,410 |
| 3 | 1.665 | 1.747 | 76,540 |
| 4 | 1.690 | 1.758 | 76,840 |
| 5 | 1.696 | 1.761 | 76,910 |
| 6 | 1.698 | 1.762 | 76,925 |
| 7 | 1.698 | 1.762 | 76,930 |

density, the volume flow rate becomes $Q = m'/\rho_{01} = 0.155$ m³/s. $N_s$ becomes 0.302, with $D_s = 8$ and $\eta_T = 0.70$. Correct gH to 45,296 m²/s², changing $N_s$ to 0.320 and $\eta_T$ to 0.71 with $D_s$ at 7.9. Considering this to be a converged value, calculate the impeller diameter as $D = 0.213$ m (compared to a turbine diameter of 0.236 m). To calculate the pressure rise and pressure ratio requires $\Delta p_T = \rho_{02}gH$, rather than gH, so one needs the outlet density for the air flow. This must also be done iteratively, since density depends on pressure or pressure ratio. Using the polytropic exponent of n = 1.67, carry out the calculations as shown in Table 7.2 to achieve a converged value of $\rho_{02}$. With $\rho_{02}$, one can calculate the velocity components for the compressor discharge if one chooses blade widths to force a constant radial velocity component. Set d/D = 0.379, then d = 0.0807 m. Using $w_1 = d/4 = 0.0202$ m, one obtains $w_2 = (\rho_{01}/\rho_{02})(d/D)w_1 = 0.0055$ m. Then,

$$C_{r1} = C_{r2} = 30.3 \text{ m/s}$$

$$U_1 = 103 \text{ m/s}$$

$$U_2 = 272 \text{ m/s}$$

$$C_{\theta 2} = 235 \text{ m/s}$$

$$C_{\theta 1} = 0.0$$

The inlet and outflow angles become

$$\beta_1 = 16.4° \quad \text{and} \quad \beta_2 = 39.2°$$

The geometric layout for this impeller is sketched in Figure 7.32.

A photo showing the complexity of the impeller geometry for radial flow compressors is given in Figure 7.33. The figure also clearly shows the axial inducer section used to maintain careful control of the inlet flow to the radial passages of the impeller.

Compressor Outlet Vector

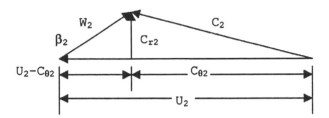

Compressor Inlet Vector

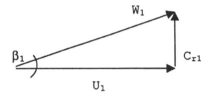

Blade Shapes (with Tandem Half-Blades)

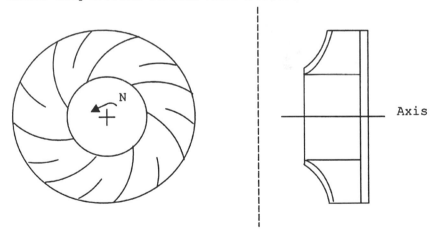

**FIGURE 7.32**  Geometry developed for the compressor impeller.

**FIGURE 7.33**  The complex blading shapes for a typical radial compressor impeller. Note the fully developed axial inducer section at the impeller inlet. (Barron Industries.)

## 7.7  DIFFUSERS AND VOLUTES

When the designers or the end users of a fan, blower, compressor, or turbine must decide on the overall layout and sizing of the machine, they must supply or receive a certain volume flow or mass flow of fluid at a specified pressure. Although one frequently works with the total pressure rise or drop through a machine, the pressure of most general interest is the static pressure at the point where the turbomachine is to be connected or inserted into the user's system. Knowing the total pressure of the machine and its components may not be adequate information unless the equipment happens to match the user's ductwork at the point of installation, or the device discharges into a large plenum, the atmosphere, or a large body of water, for example. To properly evaluate the machine, the user must know the net static pressure at which the required flow rate is delivered and the power required to produce it.

A pumping machine must then be designed or specified on the basis of the best static efficiency of a system, a combination of turbomachine and diffusing device, that can match the user's system geometry (Wright, 1984d). By itself, the turbomachine is capable of producing rather high efficiencies through careful design and control of tolerances. However, it is frequently necessary to reduce the absolute discharge velocities to an acceptable level to avoid large frictional or sudden expansion losses of energy and hence efficiency. The actual static or total pressure rise supplied by a pumping machine with its diffusion device depends on both the diffuser design and the static to total pressure ratio at the machine discharge (into the diffuser). For axial flow machines, one must then make a careful examination of the performance capabilities of conical or annular diffusers and their performance limits. For centrifugal machines, one must develop the design and performance characteristics of the scroll diffuser or spiral volute as the equivalent flow discharge device. One must also be concerned with the influence of downstream diffusion on the overall output capabilities of turbines—axial and centrifugal, gas and hydraulic.

For the axial case, one is concerned with the ratio of the net outlet static pressure to the total pressure delivered at the discharge flange of the machine. This is defined as an efficiency ratio $\eta'$ given by

$$\eta' = \frac{\eta_s}{\eta_T} = \frac{\psi_s}{\psi_T} \tag{7.19}$$

where $\psi_s$ is written as

$$\psi_s = \psi_T - k'\phi^2 \tag{7.20}$$

$k'$ is a loss coefficient combining a residual velocity pressure at the diffuser discharge and the frictional losses within the diffuser. $\psi$ and $\phi$ are the work and flow coefficients, respectively, developed earlier in the chapter. Combining these equations yields

$$\eta' = 1 - \frac{k'\phi^2}{\psi_T} \tag{7.21}$$

One can develop a functional form for $k'$ in terms of the diffuser geometric parameters.

The range of diffuser configurations is restricted to conventional annular diffusers with conical outer walls and cylindrical inner walls; as well, no bleeding of the viscous boundary layers is allowed—to prevent stall or increase permissible range of diffusion. Although complex wall shapes and wall-bleed schemes can provide significant performance gain or size and weight reductions for some applications, they will generally not provide cost-effective or efficient diffusion for practical situations such as forced or induced draft, ventilation, or heat exchange. For example, a typical boundary layer control scheme, characterized by very rapid area increase

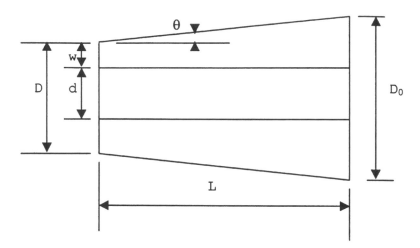

**FIGURE 7.34** Nomenclature and geometry for a conical-annular diffuser typically used in with axial flow machines.

with wall bleed to prevent stall, will require about 8 to 10% of the through-flow of the turbomachine to be removed from the walls (Yang, 1975). For most applications of practical interest, a penalty of eight to ten points in static efficiency would be incurred unless the bleed fluid were required for some secondary purpose. A cusp diffuser (Adkins, 1975), which relies on mass bleed to stabilize a standing vortex to turn the flow through a very rapid expansion, again requires an 8 to 10% bleed rate for annular configurations. The same restrictions as to applicability must be considered. Although these schemes and others involving aspiration, wall-blowing, vortex generation, and the like might have useful applications, they will not be further considered in this examination of performance.

The coefficient k′ can be defined by studying the fluid mechanics of the decelerating flow in a rigorous analytical fashion (Weber, 1978; Papailiou, 1975), or by relying on experimental information (Howard et al., 1967; Sovran and Klomp, 1967; McDonald, 1965; Adenubi, 1975; Smith, 1976; Idel'Chik, 1966). Examination of the literature yields several examples of experimentally derived correlations for predicting diffuser performance and first stall as functions of diffuser geometry. Figure 7.34 defines the geometry for the axial-annular diffusers being considered here, while Figure 7.35 shows a summary of the influence of geometry on the optimum level of static pressure recovery with minimum diffuser size or length. Also shown in the figure is the limit line for the rate of area increase, above which the flow in the diffuser becomes unstable (the "first stall" condition). Here, L/w is the ratio of diffuser length to inlet annulus height (Dr), and AR is the outlet to inlet annular area ratio. The optimal area ratio in terms of L/w as shown can be expressed as

$$AR \cong 1.03 + 1.85\left(\frac{L}{w}\right) - 0.004\left(\frac{L}{w}\right)^2 \tag{7.22}$$

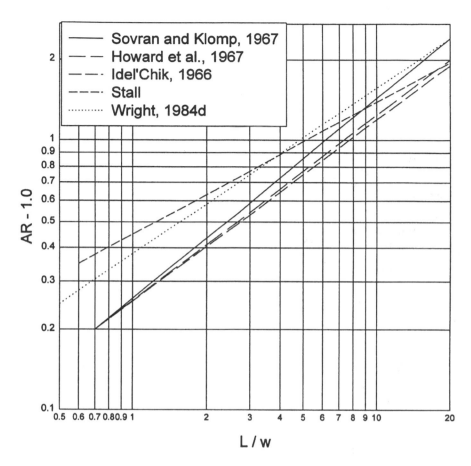

**FIGURE 7.35** Summary of the influence of geometry on optimum static pressure recovery with minimum diffuser size.

The corresponding values of $k'$ are shown in Figure 7.36 and can be expressed as

$$k' \cong 0.127 + \frac{1.745}{\left[ \left( \dfrac{L}{w} \right) + 2 \right]} \tag{7.23}$$

In a given application, one can use the relationship between $k'$ and $L/w$ with the earlier equation for $\eta'$ to make constraints between $L/w$ and the specific diameter of the machine, $D_s$. This can be done by recognizing that $\eta'$ can be rearranged to yield

$$D_s = \left( \frac{k'}{(1-\eta')} \right)^{1/4} \tag{7.24}$$

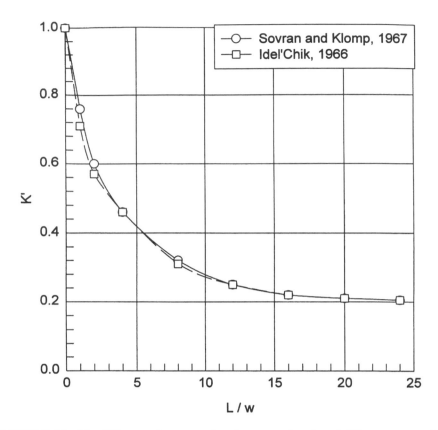

**FIGURE 7.36**  The diffuser static pressure loss factor as a function of diffuser geometry for conical-annular diffusers.

Consider an example where an axial fan is capable of a total efficiency of $\eta_T = 0.88$, and an efficiency ratio or $\eta'$ of 0.9, or $\eta_s = 0.792$, is required. If constrained to a value of L/w = 3.0, then k′ = 0.476 and the fan must have a specific speed no less than $D_s = 1.477$. If the length ratio is relaxed to allow L/w = 6, then k′ = 0.345 and $D_s$ must be at least 1.36. By doubling the diffuser length ratio, fan size is reduced by about 8%. Actual diffuser length is up 84%. This might represent a very favorable tradeoff in the cost of rotating equipment versus stationary equipment, provided there is room to install the longer diffuser.

If, on the other hand, the fan size and specific diameter are constrained, one must settle for either a fixed value of $\eta'$ if L/w is constrained, or an imposed value of L/w if $\eta'$ is fixed. If one is forced to use a value of $D_s = 1.25$ and $\eta'$ must be at least 0.92, then the required value of L/w is 23.5, which is a very long diffuser. If one retains the requirement for $D_s = 1.25$ and restricts L/w to be no greater than 10.0, the resulting value of $\eta'$ becomes $\eta' = 0.89$. Clearly, it pays to examine what one is asking for. In this case, a huge reduction in diffuser length requirement results in a static efficiency reduction of only 3%. The decrease may be the right answer in a space-constrained or initial cost-constrained design problem.

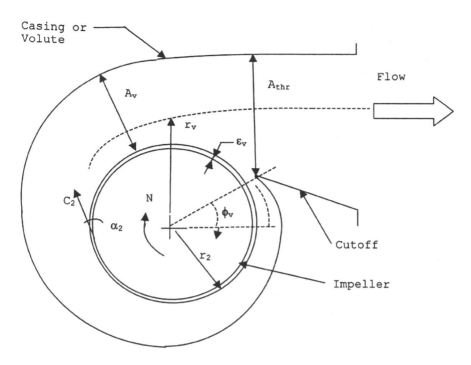

**FIGURE 7.37**   Geometry and nomenclature for a vaneless volute diffuser.

## 7.8   VOLUTE DIFFUSERS

For the centrifugal or radial flow case, one must generate information comparable to the previous section for spiral volutes. These devices are variously called housings, scrolls, collectors, diffusers, volutes, or casings. The choice depends on the type of turbomachine being discussed (pump, compressor, etc.) and, to a degree, on the point of emphasis being made. While all of these machines collect and direct the flow to a discharge station, not all of them diffuse the flow in the process. In a general way, one can consider the layout and performance of diffusing collectors and refer to them as volutes.

A prominent difference that exists from machine to machine is the existence or absence of straightening vanes in the diffuser. These vanes are primarily in place to recover the swirl energy in the discharge flow of highly loaded or high-speed machines and convert it to a useable static pressure rise. For more lightly loaded machines, the swirl recovery task is included in the collection process. The total energy of these flows is more equally distributed in static pressure rise through the impeller, velocity head associated with the through-flow, and the velocity head associated with the swirl component. Figure 7.37 defines the geometry of a vaneless volute diffuser in terms of the dominant features. The cross-section of the device can be circular, square, or more or less rectangular and varies with application.

The geometry is generic to centrifugal devices and will be modified to include diffusion vanes in the following discussion. Here, the impeller radius is given as $r_2$, and a radius characteristic of the volute is given as $r_v$. The volute radius is a function of the angular position in the volute and is usually linear between the volute "entrance and exit." The entrance/exit points as seen in Figure 7.37 relate to the "cutoff" of the volute. Various terms are used for this device, including tongue, cutwater, and splitter. The location of the cutoff relative to the outlet perimeter of the impeller requires a clearance gap, $\varepsilon_c$, between the stationary volute and rotating impeller. In general, for high-pressure and high-efficiency applications, this clearance will be designed as a nearly minimal value subject to fit-up and manufacturing tolerances required to avoid a "rub." Using the smallest possible clearances can result in serious acoustic pulsations in fans, blowers, and compressors and potentially destructive pressure pulsation in liquid pumps and high-pressure gas machines. A more conservative design choice frequently followed is to set the clearance value, $\varepsilon_c/D$, in the range of 10 to 20%. For example, most centrifugal fans of moderate pressure rise will employ a cutoff clearance ratio of about 12%, or between 9 and 15%. High-pressure blowers and compressors may use slightly smaller values, perhaps 6 to 10%, and liquid pumps may use a "reasonably safe distance" (Karassik, 1986) of 10 to 20% of impeller diameter.

Common design practice for shaping the volute employs maintaining simple conservation of angular momentum along the mean streamline of the volute (at $r = r_v$) so that

$$\frac{C_2 r_2}{C_v r_v} = \text{Constant} \tag{7.25}$$

where the constant is 1.0 for simple conservation but may be less than 1.0 if diffusion is taken within the volute. $C_2$ is the impeller discharge velocity (absolute frame of reference) and is formed with $C_{\theta 2}$ from total pressure rise and $C_{r2}$ from the volume flow rate Q. For linear volutes (the most common), the volute area is related to the throat area by $A_v = A_{thr}(\phi_v/360°)$, and the throat area is related to $C_v$ and Q according to $C_v \cong Q/A_{thr}$. The flow process in the volute is at best a non-diffusing frictional flow and may include total pressure losses from friction and diffusion. As modeled by Balje (Balje, 1981), the magnitude of the loss depends on parameters closely related to the specific speed of the machine, such as the impeller width ratio, the absolute or blade relative discharge angles, and the degree of diffusion incorporated into the volute design. Typical values (Balje, 1981; Wright, 1984c; Shepherd, 1956) can be defined in terms of a loss coefficient defined by

$$\zeta_v = \frac{\Delta H}{\left(\dfrac{C_2^{\,2}}{2g}\right)} \quad \text{or} \quad \omega_v = \frac{\Delta H}{\left(\dfrac{U_2^{\,2}}{2g}\right)} \tag{7.26}$$

For low specific diameter centrifugal machines such as fans and blowers, $0.2 \le \zeta_v \le 0.4$; for large specific diameter machines such as high-pressure pumps and gas compressors, $0.6 \le \zeta_v \le 0.8$. As pointed out by most investigators or designers in the field, the details of vaneless diffuser or volute scroll design are based on experience and craft.

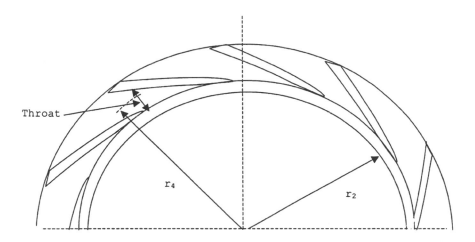

**FIGURE 7.38**   Vaned diffuser section in a pump volute.

For low specific speed centrifugal machines, it is the usual practice to include a vaned diffuser section in the volute. This is true in both high-pressure gas compressors and in high-pressure pumps. For a pump, this section consists of a number of vanes set around the impeller discharge as sketched in Figure 7.38. Following the procedure outlined by Karassik (Karassik, 1986), the flow can be considered to have several throats rather than the one seen in the vaneless volute. The volute outlet radius, $r_4$, is now smaller, and the volute velocity, $C_v$, would be commensurately larger except that a diffusing constant of about 0.8 is generally used in pump design. This vane channel may be straight or curved as sketched, but it generally results in a gradual rate of increase in the cross-sectional area $A_v$ as before. Although the sidewalls of these passages are generally designed as parallel, or even spreading apart, it would seem that a slight taper in passage width with increasing radius would provide a more conservative design. Karassik suggests a number of vanes equal to the number of impeller blades plus one.

In a high-speed compressor volute, the vane section will appear somewhat like the vane row sketched in Figure 7.39. The throat area must be carefully configured to match the relatively high-speed approach. The channels are usually straight and configured to yield outlet flow angles around 3° to 5° greater than the inlet flow angles. The static pressure achieved in these passages must generally be calculated using compressible isentropic or polytropic density calculations. Diffuser recovery values are, as usual, dependent on the inflow angle and impeller width ratio (intrinsically on the specific diameter). They may lie in the range of 70% of $C_2^2/2g$ for narrow impellers ($w_2/d = 0.02$) up to about 80 to 85% for wider impellers ($w_2/D = 0.06$) (Balje, 1981).

In the case of both pump and compressor, the overall diffuser performance must be modeled as a combination of the diffusing elements: an initial vaneless space, followed by a vaned section, followed by a final collector/diffuser section to the volute discharge.

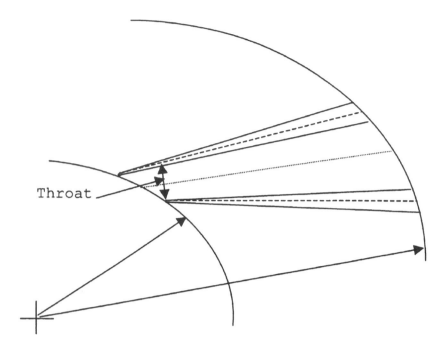

**FIGURE 7.39**    Vanes typical of a high-speed compressor.

## 7.9  SUMMARY

This chapter introduced the formal process of turbomachinery vector diagrams, using both dimensional and dimensionless forms. The dimensionless form, with work and through-flow coefficients, yields a high degree of generality and is applicable to all turbomachines and multi-stage configurations.

First, axial flow pumping machines and turbines were discussed to develop a familiarity with the terms and concepts of the interacting components, including inlet guide vanes, outlet guide vanes, and nozzles. The degree of reaction definition was developed and used to differentiate between different design approaches to turbine layout. An example of design procedure was provided in the context of a parametric layout of a hydraulic turbine.

These ideas and definitions were then extended to centrifugal and pure radial flow machines. Although it was necessary to modify the diagramming procedure slightly, the same conceptual form was applied and used to develop example layouts. An extensive example for a centrifugal fan at rather high specific speed was used to provide insight into the layout variables and procedures for radial discharge flow.

An example of a mixed-flow fan was used to illustrate techniques for diffusion levels in nearly-axial flow configurations. Another example was used to illustrate pump design, and yet another was presented for the fully compressible flow process in a radial turbocharger design problem. This last example provided an opportunity to examine the matching requirements between components of a combined system.

Finally, stationary diffusing devices were considered as the necessary adjunct equipment used to control the amount of discharge energy leaving the machine. For axial flow machines, the geometry and performance of annular diffusers was discussed in concept and by example. For the centrifugal flow machines, the volute scroll or collector was examined. This device is used to reduce the through-flow velocity and to recover the high levels of energy embodied in the swirl velocity. Layout and design techniques were examined and discussed with examples to illustrate the difference of the vaneless and vaned diffuser layouts.

## 7.10   PROBLEMS

7.1. Flow enters a centrifugal blade row (Figure P7.1), with $\rho = 1.2$ kg/m³ and mass flow rate of 1200 kg/s. If $C_1 = C_2 = 100$ m/s, calculate the power and pressure rise. Note that the vectors shown are absolute and $N = 1000$ rpm.

7.2. Repeat Problem 7.1 for an inlet absolute flow angle of 0° and +15° (relative to radii).

7.3. Determine the absolute outlet flow angles for Problem 7.2 such that the total pressure rise is zero.

7.4. A centrifugal blower is proposed with a blade exit angle $\beta_2$, and with $C_{r1} = C_{r2}$. Use the Euler equation to show that

$$\psi_{ideal} = \frac{\Delta p_T}{\left(\frac{1}{2}\rho U^2\right)} = 2\left(1 - \phi_c \cot \beta_2\right)$$

where $\phi_e = C_{r2}/U_2$.

7.5. Repeat the pump layout example (see Figure 7.30) using $Q = 600$ gpm and $H = 75$ ft.

7.6. Define the frictional effects in a fan impeller in terms of a total pressure loss, $\delta p_T$. Using the concept of a force coefficient, or "loss coefficient," one can specify $\delta p_T$ as:

$$\delta p_T = \frac{1}{2}\rho W_1{}^2 C_{wb} N_B$$

where $C_{WB}$ is a blade (or wake) loss coefficient and $N_B$ is the number of blades. Use the formulation for $\psi$ in Problem 7.4 to show that the losses can be written as

$$\delta \psi_{loss} = \frac{\delta p_T}{\left(\frac{1}{2} 2\rho U_2{}^2\right)} = \left(\left(\frac{d}{D}\right)^2 + \phi_e{}^2\right) N_B C_{wb}$$

where $d\psi_{loss}$ is the viscous correction to $\psi_{ideal}$.

7.7. Devise a means of estimating $C_{WB}$ as used in Problem 7.6. On what parameters should CWB depend? (Hint: Is the Cordier diagram of any help?)

7.8. In working with swirl recovery vanes (OGVs) behind heavily loaded rotors, the limit in performance for the stage may become the de Haller ratio in the vane-row cascade.
  (a) For example, if $C_{\theta 1} = C_{x1}$ entering the vane row and $C_{x2} = C_{x1}$, show that the inflow angle to the vane is 45° and the de Haller ratio becomes $W_2/W_1 = C_2/C_1 = 0.707$.
  (b) For $C_{\theta 1} > C_{x1}$ (very heavy loading), the approach angle to the vanes, β, becomes greater than 45° and $W_2/W_1 < 0.707$ until, if β is further increased, $W_2/W_1$ becomes unacceptably small. Show that, for $W_2/W_1 = 0.64$ as a lower limit, then $β_1$ must be less than 50°.

7.9. One way to alleviate the vane overloading described in Problem 7.8 is to allow residual swirl to exit the vane row. That is, $C_{\theta 2}$ may be non-zero and $β_2 = \tan^{-1}(C_{\theta 2}/C_{x2}) > 0$.
  (a) Show that to maintain a given level of $(W_2/W_1)$, one must constrain $β_2$ to:

$$\beta_2 = \cos^{-1}\left\{\frac{Cos\beta_1}{\left(\dfrac{W_2}{W_1}\right)}\right\}$$

  (b) If $β_2$ is allowed to be greater than zero as specified in part (a), what penalty in performance will result (qualitatively)?
  (c) For the example above with $C_{x2} = C_{x1}$ and $W_2/W_1$ constrained to 0.707, develop a table and curve of $β_2$ versus $β_1$ for 45° < $β_1$ < 70°.

7.10. Derive a quantitative expression for the reduction of static pressure rise and static efficiency associated with the vanes of Problem 7.9, which allow residual swirl in the flow discharge. (Hint: Account for the velocity pressure of a residual swirl $(\rho C_{\theta 2}/2)$ as a decrement to total pressure along with the axial velocity pressure $(\rho C_x^2/2)$.)

7.11. For the fan of Problem 6.16, define the velocity triangles in both the absolute and relative reference frames. Do so for both blades and vanes at hub, mean, and tip radial stations. Calculate the degree of reaction for all stations.

7.12. A hydraulic turbine-generator set is supplied with water at H = 67 ft at a rate of Q = 25,000 ft³/s. The turbine drives a 72-pole generator at N = 100 rpm. Calculate the non-dimensional performance parameters $N_s$, $D_s$, φ, and ψ, and estimate the efficiency. Develop a velocity diagram for the mean radial station of the turbine.

7.13. A fan specification calls for 7 m³/s of air at a pressure rise of 1240 Pa with a density of 1.21 kg/m³. Using a 4-pole motor as a cost constraint,

lay out the fan blade relative velocity vectors at the hub, mean, and tip
radial stations.

7.14. A small fan is required to supply 0.1 m³/s of air with a total pressure rise
of 2.0 kPa in air with $\rho = 1.10$ kg/m³. Lay out a centrifugal impeller, with
a single-inlet, single-width configuration. Construct the outer shape of a
volute scroll to work with the impeller layout. (Hint: Neglect compressibility.)

7.15. Select the layout parameters for an annular diffuser for the fan of Problem
7.13. Choose a length/diameter ratio that will provide optimal pressure
recovery, and estimate the diffusion losses (total pressure) and the resulting static efficiency.

7.16. For the small centrifugal blower analyzed in Problem 7.14, select a reasonable cutoff location and shape, and estimate the total pressure losses
in the volute. Compare these total pressure losses to the total pressure rise
through the impeller.

7.17. In Problem 7.14, the performance requirements for a centrifugal fan were
given as 0.1 m³/s of air at a pressure rise of 2.0 kPa with $\rho = 1.10$ kg/m³.
These requirements can be imposed on smaller high-speed axial fans if
one includes a "mixed flow" character by allowing the hub and tip radius
to increase between the inlet and outlet stations of the fan. Begin an
analysis with $N_s = 3.0$ and $D_s = 1.5$, using $d/D = 0.55$, and calculate
$(W_2/W_1)_{hub}$. Then, hold N and $d/D$ constant and allow $D_s$ at the exit plane
of the fan to increase to values greater than 1.5, recalculating the hub de
Haller ratio. Sketch several such flow paths and compare them to the
original and the one that allows the de Haller ratio to approach 0.7 at the
hub station.

7.18. Following the concept of Problem 7.17, let the flow path be conical as
sketched in Figure 7.24. At the entrance point of the blade row, let $D = 0.1$,
$d/D = 0.5$ m, and let the length, L, along the axis be 0.1 m. With $Q = 0.2$
m³/s, $N = 18,000$ rpm, and with $C_m$ held constant, calculate the maximum
ideal pressure rise possible, using $\alpha_{fp} = 45°$. Constrain the pressure so
that $w_2/w_1 = 0.72$ on the mean streamline.

7.19. Examine the influence of the narrowing of the impeller of Problem 6.19
on the shape and efficiency of a volute scroll designed to accompany the
design variations. Sketch the volute shapes and discuss their differences.

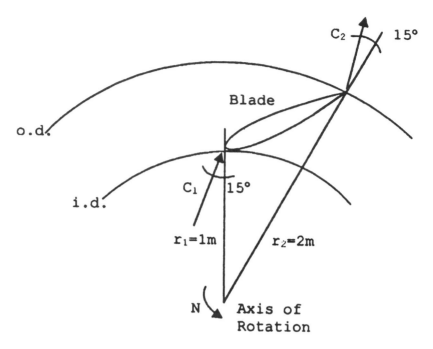

**FIGURE P7.1**   A centrifugal fan cascade.

# 8 Cascade Analysis

## 8.1 AXIAL FLOW CASCADES

In prior analyses of the flow through blade rows with energy addition or extraction, it was implicitly assumed that the velocity vectors along a blade or vane were constrained to lie along the mean camber line of the blade or vane. Although this is true exactly at the surface, the flow is not generally constrained completely by the blade shape through the entire flow channel. Figure 8.1 illustrates the problem: If the channel width is of the same scale as the channel length, then the flow will be fairly closely guided by the mean-line shape. As the width decreases relative to the length, all of the flow becomes aligned with the ideal exit angle. However, for large widths, the flow is not well constrained, and the average outflow angle will deviate significantly from the ideal or mean camber angle as shown in Figure 8.2. $\beta_2$ is defined as the mean exit angle or flow angle, while for distinction $\beta_2^*$ is the angle associated with the blade shape. In general, $\beta_2^* < \beta_2$ such that the ideal value of $\theta_{fl}$ considered thus far is now replaced by the effective value $\theta_{fl}^* = (\beta_1 - \beta_2) < \theta_{fl}$. As a result, $C_{\theta 2}$ is smaller than previously assumed. That is, we do not do as much work on the fluid as we would expect from the simple calculation we have been using so far.

To increase the accuracy of the predictions and design calculations, one needs to formulate a quantitative model of this flow behavior. The failure to achieve the expected level of flow turning, called the "deviation" of the flow vector from the ideal angle, must be modeled as $\delta$, where $\delta$ is a function of the geometric and velocity properties of the blade or vane cascade.

Here, the term "cascade" refers to the blade-to-blade relationship as illustrated in Figure 8.3a. Figure 8.3b shows the wealth of geometric variables at hand and provides a hint of the possible complexity of the modeling problem. It seemed to the early investigators (see, for example, Howell, 1945) that if the incidence factors could be fixed at the leading edge tangency condition (the simple assumption in earlier chapters), then the vector field could be resolved by expressing the trailing edge deviation as a simple function of the channel proportions. Thus, $\delta$ was reasoned to depend on the amount of turning required ($\phi_{fl}$ or $\phi_c$) and the solidity parameter expressed as the ratio of the blade chord to the blade-to-blade spacing. That is, they expected $\delta$ to be a function of $\phi_c$ and $\sigma = c/s$, where $s = 2\pi r/N_B$ (see Figure 8.3b). Howell's correlation of the limited experimental data available in 1945 suggested an algorithm as

$$\delta = \frac{0.26\phi_c}{\sigma^{1/2}} \tag{8.1}$$

This is a very simple form that provided a means for design guidance (in 1945) and was quickly supplanted by increasingly complex correlations.

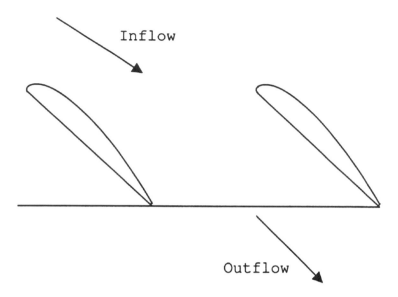

**FIGURE 8.1**    Blade channel flow.

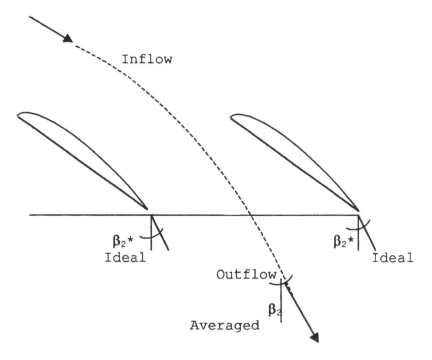

**FIGURE 8.2**    Concept of averaged flow direction.

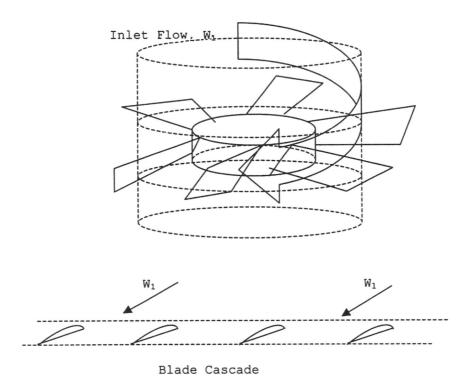

Blade Cascade

**FIGURE 8.3a**   Concept of a cascade: the 2-D linear analogy of blade-to-blade proximity.

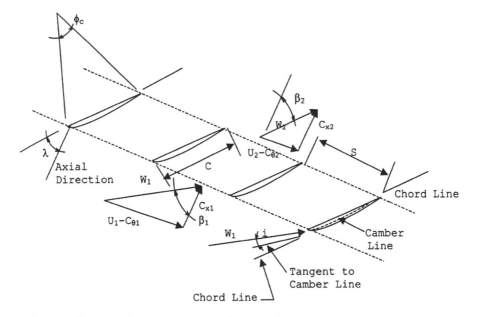

**FIGURE 8.3b**   Cascade geometry, nomenclature, and vectors.

**TABLE 8.1**
**Deviation Angle Values, $\delta$, with $\phi_c = 20°$, Using the Simple Howell Rule as a Function of $\sigma$ and $\beta_1$**

| $\beta_1/\sigma$ | 0.5 | 1.0 | 1.5 | 2.0 |
|---|---|---|---|---|
| 30 | 9.9 | 7.0 | 5.7 | 4.9 |
| 45 | 9.1 | 6.4 | 5.2 | 4.5 |
| 60 | 8.2 | 5.8 | 4.7 | 1.4 |

The value 0.26 was soon recognized to be a function of the cascade inlet flow angle $\beta_1$, and the model was rewritten as

$$\delta = \frac{m\phi_c}{\sigma^{1/2}}$$
(8.2)

with m given by

$$m = 0.41 - 0.2\left(\frac{\beta_1}{100}\right)$$
(8.3)

(Note that Howell's original form was written in terms of the complement to the angle $\beta_1$, $\alpha = 90 - \beta_1$, so that m = 0.23 + 0.1 ($\alpha/50$)). This is still a rather simple rule and is referred to as the Simple Howell Rule for circular arc cambered blades. For a typical cascade camber angle of, say, 20° at a value of $\beta_1 = 45°$ and a solidity of $\sigma = 1.0$, the older form of Howell's rule gives $\delta = 5.2°$, while the Simple Howell Rule gives $\delta = 6.4°$. Clearly, neither 5.2 nor 6.4 is negligible compared to 20.0. The early rules thus provided a needed measure of conservatism in cascade layout for axial compressors and fans. The rule in either form clearly shows the expectation of large deviation angles at low solidity and the ability to reduce deviation through increased solidity. The influence of inflow angle $\beta_1$, and the solidity $\sigma$, on $\delta$ can be seen in Table 8.1 where both parameters are varied. The range of deviation is seen to be substantial compared to the simplistic assumption of total guidance of the flow. The table values are all directly proportional to the camber angle and range from about 20 to 50% of $\phi_c$.

Following the earlier work of Howell and many others, the NACA staff at Langley Aeronautical Laboratory (now the NASA Langley Research Center) and the NASA Lewis Research Center undertook an extensive series of wind tunnel tests to determine the performance of compressor or fan blades in typical cascade configurations. The work was carried out in the late 1940s and continued through most of the 1950s. Tests were run using a particular family of airfoils called the NACA 65-Series Compressor Blade Section, based on the original NACA 65-Series airfoil used on aircraft built during World War II, perhaps most notably on the P-51 fighter aircraft. A large array of experimental configurations was examined with camber angles ($\phi_c$) varying from 0° to about 70°, solidity ($\sigma$) varying from 0.5 to 1.75, and

**TABLE 8.2**
**Cascade Combinations Tested by NACA**
**(Emery et al., 1958)**

| σ | $\beta_1 = 30°$ | $\beta_1 = 45°$ | $\beta_1 = 60°$ | $\beta_1 = 70°$ |
|---|---|---|---|---|
| | | 65-410 | 65-410 | |
| 0.50 | | 65-(12)10 | 65-(12)10 | |
| | | 65-(18)10 | 65-(18)10 | |
| | | 65-410 | 65-410 | |
| 0.75 | | 65-(12)10 | 65-(12)10 | |
| | | 65-(18)10 | 65-(18)10 | |
| | 65-010 | 65-010 | 65-010 | 65-010 |
| | 65-410 | 65-410 | 65-410 | 65-410 |
| | 65-810 | 65-810 | 65-810 | 65-810 |
| | 65-(12)10 | 65-(12)10 | 65-(12)10 | 65-(12)10 |
| 1.00 | 65-(15)10 | 65-(15)10 | 65-(15)10 | 65-(15)10 |
| | 65-(18)10 | 65-(18)10 | 65-(18)10 | |
| | | 65-(21)10 | 65-(21)10 | |
| | | 65-(24)10 | | |
| | | 65-(27)10 | | |
| | 65-410 | 65-410 | 65-410 | 65-410 |
| 1.25 | 65-(12)10 | 65-(12)10 | 65-(12)10 | 65-810 |
| | 65-(18)10 | 65-(18)10 | 65-(18)10 | 65-(12)10 |
| | | | | 65-(15)10 |
| | 65-010 | 65-010 | 65-010 | 65-010 |
| | 65-410 | 65-410 | 65-410 | 65-410 |
| | 65-810 | 65-810 | 65-810 | 65-810 |
| 1.50 | 65-(12)10 | 65-(12)10 | 65-(12)10 | 65-(12)10 |
| | 65-(15)10 | 65-(15)10 | 65-(15)10 | 65-(15)10 |
| | 65-(18)10 | 65-(18)10 | 65-(18)10 | |
| | | 65-(21)10 | 65-(21)10 | |
| | | 65-(24)10 | 65-(24)10 | |

inlet flow angle ($\beta_1$) varying from 30° to 70°. The tested values are summarized in Table 8.2 (taken from Emery et al., 1958). The NACA notation, for example the 65-010, refers to the basic airfoil properties. Here, 65 is the "series," the first digit beyond the dash referring to the Design Lift Coefficient $C_{lo}$, 0 being zero, and the last two digits referring to the blade relative thickness in percent of chord, t/c. For the 65-010, the t/c value is 0.10, or 10%. A 65-810 has a $C_{lo}$ value of 0.8, equivalent to $\phi_c = 20.0°$ of circular arc camber. (The equivalency is $\phi_c = 25\ C_{lo}$). All values tested in the initial work were for 10% thick airfoils. Table 8.3 also includes the basic information needed to generate the shapes of the airfoils considered here. Figure 8.4 shows a schematic view of the cascade wind tunnel used for the tests and the shapes of the airfoil sections used in the blade cascades.

The test results are covered in part by Emery et al., and they are perhaps best summarized and analyzed in the NASA Special Publication Number 36, Aerodynamic

**TABLE 8.3**
**NACA 65 Series Blades (Emery et al., 1958)**

NACA 65-010 BASIC THICKNESS FORMS

Ordinates, +y                                                    a = 1.0 mean camber line

| Station, x | 65(212)-010 airfoil combined with y = 0.0015x | Derived 65-010 airfoil | Station, x | Ordinates y | Slope, dy/dx |
|---|---|---|---|---|---|
| 0 | 0 | 0 | 0 | 0 | — |
| 0.5 | 0.752 | 0.772 | 0.5 | 0.250 | 0.42120 |
| 0.75 | 0.890 | 0.932 | 0.75 | 0.350 | 0.38875 |
| 1.25 | 1.124 | 1.169 | 1.25 | 0.535 | 0.34770 |
| 2.5 | 1.571 | 1.574 | 2.5 | 0.930 | 0.29155 |
| 5.0 | 2.222 | 2.177 | 5.0 | 1.580 | 0.23430 |
| 7.5 | 2.709 | 2.647 | 7.5 | 2.120 | 0.19995 |
| 10 | 3.111 | 3.040 | 10 | 2.585 | 0.17485 |
| 15 | 3.746 | 3.666 | 15 | 3.365 | 0.13805 |
| 20 | 4.218 | 4.143 | 20 | 3.980 | 0.11030 |
| 25 | 4.570 | 4.503 | 25 | 4.475 | 0.08745 |
| 30 | 4.824 | 4.760 | 30 | 4.860 | 0.06745 |
| 35 | 4.982 | 4.924 | 35 | 5.150 | 0.04925 |
| 40 | 5.057 | 4.996 | 40 | 5.355 | 0.03225 |
| 45 | 5.029 | 4.963 | 45 | 5.475 | 0.01595 |
| 50 | 4.870 | 4.812 | 50 | 5.515 | 0 |
| 55 | 4.570 | 4.530 | 55 | 5.475 | −0.01595 |
| 60 | 4.151 | 4.146 | 60 | 5.355 | −0.03225 |
| 65 | 3.627 | 3.682 | 65 | 5.150 | −0.04925 |
| 70 | 3.038 | 3.156 | 70 | 4.860 | −0.06745 |
| 75 | 2.451 | 2.584 | 75 | 4.475 | −0.08745 |
| 80 | 1.847 | 1.987 | 80 | 3.980 | −0.11030 |
| 85 | 1.251 | 1.385 | 85 | 3.365 | −0.13805 |
| 90 | 0.749 | 0.810 | 90 | 2.585 | −0.17485 |
| 95 | 0.354 | 0.306 | 95 | 1.580 | −0.23430 |
| 100 | 0.150 | 0 | 100 | 0 | — |
| L. E. radius | 0.666 | 0.687 | | | |

Design of Axial-Flow Compressors (Johnsen and Bullock, 1965). The chapter on "Experimental Flow in Two-Dimensional Cascades" provides an excellent summary and review, and presents correlations for the behavior of fluid turning or deviation, optimal leading edge incidence, and total pressure losses through the cascades tested. The observed data is summarized in terms of a leading-edge incidence angle, i, in degrees and the cascade deviation angle, $\delta$, as used by Howell et al. Incidence is presented as a linear function of camber in the form

$$i = i_o + n\phi_c \tag{8.4}$$

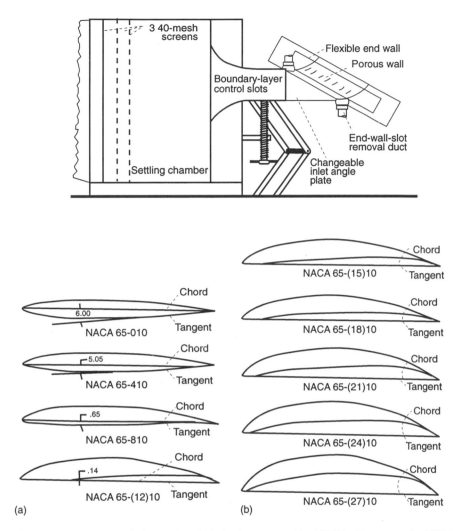

**FIGURE 8.4**   Cascade wind tunnel and blade shapes tested by NACA. (Emery et al., 1958.)

where $i_o$ is the incidence angle required for a zero-camber blade profile and n is a slope factor relating incidence to camber. Both $i_o$ and n are complicated functions related to the cascade variables $\beta_1$, t/c, and $\sigma$. $i_o$ is expressed as

$$i_o = \left(K_t\right)_{sh}\left(K_t\right)_i\left(i_o\right)_{10}\tag{8.5}$$

This is a messy function intended to be very general. The "t" subscript refers to variation with relative thickness of the blade and the "sh" subscript refers to behavior using shapes other than the 65-Series airfoils. In the limit case for very thin blades (t approaching zero), the value of $(K_t)_i$ goes to zero so that $i_o$ goes to zero. For 10% thick 65-Series blades, the function reduces to

$$i_o = \left(i_o\right)_{10} \tag{8.6}$$

reducing the problem of determining $(i_o)_{10}$ and $n$ as functions of $\beta_1$ and $\sigma$. The original correlations were presented as graphs of these values versus $\beta_1$ with $\sigma$ as a parameter. More recent efforts have reduced these extensive graphs to a set of curve fits (Wilson, 1983; Wright, 1985) in this form or one slightly modified for convenience. For example, the use of $i = i_o + n\phi_c$ with $n = n_o/\sigma^c$ with $n_o$ and $c$ as functions of $\beta_1$ and $\sigma$ yields a form very similar to the original correlations of Howell. That is,

$$i_o = i_o + \left(\frac{n_o}{\sigma^c}\right)\phi_c \tag{8.7}$$

For small values of $t/c$, the formula for $i$ reduces to

$$i = \left(\frac{n_o}{\sigma^c}\right)\phi_c, \qquad \frac{t}{c} \to 0 \tag{8.8}$$

which is virtually the same form as the simple Howell equation, but with the square root of $\sigma$ replaced by a variable exponent $c$, which can be a function of $\sigma$ and $\beta_1$.

For the $i_o$ function, a very simple curve fit results in the approximate expression

$$i_o = \left[8.0\left(\frac{\beta_1}{100}\right) - 1.10\left(\frac{\beta_1}{100}\right)^2\right]\sigma \tag{8.9}$$

for 10% thick 65-Series blades. The values of $n_o$ can be approximated by

$$n_o = -\left[0.0201 + 0.3477\left(\frac{\beta_1}{100}\right) - 0.5875\left(\frac{\beta_1}{100}\right)^2 + 1.0625\left(\frac{\beta_1}{100}\right)^3\right] \tag{8.10}$$

and the exponent $c$ becomes

$$c = 1.875\left[1.0 - \left(\frac{\beta_1}{100}\right)\right]\sigma \tag{8.11}$$

for $\sigma \le 1.0$, and

$$c = 1.875\left[1.0 - \left(\frac{\beta_1}{100}\right)\right] \tag{8.12}$$

for $\sigma > 1.0$. This group of equations becomes the algorithm for more formal, careful, and accurate treatment for establishing the optimal leading-edge incidence angles for blade layout. They replace the rather simplistic tangency condition used earlier, since that condition is reasonably accurate for only high solidity cascades of very thin blades.

To complete the more accurate treatment of the blade cascade, one needs to replace the Howell algorithm for flow deviation with the newer forms given in NASA SP-36. Again, the correlation was assumed to be a linear function of camber angle, whose coefficients are function of $\sigma$ and $\beta_1$. That is,

$$\delta = \delta_o + m\phi_c \tag{8.13}$$

The expression for $\delta_o$ is dependent on thickness and shape as was the form for $i_o$ so that

$$\delta = \left(K_t\right)_{sh}\left(K_t\right)_\delta \left(\delta_o\right)_{10} \tag{8.14}$$

As before, $(K_t)_{sh}$ goes to 1.0 for 65-Series blade shapes and $(K_t)_\delta$ goes to 1.0 for 10% thick blades, or to zero for very thin blades. Using the Howell form for the slope factor m as $m = m_o/\sigma^b$, the algorithm can be reduced to

$$\delta = \delta_o + \left(\frac{m_o}{\sigma^b}\right)\phi_c \tag{8.15}$$

or, for very thin blades,

$$\delta = \left(\frac{m_o}{\sigma^b}\right)\phi_c, \qquad \frac{t}{c} \to 0 \tag{8.16}$$

For the 10% thick 65-Series airfoils, the values of $\delta_o$, $m_o$, and b can be approximated as

$$\delta_o = 5.0\, \sigma^{0.8}\left(\frac{\beta_1}{100}\right)^2 \tag{8.17}$$

$$m_o = 0.170 - 0.0514\left(\frac{\beta_1}{100}\right) + 0.3592\left(\frac{\beta_1}{100}\right)^2 \tag{8.18}$$

$$b = 0.965 - 0.0200\left(\frac{\beta_1}{100}\right) + 0.1249\left(\frac{\beta_1}{100}\right)^2 - 0.9720\left(\frac{\beta_1}{100}\right)^3, \qquad \text{for } \sigma > 1.0$$

and

$$b = \left(0.965 - 0.0200\left(\frac{\beta_1}{100}\right) + 0.1249\left(\frac{\beta_1}{100}\right)^2 - 0.9720\left(\frac{\beta_1}{100}\right)^3\right)\sigma, \qquad \text{for } \sigma < 1 \tag{8.19}$$

This set of equations allows an improved estimate for the deviation angle in preparing blade layouts.

Now, one needs to assemble all of this accurate detail for blade layout into a rational scheme for choosing blade cascade parameters. The basic object of the layout process is, of course, to pick a geometry that will achieve the required head rise or fluid turning, $C_{\theta 2}$ or $\theta_{fl}$ at a given mass or volume flow rate. Reference to Figure 8.3 shows that the geometry to achieve the fluid turning becomes

$$\theta_{fl} = \phi_c + i - \delta \tag{8.20}$$

Bringing in the equations just developed for i and $\delta$, this may be rewritten as

$$\theta_{fl} = \phi_c + i_o + \left(\frac{n_o}{\sigma^c}\right)\phi_c - \left[\delta_o + \left(\frac{m_o}{\sigma^b}\right)\phi_c\right] \tag{8.21}$$

The amount of camber required to achieve the required value of $\theta_{fl}$ is

$$\phi_c = \frac{\left(\theta_{fl} - i_o + \delta_o\right)}{\left(1 + \dfrac{n_o}{\sigma^c} - \dfrac{m_o}{\sigma^b}\right)} \tag{8.22}$$

The correlations for $m_o$, $n_o$, b, and c are shown in Figures 8.5a and 8.5b.

Take a look at the earlier, simple example of deviation using the Howell correlation for a cascade with $\beta_1 = 45°$ and various values of $\sigma$. If the required fluid turning was 20°, then to meet that target, one requires

$$\phi_c = \theta_{fl} - i + \delta \tag{8.23}$$

For the leading edge tangency assumption (i = 0) and Howell's original correlation for $\delta$, then

$$\phi_c = \frac{\theta_{fl}}{\left[1.0 - \left(\dfrac{0.26}{\sigma^{1/2}}\right)\right]} \tag{8.24}$$

For $\sigma = 1.0$, $\phi_c = 26.3°$; for $\sigma = 0.5$, $\phi_c = 31.6°$; etc. One can use the more complex and more accurate form for i and $\delta$ to recalculate the camber and compare this to the Howell form just used. Using 10% thick 65-Series airfoils, one calculates (for $\sigma = 1.0$, $\theta_{fl} = 20°$ and $\beta_1 = 45°$)

$$i_o = 3.37; \quad \delta_o = 1.01; \quad b = 0.894$$

$$c = 1.031; \quad m_o = 0.2196; \quad n_o = -0.1344$$

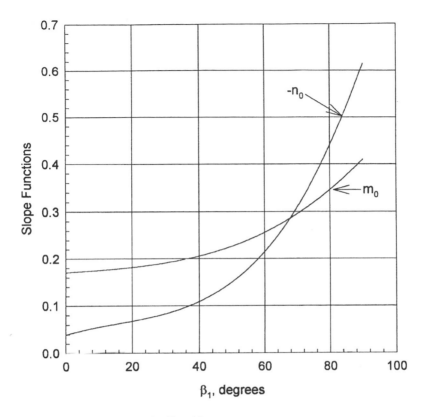

**FIGURE 8.5a**  Slope functions for δ and i.

Thus, the camber must be $\phi_c = 27.3°$. That is extremely close to the Howell value, differing by only 1°. If $\sigma = 0.5$, the complicated method gives $\phi_c = 34.6°$; this time, there is a somewhat larger difference from the Howell-based result, a difference of just over 3°.

As an example, examine the design requirements for an axial fan mean blade station where $Q = 10,000$ cfm and $\Delta p_T = 1.8$ InWG at standard density ($\rho g = 0.0748$ lbf/ft³). Calculation of $N_s$ and $D_s$ for various choices suggests a vane axial fan with $N = 1423$ rpm, $D = 2.5$ ft, and $\eta_T = 0.89$. Figure 6.10 suggests $d/D = 0.6$ or $d = 1.5$ ft and $r_m = (d + D)/4 = 1$ ft. One can calculate $U_m = 149$ ft/s and $C_x = Q/A_{ann} = 53$ ft/s. Now, using the Euler equation, $C_{\theta2}$ is calculated as

$$C_{\theta2} = \frac{\left(\dfrac{\Delta p_T}{\rho}\right)}{(U_m \eta_T)} = 30 \text{ ft/s}$$

This gives

$$\beta_1 = 70.4°; \quad \beta_2 = 66.0°; \quad \theta_{fl} = 4.4°$$

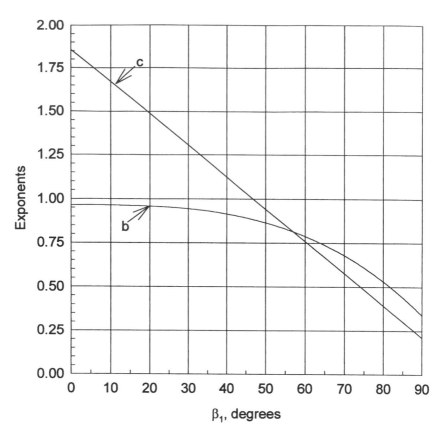

**FIGURE 8.5b**  Exponents for solidity in deviation and incidence angle correlations.

These three parameters provide the information necessary to lay out the mean blade station. However, one does have to select a value of $\sigma$ first. The de Haller ratio with the given velocity components is $W_2/W_1 = 130.3/158.1 = 0.82$. This is rather conservative, so choose a moderate value of solidity of $\sigma = 1.0$. Now proceed to calculate required values for the camber angle, $\phi_c$, and the blade setting angle, $\lambda$. For simplicity (perhaps for low cost), use a stamped sheet metal blade whose t/C value is very small, about 0.005. This choice allows one to set $i_o$ and $\delta_o$ to zero and work with

$$i = \left(\frac{n_o}{\sigma^c}\right)\phi_c \quad \text{and} \quad \delta = \left(\frac{m_o}{\sigma^b}\right)\phi_c \qquad (8.25)$$

Then calculate

$$m_o = 0.312; \quad b = 0.674; \quad n_o = -0.340;$$

$$c = 0.555; \quad i_o = 0; \quad \delta_o = 0$$

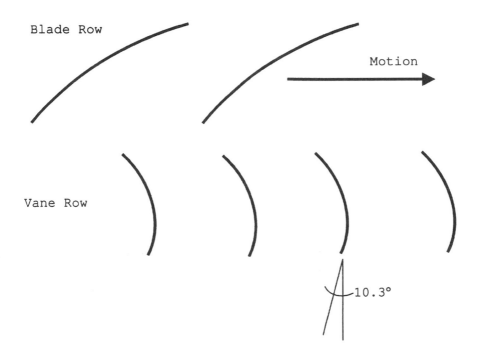

**FIGURE 8.6**  Layout for the example fan blade and vane rows.

so

$$\phi_c = 12.6°$$

Even with a moderately high level of solidity, one observes significant effects due to deviation and optimal incidence setting (–4.28° and 3.93°, respectively) with 12.6° of circular arc camber required to achieve only 4.4° of fluid turning. The blade pitch or setting angle is calculated as

$$\lambda = \beta_1 - i - \left(\frac{\phi_c}{2}\right) = 68.4° \qquad\qquad (8.26)$$

The layout for this blade mean station is shown Figure 8.6 (the vanes, to be developed, are also shown).

To extend this example, select some number of blades to put some precise geometry down on paper. Use $N_B = 8$ and calculate the blade chord and spacing from the solidity. Here, $\sigma_m = N_B C_m / (2\pi r_m)$ so that $C_m = (2\pi r_m \sigma_m)/N_B = 0.785$ ft. The blade length, $L_B$, from hub to tip is $(D - d)/2 = 1$ ft so that the blade is nearly square in appearance, with an aspect ratio of $L_B/C_m = 1.274$. This is a somewhat stubby blade, as aspect ratios of 2 to 3 are not unusual.

**TABLE 8.4**

**Effect of Blade Number on Pressure Rise**

| $N_B$ | $\sigma$ | $\delta$ | $\theta_{fl}$ | $\beta_2$ | $C_{\theta 2}$ | $\Delta p_T$ |
|----|------|------|------|------|------|-------|
| 2  | 0.25 | 10.0 | -1.7 | 72.1 | -15  | -0.89 |
| 4  | 0.50 | 6.27 | 2.05 | 68.2 | 15.5 | 0.92  |
| 6  | 0.75 | 4.77 | 3.55 | 66.8 | 24.8 | 1.49  |
| 8  | 1.00 | 3.93 | 4.40 | 66.0 | 30.0 | 1.78  |
| 10 | 1.25 | 3.38 | 4.94 | 65.4 | 32.9 | 1.95  |
| 12 | 1.50 | 3.00 | 5.33 | 65.0 | 35.3 | 2.09  |
| 14 | 1.75 | 2.69 | 5.62 | 64.8 | 36.5 | 2.17  |
| 16 | 2.00 | 2.46 | 5.85 | 64.5 | 37.7 | 2.24  |

What if one decides to use the same hub and blade designs but wants to modify the fan performance by selecting fewer or greater numbers of blades (of the same size and shape)? For example, one might market fans with $N_B$ = 4, 6, 8, 10, 12, 14, and 16 blades to amortize the engineering, development, and tooling costs over a broader range of products. The corresponding values of mean station solidity are $\sigma_m$ = 0.5, 0.75, 1.0, 1.25, 1.50, 1.75, and 2.0. Intuitively, one should expect that removing blades from the fan (e.g., from eight to four blades) will reduce the amount of work done on the fluid and hence the pressure rise at Q = 10,000 cfm will be reduced. On the other hand, if one adds blades (e.g., from 8 to 16), the work and pressure rise should be increased. Take care in this process to keep the blade setting angle and the flow rate the same, so that $\lambda$, $\phi_c$, $\beta_1$, and i are unchanged. Only the value of $\delta$ changes with the change in $\sigma$. $\theta_{fl}$ is modified to

$$\theta_{fl} = \phi_c + i - \delta \qquad (8.27)$$

Using the unchanged values of $m_o$, b, I, and $\phi_c$ yields $\theta_{fl} = 8.32° - 3.93°/\sigma^{0.674}$. Use the modified value of turning to revise the pressure rise value. That is,

$$\beta_2 = \beta_1 - \theta_{fl}$$

$$C_{\theta 2} = U_m - C_x \tan \beta_2$$

$$\Delta p_T = \eta_T \rho U_m C_{\theta 2}$$

Table 8.4 summarizes the calculations.

The behavior of the pressure rise with solidity is shown in Figure 8.7. Several observations are made. First, the pressure rise does indeed increase with $\sigma$ or $N_B$, but the relationship is decidedly nonlinear. Using half of the eight blades of the initial design layout yields 52% of the design pressure rise. That is, half the solidity gives half the pressure rise. However, using 16 blades instead of 8 gives only 126% of the design pressure rise. Doubling solidity does not come close to doubling the

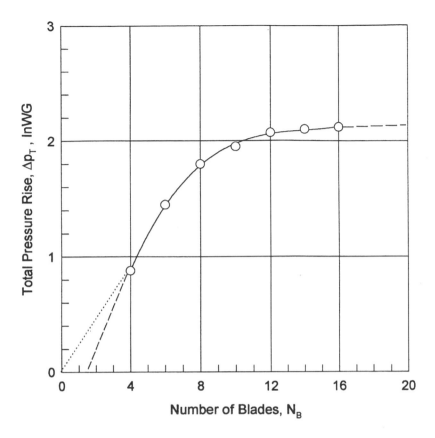

**FIGURE 8.7** Variation of total pressure rise with the number of blades. All other parameters are held constant.

pressure rise. The overall rate of change appears to be asymptotic as solidity becomes large and is a clear reflection of the $1/\sigma^c$ behavior of $\delta$ in the deviation angle algorithm. $\delta$ is slowly approaching zero, so the pressure rise is gradually reaching an upper limit. According to the algorithm, when $\delta$ approaches zero, $\theta_{fl}$ nears the limit of $\phi_c + i = 8.82°$. This gives $\Delta p_T = 2.90$ InWG. Of course, this value requires that the assumptions prevail at infinite solidity; it has already been noted that a practical upper limit for solidity is about $\sigma = 2.0$ or less because of flow channel blockage due to blade thickness and viscous boundary layer thickness. The 16-blade case is therefore a probable upper limit of performance for this design.

Referring again to Table 8.4, one sees that the values for swirl velocity, fluid turning, and pressure rise all become negative for a 2-blade configuration. This seems to imply that at very low solidity, the fan is operating as a turbine and is capturing energy from the airflow. However, the very great probability is that the algorithms for incidence and deviation are breaking down because of the paucity of experimental performance data for really low values of solidity (see Table 8.3). Rather than extrapolate these correlations toward zero solidity, first recognize the singular nature of the limit both physically and mathematically. In the curve-fit correlations, the use of the form $1/\sigma^a$ yields non-finite results as $\sigma$ becomes vanishingly small. In the physical situation, if performance requirements are held at a fixed level, one is forced to use very high values of camber angle or blade incidence to achieve any fluid turning at all as blades become very narrow in proportion to the blade spacing. The answer to this difficulty lies in establishing an analytical limit based on airfoil theory for "isolated airfoils" (not in cascade) and then devising a means of connecting known behavior of moderately low solidity cascades to this more reliable limit. Recent work in the area of low solidity cascades (Myers and Wright, 1993) has provided some guidance and a means of circumventing the difficulty. Growing out of that work, a very simple method of computation has been developed. The approach relies on the classical, isolated airfoil theory as shown in many references (e.g., Mellor, 1959). Mellor defined the limiting ability of a cascade to turn the flow in terms of a lift coefficient, which is a function of the cascade and flow variables. This value $C_{Lm}$, from theoretical considerations, is

$$C_{Lm} = \left(\frac{2}{\sigma}\right)\cos^2\beta_m\left(\tan\beta_1 - \tan\beta_2\right) \tag{8.28}$$

where $\beta_m$ is a mean value of $\beta$ defined by

$$\beta_m = \tan^{-1}\left[\frac{\left(\tan\beta_1 - \tan\beta_2\right)}{2}\right] \tag{8.29}$$

The camber value for isolated or very low solidity blades is established on the basis of the lift coefficient at an arbitrarily low solidity. The actual value is based on an interpolation between this "isolated" value and the experimentally correlated value at $\sigma = 0.6$. When dealing with a $\sigma$ value less than 0.6, the two controlling values must be established: at $\sigma = 0.6$, calculate $\phi_{c\,0.6}$ with the existing algorithms; for the low value, set $\sigma$ to 0.05 to get $\phi_{c\,isol} = 36.4\,C_{Lm} = 3.64\cos^2\beta_m\,(\tan\beta_1 - \tan\beta_2)$. The interpolated value is then given by

$$\phi_c = \left(1.5\phi_{c0.2} - 0.5\phi_{c0.6}\right) + \left(\frac{\sigma}{0.4}\right)\left(\phi_{c\,0.6} - \phi_{c\,0.2}\right) \tag{8.30}$$

Similarly, the incidence angle is adjusted to

$$i = -\frac{\phi_{c\,02}}{2} + \left(\frac{\sigma}{0.4}\right)\left(i_{0.6} + \frac{\phi_{c\,02}}{2}\right) \tag{8.31}$$

These equations substitute for the low range values such as those one would have obtained in the earlier example if one had tried to use the existing algorithms with very low solidity.

For the post-design analysis carried out with the existing blade shape and hub, one needs a slightly different kind of limit in terms of the amount of flow turning expected. Here, one can go to the actual limit and state clearly that the amount of turning for a zero-solidity cascade is zero. In terms of the deviation angle $\delta$, $\delta$ is given by $\theta_{fl} = 0$ or

$$\delta_{isol} = (\phi_c + i) = \frac{\phi_c}{2} + \beta_1 - \lambda \qquad (8.32)$$

Then calculate $\delta$ using $\sigma = 0.6$ and interpolate for the actual low value of $\sigma$ according to

$$\delta = \left(\frac{\sigma}{0.6}\right)(\delta_{0.6} - \delta_{isol}) + \delta_{isol} \qquad (8.33)$$

Now, re-examine the low solidity performance of the eight-blade fan with only two blades. The use of $\sigma = 0.25$ and the cascade algorithms gave a deviation angle of $10°$, resulting in a negative turning angle and pressure rise. Using the zero solidity limit pressure, since $\sigma$ is less that 0.6, one calculates $\delta_{isol} = 8.23°$. Then, the end value at $\sigma = 0.6$ is given by $\delta_{0.6} = 3.93°/\sigma^{0.674} = 5.55°$. Interpolation at $\sigma = 0.25$ is done according to

$$\delta = \left(\frac{0.25}{0.6}\right)(5.55° - 8.23) + 8.23° = 7.4°$$

Following through the remaining calculations, the turning angle is a positive $1.12°$, $\beta_2$ is $69.3°$, $C_{\theta2}$ is 8.88 ft/s, and $\Delta p_T$ is 0.527 InWG, a nice, satisfying, positive value. Carrying out the same calculations for a four-blade configuration yields a pressure rise of 0.97 InWG, only slightly higher than before. The new values are shown on Figure 8.7 to form the dashed line to the limit of zero pressure rise for zero blades.

The methods developed here can also be used to lay out a vane section for the vane axial fan. Again, working at the mean radial station of the fan, determine the inflow properties for the vane row directly from the outflow of the blade row. The blade relative velocity components must be converted velocities in the absolute non-rotating reference frame of the stationary vanes. The discharge velocity requirements for the vane row are to simply provide a pure axial flow at the outlet (complete swirl recovery). The inlet angle for the vane, $\beta_1$, is set by the value of $C_x$ and $C_{\theta2}$ from the blade row. These were $C_x = 53$ ft/s and $C_{\theta2} = 30$ ft/s, so that $C_1 = 60.9$ ft/s. At the discharge, the value of $C_2$ is just $C_x$, so $C_2 = 53$ ft/s. The de Haller ratio for the vanes is

$$\frac{W_2}{W_1} = \frac{C_2}{C_1} = 0.87$$

This is a very conservative value; thus, one can assign a solidity value to set up the vane mean station cascade of, say, $\sigma = 0.75$. The inflow angle is

$$\beta_1 = \tan^{-1}\left(\frac{C_{\theta 2}}{C_x}\right) = 29.5°$$

The fluid turning of the vane must drive the outflow angle to zero, so $\theta_{fl} = \beta_1 = 29.5°$ and $\beta_2 = 0°$. Using a thin airfoil section, calculate

$$m_o = 0.186; \quad b = 0.709; \quad n_o = -0.0908$$

$$c = 0.991; \quad i_o = 0; \quad \delta_o = 0$$

to yield

$$\phi_c = 45.3°$$

Since $N_B = 8$ for this design, use $N_V = 17$. Choose the number of vanes primarily to avoid a number that is evenly divisible by a multiple of the number of blades so as to avoid the natural resonances of the acoustic frequencies of the two rows. Secondarily, try to keep the aspect ratio of the vane ($L_V/C_{mV}$) between 1 and 6. With 17 vanes, the mean chord (based on a vane row solidity of 0.75) is 0.227 ft, with a height of 1 ft. That gives an aspect ratio of about 4.4. Seven vanes would require a chord of 0.551 ft with an aspect ratio of 1.8; 23 vanes would give a chord of 0.168 ft and an aspect ratio of nearly 6.0. Either the 7-vane or 17-vane configuration looks all right, so one can stay with $N_V = 17$. Values of i and $\delta$ for the thin-vane mean station are calculated as

$$i = \left(\frac{n_o}{\sigma^c}\right)\phi_c = -5.47° \quad \text{and} \quad \delta = \left(\frac{m_o}{\sigma^b}\right)\phi_c = 10.33°$$

Finally,

$$\lambda = \beta_1 - i - \frac{\phi_c}{2} = 12.4°$$

The mean station blade and vane shapes and orientations for this preliminary design layout are shown in Figure 8.6. Note that the vane appears to turn the flow well past the axial flow direction. This appearance led to the term "over-turning" commonly seen in the cascade literature years ago. It is clearly a misnomer since the extra 10.33° simply accounts for the deviation and corrects the average velocity across the channel to be purely axial.

## 8.2 RADIAL FLOW CASCADES

As found for the flow through axial cascades, incomplete guidance of the fluid leads to a reduction in $C_{\theta 2}$ compared to an ideal value based purely on camber line geometry of the blades. In effect, the value of $C_{\theta 2}$ is always less than the ideal value. One can define $\mu_E$ as

$$\mu_E = \frac{C_{\theta 2}'}{C_{\theta 2}} \tag{8.34}$$

where $C_{\theta 2}'$ is the adjusted value and $C_{\theta 2}$ is the camber line calculation. This ratio is called the slip coefficient (Csanady, 1964; Shepherd, 1956). As in axial flows, there are many schemes, mostly based on numerical flow solutions and geometric arguments, to estimate $\mu_E$. The more commonly used methods used to calculate slip include the method of Busemann (Busemann, 1928), which is used in initial design and layout for blades with both large and moderate discharge angles, in the range of $45° \le \beta_2 \le 90°$ (see Wright, 1978). Another method used extensively in compressor layout is the method by Stanitz (Stanitz, 1951). The method presented by Balje (Balje, 1981) is interesting in the context of its great simplicity. The method developed by Stodola (Stodola, 1927) is perhaps the oldest in common use and is the one recommended here as a preferred choice in preliminary design blade layout for centrifugal machines. The Stodola formula for slip is

$$\mu_E = 1 - \frac{\left(\dfrac{\pi \sin \beta_2}{N_B}\right)}{\left(1 - \dfrac{C_{m2} \cot \beta_2}{U_2}\right)} \tag{8.35}$$

For moderate values of $C_{m2}/U_2$, the formula is usually simplified to

$$\mu_E = 1 - \left(\frac{\pi \sin \beta_2}{N_B}\right) \tag{8.36}$$

This form is claimed to be most relevant for blades that are very flat in the trailing edge region. $C_{m2}$ can be replaced in most cases by $C_{r2}$. A more complicated form of the Stodola formula includes a blade trailing face viscous blockage factor, $F_{vb}$, of the order of about 0.7 to 0.8, which acts to modify $\mu_E$ according to

$$\mu_E = 1 - \left(\frac{\pi \sin \beta_2}{N_B}\right) F_{vb} \tag{8.37}$$

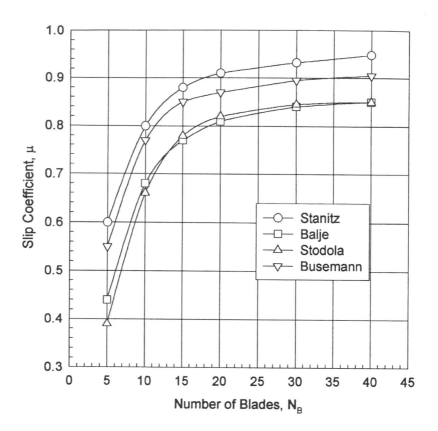

**FIGURE 8.8a**  Slip coefficient formula comparisons for radial blades.

For $F_{vb} = 0.75$, $\beta_2 = 30°$, and $N_B = 20$, for example, $\mu_E = 0.941$ compared to 0.921 for the unmodified Stodola value. The refinement, which is unconservative, is probably not justified in light of uncertainty in the proper value to use for $F_{vb}$ (see Balje, 1984).

Figure 8.8a provides a comparison study of the different methods for calculating $\mu_E$ for radial bladed or radial tipped blowers ($\beta_2 = 90°$). As seen, the methods by Stodola, Busemann, and Balje are in substantial agreement, while the Stanitz method provides the highest estimates for $\mu_E$ by 5% or so, and the Balje method is moderately conservative at blade numbers greater than ten. As seen in Figures 8.8b and 8.8c, the methods of Busemann and Stodola are in excellent agreement for the lower range of blade angles when the blade number is greater than about eight. The Stodola formula in its simpler form is thus considered to be very easy to use, and accuracy is adequate for use in preliminary blade layout for centrifugal machines. Clearly, refinement of a final choice of layout, by laboratory development or detailed numerical flowfield analysis, would be required in an advanced phase of design.

Consider the layout of the trailing edge angle of the blade of a water pump. Start with performance requirement of 30 m of head at a flow rate of 0.015 m³/s. Selecting a rotating speed of 3000 rpm yields a specific speed of $N_s = 0.542$. Cordier analysis sets $D_s = 4.9$, $D = 0.145$ m, and $\eta_T = 0.87$. d/D should be about 0.383 based on

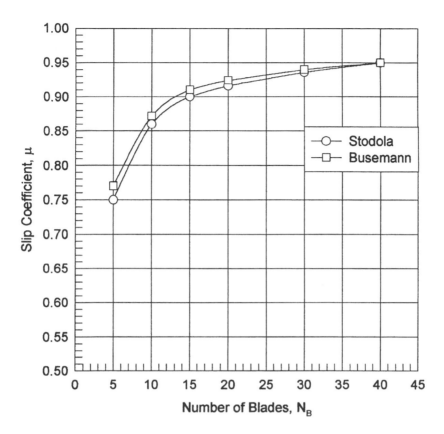

**FIGURE 8.8b**   Slip coefficient comparison with $\beta_2 = 20°$.

Equation (6.41). The Euler equation with $\eta_T = 0.87$ provides a requirement on $C_{\theta 2}$ of 14.9 m/s. If $C_{r1}$ is set to match the throat velocity, then $C_{r1} = Q/(\pi d^2/4) = 6.09$ m/s. Then choose a reasonable value of $C_{r2} = 4.0$ m/s and calculate a conservative de Haller ratio of about $W_2/W_1 = 0.8$. The values of $C_{r2}$ and $C_{r1}$ will size the inlet and outlet widths of the impeller. The results give

$$U_1 = 8.79 \ \text{m/s}; \quad U_2 = 22.8 \ \text{m/s}$$

$$C_{r1} = 6.09 \ \text{m/s}; \quad C_{r2} = 4.0 \ \text{m/s}; \quad C_{\theta 2} = 14.9 \ \text{m/s}$$

where $C_{\theta 1}$ has been set to zero.

Based on a zero slip calculation (ideal Euler equation), the trailing edge mean camber line would lie at an angle of

$$\beta_2 = \tan^{-1}\left(\frac{C_{r2}}{(U_2 - C_{\theta 2})}\right) = 26.9°$$

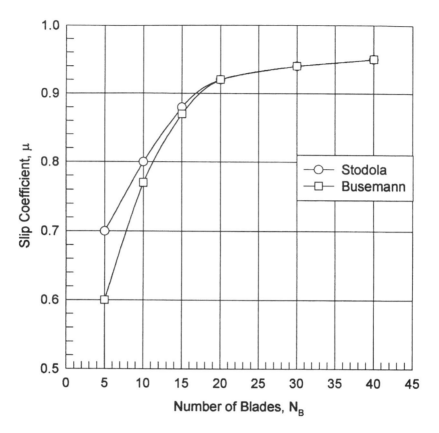

**FIGURE 8.8c**   Slip coefficient comparison for $\beta_2 = 40°$.

above the tangent to the impeller at the trailing edge.

However, it is necessary to correct this really approximate setting by the estimated slip coefficient. Thus, the actual angle will depend on the number of blades
one decides to use. Using

$$\mu_E = 1 - \frac{\pi \sin \beta_2}{N_B}$$

correct the value of $C_{\theta2}$ with $C_{\theta2}/\mu_E$ and replace the ideal $\beta_2$ with

$$\beta_2{}^* = \tan^{-1}\left[\frac{C_{r2}}{\left(U_2 - \dfrac{C_{\theta2}}{\mu_E}\right)}\right]$$

**TABLE 8.5**
**Variation of Blade Angle with Blade Number**

| $N_B$ | $\mu_E$ | $\beta_2^*$ | $\beta_2$ |
|---|---|---|---|
| 6 | 0.763 | 50.6 | 26.9 |
| 10 | 0.858 | 36.4 | 26.9 |
| 15 | 0.905 | 32.2 | 26.9 |
| 20 | 0.929 | 30.6 | 26.9 |
| 25 | 0.943 | 29.7 | 26.9 |
| 30 | 0.953 | 29.2 | 26.9 |
| 35 | 0.959 | 28.8 | 26.9 |
| 40 | 0.965 | 28.6 | 26.9 |

If using ten blades, then $\mu_E = 0.858$ and $\beta_2 \cong 36.4°$. That is a substantial change—9.5°—from 26.9° and is clearly non-negligible. Higher blade numbers will reduce the disparity and the required angle for the blade. Table 8.5 gives the values for $N_B$ ranging to about 40 blades. The discharge fluid angle, $\beta_2$, is constant to meet the required head rise and is included in the table for emphasis.

The behavior of slip can be illustrated further by examining the pressure rise performance of a centrifugal machine as one adds or omits blades. Once the blade setting angle, $\beta_2^*$, is fixed to some value, one can recalculate the $C_{\theta 2}$ based on a different number of blades. That is, the ratio of the new value of $C_{\theta 2\,new}$ to the original value $C_{\theta 2\,orig}$ will be

$$\frac{C_{\theta 2\,new}}{C_{\theta 2\,orig}} = \frac{\mu_{E\,new}}{\mu_{E\,orig}} = \frac{\left(1 - \dfrac{\pi \sin \beta_2^*}{N_{B\,new}}\right)}{\left(1 - \dfrac{\pi \sin \beta_2^*}{N_{B\,orig}}\right)}$$

Following the previous example and using $N_{B\,orig} = 40$ with $\beta_2^* = 28.6°$ reduces the above equation to

$$\frac{H_{new}}{H_{orig}} = 1.039 - \frac{1.543}{N_{B\,new}}$$

assuming no change in efficiency. Table 8.6 gives the ratios for blade numbers beginning at 6 and increasing to 40. Using the velocities from the previous work to calculate $H = U_2 C_{\theta 2}/g$ allows inclusion of real head numbers in the last column ($U_2$ was 22.8 m/s and the design value for $C_{\theta 2}$ was 14.9 m/s).

An interpretation of the results shown in Tables 8.5 and 8.6 indicates that the design should probably not be executed with more than 12 to 15 blades. The gains above the 90% level achieved by doubling or tripling the numbers of blades do not

**TABLE 8.6**
**Head Rise Variation with Blade Number**

| $N_B$ | $H_{new}/H_{orig}$ | H (m) |
|---|---|---|
| 6 | 0.786 | 23.6 |
| 8 | 0.850 | 25.5 |
| 10 | 0.889 | 26.7 |
| 12 | 0.914 | 27.4 |
| 15 | 0.940 | 28.2 |
| 20 | 0.966 | 29.0 |
| 25 | 0.989 | 28.4 |
| 30 | 0.992 | 29.8 |
| 35 | 0.999 | 30.0 |
| 40 | 1.000 | 30.0 |

justify the extra cost and complexity of the impeller layout for the higher blade numbers.

The reader has probably noted that no presentation is given here on choosing an optimal leading edge incidence for the centrifugal cascade. No reliable analysis to predict such values is known to the author. It is recommended that the leading edge of the blade be matched to the blade relative velocity vector at the blade row entrance, based on uniform radial velocity into the cascade. Although this uniformity of flow is a relatively simplistic approach, refinement beyond this assumption requires a detailed knowledge of the three-dimensionality of the flow passing into the impeller inlet and approaching the blade leading edge. A rudimentary preliminary design analysis can be carried out with the simple assumption stated here. In a later chapter, this restriction is alleviated somewhat by the introduction of approximate quasi-three-dimensional analysis for estimating the distribution of the flow across the blade inlet (in the axial direction). As was stated for the axial flow analysis above, the geometry developed for the centrifugal impeller using these preliminary design techniques should be refined by rigorous numerical analysis of the impeller flowfield or through an experimental development program—or both.

## 8.3  SELECTION OF SOLIDITY FOR AXIAL CASCADES

As in the development of the modified Howell rule for axial cascades, deviation is a function $\beta_1$, $\sigma$, and $\phi_c$. The rule seems to imply that a low value of $\sigma$ can simply be compensated for by increasing the amount of $\phi_c$ to meet the requirements on $\theta_{fl}$. It is really not that simple because the details for estimating the allowable diffusion on a blade suction surface depend, in part, on the solidity of the cascade. From the de Haller ratio and the diffusion analogy, the rule was established that $W_2/W_1 > 0.72$ would yield an acceptable blade loading level. Although it was indicated that increasing solidity could allow some adjustment of the minimum $W_2/W_1$ value, the idea was not developed further.

In the NASA Special Publication 36 (Johnson and Bullock, 1965), this concept is developed in terms of a more realistic estimate of relative velocity, W. If one denotes the maximum value of W on the low-pressure or suction surface of a blade (or vane) as $W_p$ (p for "peak") and defines a more detailed velocity parameter as $W_2/W_p$, then an alternate form can be written as

$$D_p = \frac{\left(W_p - W_2\right)}{W_p} = 1 - \frac{W_2}{W_p} \tag{8.38}$$

and called the NACA local diffusion parameter. The concept of the blade surface velocity leading to this parameter is illustrated in the velocity distribution shown in Figure 8.9. NACA used theoretical potential flow velocity distributions on the airfoil surface to estimate the boundary layer property $\theta^*/C$ as a function of $D_p$. $\theta^*/c$ is the ratio of the momentum deficit thickness of the boundary layer at the trailing edge of the blade suction surface to the blade chord length. $\theta^*/C$ serves as a measure of blade wake thickness and momentum loss (hence drag) due to viscous effects on the blade flow. The calculated data from these airfoils is summarized in Figure 8.10 as $\theta^*/C$ versus $D_p$. As seen in the figure, $\theta^*/C$ increases monotonically as $D_p$ increases. When $D_p$ reaches a value of slightly above 0.5 (or when $W_2/W_p$ becomes less than 0.5), the magnitude of $\theta^*/C$ begins to increase rapidly. This sudden change indicates the development of flow separation of the flow from the blade suction surface. The blade cascade begins to stall and drag or total pressure loss increases rapidly as well. The concept is a clear and useful illustration of the viscous flow behavior, but calculation of $D_p$ and $\theta^*$ are non-trivial. Seeking a simpler approach, Leiblien (Leiblien, 1956) developed an approximate form for $W_p$ such that

$$W_p \cong W_1 + \frac{\Delta W_\theta}{2\sigma} = W_1 + \frac{\Delta C_\theta}{2\sigma} \tag{8.39}$$

where $\Delta W_\theta/2$ is the mean circulation velocity, additive on the suction surface and subtractive on the pressure side. The magnitude of $\Delta W_\theta$ can be extracted from the velocity diagram for the blades as shown in Figure 8.11. This is, qualitatively speaking, the source of the difference of pressure between the two surfaces illustrated in Figure 8.9. Leiblien developed a new form for the diffusion factor as

$$D_L = 1 - \frac{W_2}{W_1} + \frac{\Delta W_\theta}{2 W_1 \sigma} \tag{8.40}$$

for which an experimental correlation of existing, extensive cascade data for $\theta^*/C$ was undertaken. Results are presented in Figure 8.12 and indicate excellent collapse of the data. Again, one observes the gradual increase of $\theta^*/C$ as $D_L$ increases up to about $D_L = 0.6$, where $\theta^*/c$ has increased by a factor of about 4. Beyond $D_L = 0.6$, the data for $\theta^*/C$ shows a very rapid increase and greater scatter, indicating again the development of blade stall. Leiblien's results can be used as a solid guideline to

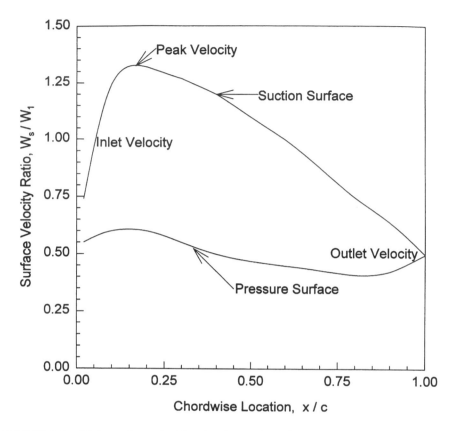

**FIGURE 8.9**   Blade surface velocities showing peak suction surface velocity for local diffusion model.

what constitutes an acceptable level of blade loading; namely, $D_L$ should clearly be less than 0.6, and for reasonably conservative design practice, the value of $D_L$ should not exceed the range 0.4 to 0.55. For relatively high levels of blade solidity, this result is seen to reduce to the de Haller ratio as a criterion of blade loading limitation.

If one establishes the level of $D_L$ that can be allowed as a design limit, $D_{L\,max}$, the required minimum value of solidity can be written as

$$\sigma_{min} = \frac{\left(\dfrac{\Delta C_\theta}{2W_1}\right)}{\left(\left(\dfrac{W_2}{W_1}\right) - \left(1 - D_{L\,max}\right)\right)} \qquad (8.41)$$

The behavior of minimum solidity can be illustrated by an example of an axial flow mean station with $C_{\theta 2} = 12$ m/s, $U_2 = U_1 = 35$ m/s, and $C_x = 10$ m/s. For these

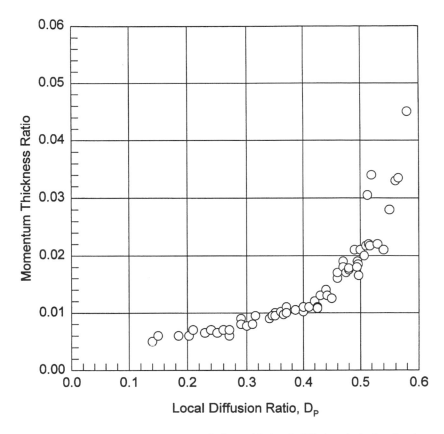

**FIGURE 8.10**   Momentum thickness variation with local diffusion (calculated values of velocity and momentum loss).

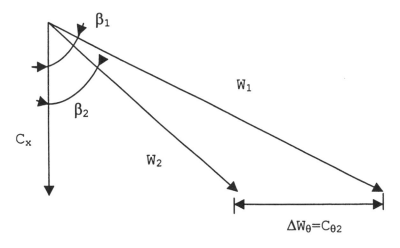

**FIGURE 8.11**   Velocities used in Leiblien's diffusion model.

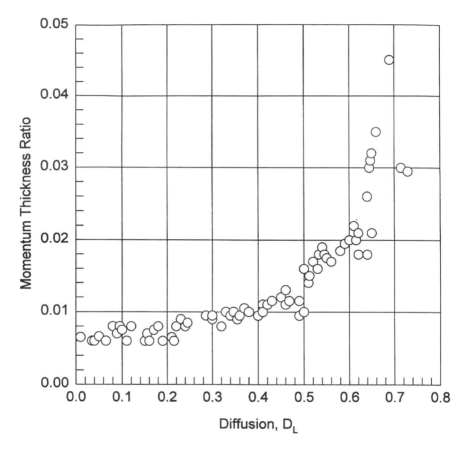

**FIGURE 8.12** Momentum loss variation with the Leiblien diffusion factor (correlation of experimental data).

values, $W_1 = 36.4$ m/s and $W_2 = 25.1$. Then, setting $D_{L\ max} = 0.5$ yields, with $W_2/W_1 = 0.690$,

$$\sigma_{min} = \frac{\left(\dfrac{12}{36.4}\right)}{\left[\left(\dfrac{25.1}{2 \times 36.4}\right) - (1 - 0.5)\right]} = 0.870$$

This is a moderate value of solidity, a little on the low side, which is consistent with a fairly conservative value of the de Haller ratio.

Choosing $D_{L\ max}$ as 0.45 would provide a more conservative level of design by allowing a more substantial margin away from stall conditions. $\sigma_{min}$ can then be calculated as a function of $W_1$, $W_2$, and $\Delta C_\theta$ or by using the alternate definition for $D_{L\ max}$ directly in terms of $\beta_1$ and $\beta_2$

$$\sigma_{min} = \left(\frac{\cos\beta_1}{2}\right)\frac{\left(\tan\beta_1 - \tan\beta_2\right)}{\left[\left(\frac{\cos\beta_1}{\cos\beta_2}\right) - \left(1 - D_{L\,max}\right)\right]} \qquad (8.42)$$

Figure 8.13 shows typical results for solidity as a function of outlet flow angle with inlet flow angle as parameter. These results show that, for large $\beta_1$, the value of $\beta_2$ does not vary much from $\beta_1$. That is, $\theta_{fl} = \beta_1 - \beta_2$ is restricted to small values as constrained by reasonable diffusion levels, even if very high values of $\sigma$ are used. On the other hand, at smaller $\beta_1$, the diffusion level is not nearly as significant a constraint, even for very large values of $\theta_{fl}$ (small $\beta_2$). The required minimum solidity is observed to take on much smaller values as well. Putting this behavior in perspective, note that large values of $\beta_1$ correspond to small values of $C_x/U$ ($\tan^{-1}\beta_1 = C_x/U$) or low specific speed (for an axial machine). Smaller $\beta_1$ values correspond to larger $C_x$ or higher specific speed. If the $\beta_1$ is fairly large and, in addition, the $\theta_{fl}$ is high, one is facing low flow and high pressure rise and can expect to have difficulty laying out an acceptable axial cascade to do the job. This difficulty is consistent with previous experience using the Cordier guidelines and arriving at unacceptable de Haller ratios. Here, one will be forced into using unacceptably large solidity to try to keep the diffusion level under control. The problem can be alleviated somewhat by selecting a very large hub or d/D value to force the $C_x$ higher and $\beta_1$ lower. This is always done at the expense of increased viscous losses and a high level of velocity pressure through the blade row (low static efficiency). Also note that, in the process of design, the solidity can be chosen directly, perhaps because of geometric constraints such as a linearly tapered blade. It is, however, very important to use that choice to calculate the diffusion level that results from that solidity to make sure the design will be below the stall limitation.

## 8.4 SOLIDITY IN CENTRIFUGAL CASCADES

In centrifugal cascades, the term "solidity" is not commonly used, nor is solidity a primary design variable used in blade layout or impeller design. As seen in the use of the various formulas for slip coefficient, the number of blades becomes a very important parameter in determining both outlet angle and pressure rise. The solidity of a radial flow cascade can be defined, if one so chooses, in a couple of different ways. The more common method is to simply define the ratio of the angular segment between leading and trailing edges of a blade to $360°/N_B$, the angular spacing between adjacent leading edges. The geometry is shown in Figure 8.14. A general rule in centrifugal cascade layout is to keep this angular solidity at about 1 or slightly greater. As seen in the examples, the use of a small number of blades leads to a small slip coefficient. The pressure rise is made up by raising the blade outlet angle to rather large values to overcome the inherently low turning. Although seemingly valid, substituting angle for blade density will lead to a very heavily loaded blade configuration and, hence, to unnecessarily high viscous total pressure losses. Using a greater number of blades reduces the requirement on outlet angle, distributes the

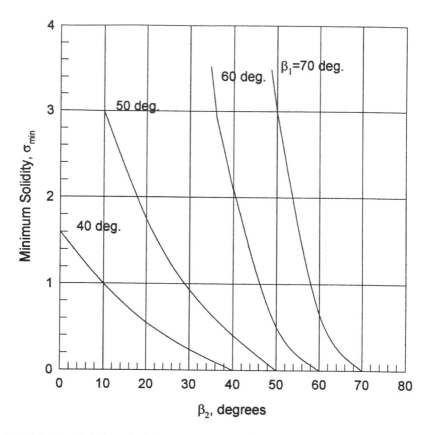

**FIGURE 8.13**   Variation of minimum solidity with $\beta_1$ and $\beta_2$.

work to a larger number of blades (larger surface area), and leads to a higher angular solidity for the cascade. Examination of typical centrifugal blade layouts will show that the slip coefficient is usually set above 0.85 so that a reasonably large number of blades is required and a fairly high angular solidity is achieved. Using the Stodola formula for illustration

$$\mu_E = 1 - \frac{\pi \sin \beta_2}{N_B}$$

Rearranging this equation yields

$$N_B = \text{INT}\left[\frac{\pi \sin \beta_2}{(1 - \mu_E)}\right] + 1 \qquad (8.43)$$

using INT for the "greatest integer" notation. Clearly, for larger outlet flow angles, the blade number increases as the sine function; and for larger slip coefficients, the

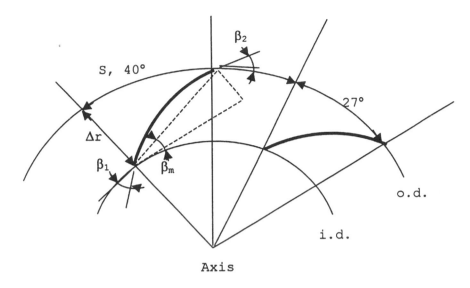

**FIGURE 8.14**  Geometry for the approximate solidity in a centrifugal cascade.

required number of blades increases strongly. Both trends drive the cascade to high angular solidity in order to control the blade loading. To that degree, the use of solidity is analogous to the axial cascade design process.

Centrifugal cascades for impellers of very high specific speed ($N_s \cong 1.0$) can be analyzed in much the same manner as an axial configuration using solidity and diffusion coefficient. One can approximate the solidity of the centrifugal cascade ($\sigma_c$) for fairly a high blade number as

$$\sigma_c \cong \frac{\left(1 - \dfrac{d}{D}\right) N_B}{\left(2\pi \sin\beta_m\right)} \tag{8.44}$$

where

$$\beta_m = \sin^{-1}\left[\frac{\left(\sin\beta_1 + \sin\beta_2\right)}{2}\right] \tag{8.45}$$

The formulation is similar to the axial solidity where the chord of the blade is divided by the arc length between adjacent trailing edges (see Figure 8.14). If one chooses to govern the blade solidity on the basis of diffusion, then again employ the Leiblien formulation in a form relevant to the centrifugal cascade

$$D_{Lc} = 1 - \frac{W_2}{W_1} + \frac{\Delta W_\theta}{\left(2W_1\sigma_c\right)} \tag{8.46}$$

**TABLE 8.7**
**Selection of Solidity for a Centrifugal Cascade**

| $N_B$ | $\mu_E$ | $\Delta W_\theta$ | $W_2/W_1$ | $\beta_2^*$ | C | $\sigma_c$ | $D_L$ |
|---|---|---|---|---|---|---|---|
| 10 | 0.892 | 135.1 | 0.62 | 21.4 | 1.40 | 1.48 | 0.638 |
| 12 | 0.915 | 131.9 | 0.636 | 20.8 | 1.40 | 1.78 | 0.575 |
| 14 | 0.923 | 130.4 | 0.645 | 20.6 | 1.40 | 2.08 | 0.550 |
| 16 | 0.933 | 129.6 | 0.650 | 20.4 | 1.40 | 2.38 | 0.504 |

If we assume that the boundary layer flow on the blade suction surface behaves the same as on the axial cascade (neglecting curvature and Coriolis effects), then we should limit the value of $D_{Lc}$ to less than 0.5 or 0.6 to avoid flow separation. This is a rather conservative constraint as illustrated in Table 8.7. The table presents calculations for a fan with Q = 150 ft³/s, $\Delta p_T$ = 60 lbf/ft², $\rho$ = 0.00233 slugs/ft³, D = 3 ft, d/D = 0.727, N = 158 radians/s (1506 rpm). Here, $N_s$ = 0.95 and $D_s$ = 3.1. $C_{\theta 1}$ is assumed to be zero so that $\Delta W_\theta$ is simply $C_{\theta 2}$. The diameter ratio is d/D = 0.727, and the widths are $w_1$ = d/4 and $w_2$ = (d/D)$w_1$. The results show a requirement for $\sigma_c \geq 1.48$ for fairly moderate values of blade number. Most centrifugal fans in this range of specific speed will have a solidity of perhaps 1.1 to 1.25, and very high total efficiencies as well, indicating very clean boundary layer flows. Twelve blades would be typical so that the diffusion constraint requires about 40% more blade length than is commonly seen. If one considers a lower specific speed, the requirements for solidity become totally unacceptable. For example, one could quadruple the pressure rise requirement for the fan and use $N_s$ = 0.40 and require $D_s$ = 6.5. The solidity requirement with 30 blades goes to about 8, which is really not workable. The de Haller ratio was only 0.53. What is happening is that the suction side of a blade in a centrifugal impeller is routinely operating with separated flow. In the axial flow machines, this was deemed unacceptable because of flow stability problems. However, the flow in the blade-to-blade channel of a centrifugal impeller with boundary layer separation is relatively benign, with a well-ordered jet-wake (Moore, 1986) flow pattern that remains stable through the channel and out the exit. Not having to preclude separated flow in the impeller channel greatly relaxes the restriction on diffusion and allows much higher blade loading with acceptable performance. Again, recourse is made to specifying an angular solidity near 1.0 with adequate blade number to keep the slip coefficient near 0.9.

## 8.5 DEVIATION IN TURBINE CASCADES

In turbine applications, one does not have to deal with diffusion limitations, but the phenomenon of deviation must be addressed to maintain design accuracy. A fairly simple method by Ainley and Mathieson (1951) (as presented by D.G. Wilson (1984)) can be used to estimate blade angle requirements in both blade and nozzle rows for axial flow turbines. It appears to be useable for hydraulic, gas, and steam turbines for Mach numbers less than 1, and it is based on geometric arguments,

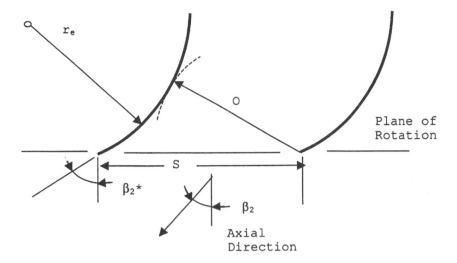

**FIGURE 8.15**  Turbine cascade parameters for the Ainley-Mathieson deviation model.

tempered by comparisons to experimental data from actual turbine testing. For purposes of illustration, this can be simplified to a mean camber line analysis. As shown in Figure 8.15, the calculation of the actual fluid outlet angle is expressed in terms of an opening ratio, o/s, and a curvature ratio $r_e/s$. The effective outlet angle or fluid angle is $\beta_2$ given by

$$\beta_2 = \left[\left(\frac{7}{6}\right)\left(\cos^{-1}\left(\frac{o}{s}\right) - 10°\right) + 4°\left(\frac{s}{r_e}\right)\right], \qquad \text{for } M_{a2} < 0.5 \qquad \textbf{(8.47)}$$

$$\beta_2 = \left[\cos^{-1}\left(\frac{o}{s}\right) - \left(\frac{s}{re}\right)^{1.787 + 4.128\left(\frac{s}{r_e}\right)}\right]\sin^{-1}\left(\frac{o}{s}\right), \qquad \text{for } M_{a2} = 1.0 \qquad \textbf{(8.48)}$$

Referring to the value for $M_{a2} < 0.5$ as $(\beta_2)_{.5}$ and the sonic value as $(\beta_2)_{1.0}$, estimates for Mach numbers between 0.5 and 1.0 can be made by

$$\beta_2 = (\beta_2)_{.5} - (2M_{a2} - 1)\left[(\beta_2)_{.5} - (\beta_2)_{1.0}\right], \qquad \text{for } 0.5 \leq M_{a2} \leq 1.0 \qquad \textbf{(8.49)}$$

These equations were developed by D.G. Wilson of MIT (1984) as curve fits to the original graphical methods developed by Ainley and Mathieson (1951). The absolute value of the calculation must be used.

For illustration, the simplifying assumption of flat blades near the trailing edge (see Figure 8.16) drives $r_e$ to very large values, reducing the flow angle equation to (for $M_{a2} < 0.5$)

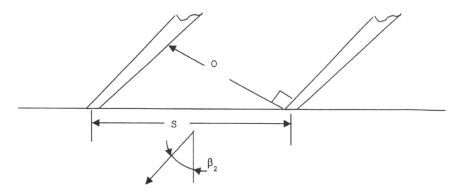

**FIGURE 8.16**  Turbine nozzle geometry in the trailing edge region.

$$\beta_2 = \left(\frac{7}{6}\right)\cos^{-1}\left(\frac{o}{s}\right) - 10° \qquad (8.50)$$

where $\cos^{-1}(o/s) = \beta_2^*$, the blade metal angle. For example, if $\beta_2^* = 50°$, then $\beta_2 = 48.3°$ or $\delta = 1.7°$. This behavior is seen as a function of $\beta_2^*$ in Figure 8.17. This method can be used to lay out inlet guide vanes for pumping machines when pre-swirl of the incoming fluid is needed ($C_{\theta 1} \neq 0.0$).

The question of choosing the correct incidence at the leading edge of a turbine takes on far less significance than was the situation for fan and blades in a pumping cascade. The incidence angle to the camber line at the leading edge can safely be set to zero (as was done by Howell for compressors) for a preliminary design study (Wilson, 1984). The remaining major variable, the cascade solidity, is not as strictly governed either, as one does not deal directly with a diffusion limitation. The choice of solidity can be adequately constrained by holding the lift coefficient for a blade to some upper limit (Zweifel, 1945). From momentum considerations (for a constant axial velocity), the lift coefficient is

$$C_L = \left(\frac{2\cos\lambda}{\sigma}\right)\cos^2\beta_2(\tan\beta_1 - \tan\beta_2) \qquad (8.51)$$

or

$$\sigma = \frac{2\cos\lambda}{(C_L)}\cos^2\beta_2(\tan\beta_1 - \tan\beta_2) \qquad (8.52)$$

Fixing $C_L$ to an upper limit yields an initial choice of solidity and hence the blade or vane chord (recall $\sigma = C/s$). Zweifel suggests $C_L$ approximately equal to 0.8, while Wilson recommends $C_L = 1.0$ as a less conservative value. Typical numbers for an inlet nozzle might be $\beta_1 = 0°$, $\beta_2 = -65°$. With $C_L = 1.0$, one obtains $\sigma = \cos(\beta_2/2)$ so that $\sigma = 0.84$ and for other values of $\beta_2$ would range from near

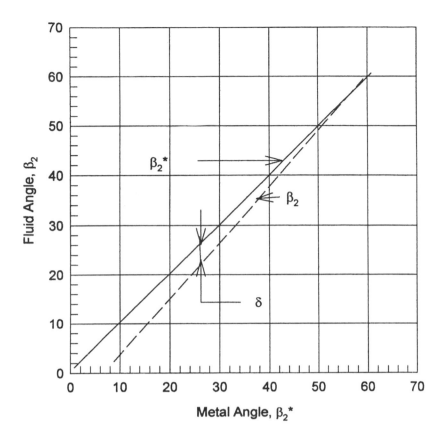

**FIGURE 8.17**   Example of deviation in a turbine nozzle.

1.0 to perhaps 0.75. For reaction blading examples, the required solidity increases substantially. With, say $\beta_1 = 30°$, $\beta_2 = -60°$, and $C_L = 1.0$, one requires $\sigma = 1.5$. Clearly, these much harder working blades need more solidity for reasonable loading.

## 8.6   LOSS SOURCES AND ASSOCIATED MAGNITUDES

In the design of axial fans and compressors, four sources of loss have been identified that significantly reduce the actual design point performance of the fan (Koch and Smith, 1976). The four sources of loss that are significant include (1) blade and vane profile losses due to surface diffusion, (2) losses associated with the end-wall boundary layers and blade end clearances, (3) losses due to shocks within the blade passages, and (4) drag losses due to part-span shrouds and supports. The net magnitude of all these losses must be considered in analyzing the overall performance of the fan design. Of the four types of losses identified, only the first two are significant in the preliminary design process of low-speed axial fans. Several different techniques are available to the designer in estimating the influence of each of these losses on the final fan design (Brown, 1972). These techniques vary in complexity and application, ranging from simple mean-line estimations to rigorous numerical solutions of shear-stress distributions. Brown has indicated that for low-speed applications, the various parameters are equal and interchangeable. The technique that will be considered in this analysis follows the NACA convention that normalizes the significant pressure losses by the rotor inlet dynamic head (Johnsen and Bullock, 1965). The use of this convention will be readily observed when performing efficiency estimations and will maintain the use of dimensionless parameters in the analysis.

Profile losses that occur in the fan design are a result of boundary-layer development along the blade surfaces. These losses for the local blade and vane elements are determined based upon the blade loading work of Lieblein (Lieblein, 1957) and expanded by Koch and Smith (Koch and Smith, 1976). This technique applies boundary-layer theory to conventional blade cascades in order to relate the profile losses with the conventional boundary-layer parameters of blade-outlet momentum thickness and trailing-edge form factor. The designer estimates Lieblein's equivalent

$$D_{eq} = \left(\frac{\cos\beta_1}{\cos\beta_2}\right)\left[1.12 + 0.0117\, i^{1.43} + 0.61\left(\frac{\cos^2\beta_1}{\sigma}\right)(\tan\beta_1 - \tan\beta_2)\right] \quad \textbf{(8.53)}$$

$$\frac{\theta^*}{C} \cong 0.00210 + 0.00533\, D_{eq} - 0.00245\, D_{eq}^{\,2} + 0.00158\, D_{eq}^{\,3} \quad \textbf{(8.54)}$$

diffusion ratio for the blade element from Equation (8.53). This parameter is then used to determine the blade-outlet momentum thickness and trailing-edge form factor from Equation (8.54) and Figure 8.18. Both of these relations are based on nominal blade chord Reynolds Number of $1.0 \times 10^6$ and an inlet Mach number of 0.05. The blade-outlet momentum-thickness and trailing-edge form factor values must be corrected to account for conditions other than those presented. These corrections are obtained from Figures 8.19 and 8.20. Figure 8.19 presents the variation of blade outlet momentum thickness and trailing edge form factor as functions of inlet Mach number and the equivalent diffusion factor. Figure 8.20 presents the variation of blade-outlet momentum thickness as a function of blade chord Reynolds Number and blade surface roughness. Once the designer has determined the boundary layer parameters, the loss in total pressure is calculated according to

$$C_{w,Bld} = \frac{\Delta p_T}{\left(\frac{\rho W_1^2}{2}\right)} = \frac{2\left(\frac{\theta_2}{c}\right)\left(\frac{\sigma}{\cos\beta_2}\right)\left(\frac{\cos\beta_1}{\cos\beta_2}\right)^2 \left(\frac{2}{3 - \frac{1}{H_{te}}}\right)}{\left[1 - \left(\frac{\theta_2}{c}\right)\left(\frac{\sigma H_{te}}{\cos\beta_2}\right)\right]^3} \quad \textbf{(8.55)}$$

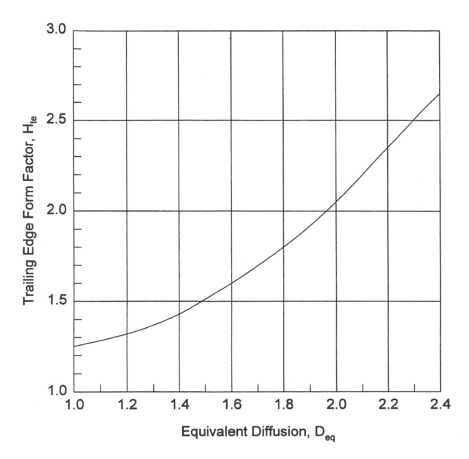

**FIGURE 8.18**   Trailing edge form factor as a function of diffusion. (Koch and Smith, 1976.)

The net pressure loss across the blade row is determined by performing a mass-averaged integration of the individual blade element pressure loss distribution across the entire blade row.

If outlet guide vanes are to be used to recover the rotor exit swirl velocity, the profile losses for the vane row are calculated in the same manner as the blade profile losses. In designs where outlet guide vanes (OGVs) are not employed, the discharge swirl velocity comprises a loss in total pressure as unrecovered work. This loss is determined using a mass-averaged integration of the tangential velocity distribution as specified for the particular design. The unrecovered work term can dominate the losses of a fairly heavily loaded design without OGVs.

A review of the literature reveals that the losses induced by the hub and tip casing boundary layers and tip clearance effects contribute to the overall reduction in fan performance (Hirsch, 1974; Horlock, 1963; Mellor and Wood, 1971). Extensive studies indicate that the size and condition of these end-wall boundary layers can lead to the inception of stall (McDougall, Cumptsy, and Hynes, 1989). Several techniques have been developed to estimate these losses (Comte, Ohayon, and

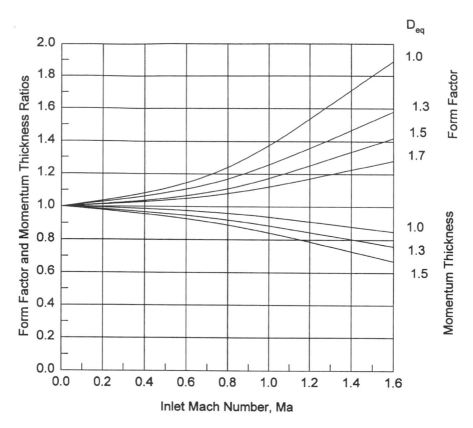

**FIGURE 8.19** Influence of Mach number on form factor and momentum thickness. (Koch and Smith, 1976.)

Papailiou, 1982; Hunter and Cumptsy, 1982; Lakshminarayana, 1970; Wright, 1984d). Boundary layer and tip clearance effects can be estimated during the preliminary design process using a semi-empirical technique that employs boundary layer theory to the hub and tip end-wall regions to calculate a combined loss parameter (Smith, 1969). With this algorithm, the loss of efficiency is estimated as a function of the boundary layer properties of displacement thickness, $\delta^*$, and tangential force thickness, $v^*$, at the end walls. The displacement thickness of the boundary layer is a measure of the amount of reduction of the mass flow (assuming a fixed core velocity in the passage), while the tangential force thickness is the amount that the blade tangential force on the fluid is reduced. In Smith's correlation, the displacement is normalized by the staggered spacing, g, between adjacent airfoils and has been related to tip clearance and fan maximum pressure rise using a semi-empirical approach. The tangential force thickness is assumed to be a fixed fraction of the displacement thickness. Koch and Smith (1976) show that by combining the effects of the hub and tip boundary layers, the parameters can be related to efficiency reduction by

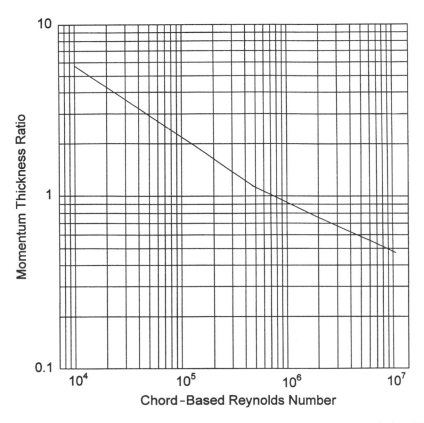

**FIGURE 8.20**   Influence of Reynolds Number on momentum thickness. (Koch and Smith, 1976.)

$$\eta = \eta_{fs} \frac{\left[1-\left(\dfrac{2\delta^{*}}{g}\right)\left(\dfrac{g}{h}\right)\right]}{\left[1-\left(\dfrac{2v^{*}}{2\delta^{*}}\right)\left(\dfrac{2\delta^{*}}{g}\right)\left(\dfrac{g}{h}\right)\right]}$$  (8.56)

where $\eta_{fs}$ is the efficiency of the fan in the absence of end-wall viscous effects.

   To determine the displacement thickness in the presence of finite tip clearance, Koch and Smith recommend

$$\frac{2\delta^{*}}{g} = \left(\frac{2\delta^{*}}{g}\right)_{\varepsilon=0} + \left(\frac{2\varepsilon}{g}\right)\frac{\Delta p_{T}}{\Delta p_{T\,max}}$$  (8.57)

where $\Delta p_{T}$ is the total pressure rise of the fan. Variation of $v^{*}$ and $\delta^{*}$ is shown in Figures 8.21 and 8.22 as a function of maximum pressure rise. The reduction in efficiency can also be estimated from

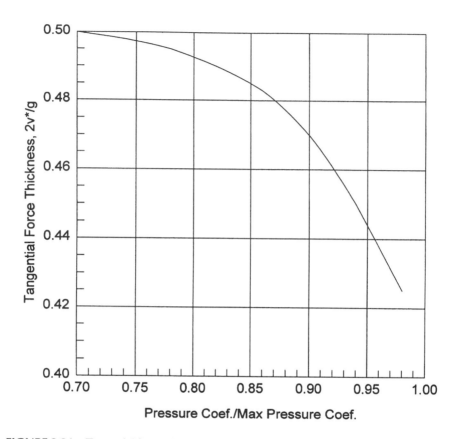

**FIGURE 8.21**   Tangential force thickness as a function of pressure rise. (Koch and Smith, 1976.)

$$C_{w,end} = \psi_{T,i}\left[1.0 - \frac{1.0 - \left(\frac{2\delta^*}{g}\right)\left(\frac{g}{h}\right)}{1.0 - \left(\frac{2v^*}{2\delta^*}\right)\left(\frac{2\delta^*}{g}\right)\left(\frac{g}{h}\right)}\right] \qquad (8.58)$$

Since for a fixed core flow the wall boundary layers reduce mass flow, the correlation developed by Smith can be used to correct the volume flow rate by

$$\phi = \phi_{fs}\left[1.0 - \frac{\left(\frac{2\delta^*}{g}\right)}{\left(\frac{g}{h}\right)}\right] \qquad (8.59)$$

During the preliminary design process, the information needed to do a rigorous estimate of the maximum pressure rise is not generally available. Koch (1981)

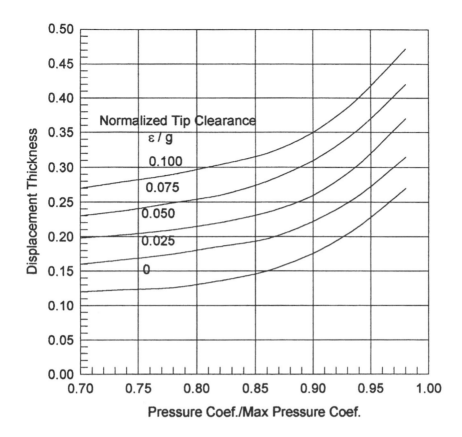

**FIGURE 8.22**  Displacement thickness as a function of pressure rise. (Koch and Smith, 1976.)

developed a technique to estimate the maximum pressure rise by employing a planar diffuser analogy to the ratio of design to maximum pressure rise. Using the stall relations developed for these diffusers (Sovran and Klomp, 1967), maximum pressure is estimated from Figure 8.23. This value is then used to estimate losses associated with end-wall viscous effects.

Having estimated these losses, efficiency is quantified by the ratio of the reduced pressure rise (ideal minus losses) to the ideal value. That is,

$$\psi_T = \psi_{T,i} - C_{w,\,Blade} - C_{w,\,vane} - C_{w,\,swirl} - C_{w,\,end} \tag{8.60}$$

Then total efficiency becomes

$$\eta_T = 1.0 - \frac{\left(C_{w,\,Blade} + C_{w,\,vane} + C_{w,\,swirl} + C_{w,\,end}\right)}{\psi_{T,I}} \tag{8.61}$$

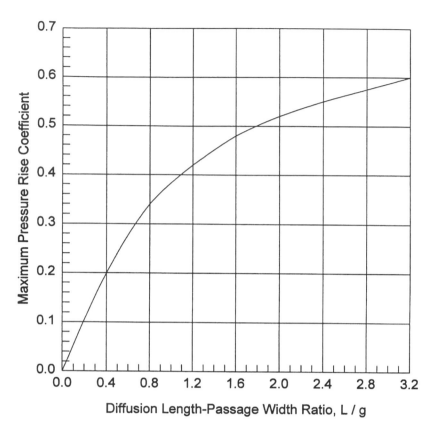

**FIGURE 8.23** Maximum pressure rise as a function of passage length-width ratio. (Koch and Smith, 1976.)

Static efficiency is estimated, then, by

$$\eta_s = \eta_T - \left(\phi^2 + \theta^2\right) \qquad\qquad (8.62)$$

where $(\phi^2 + \theta^2)$ are the non-dimensional axial and residual velocity pressures, respectively, averaged across the annulus. For full OGV swirl recovery, the value of $\theta^2$ is, of course, zero.

## 8.7   SUMMARY

The study of the detailed flow in blade cascades was introduced to provide a greater depth of understanding of the flow behavior. Employing the relationships for prediction of cascade flows also leads to significant improvement in the accuracy with which one can lay out blade and vane rows for a given specified performance. Although the final design and analysis is being done now with computational fluid mechanical analysis techniques, the established cascade methods can still provide good preliminary design and tradeoff information prior to refinement and final optimization.

The information on performance of cascades in axial flow is extensive. Most of the information is based on the studies at the Langley and Lewis research centers. The information reviewed in this chapter was initially centered on determining the flow deviation from the prescribed geometry as a function of the cascade variables, solidity, camber, and pitch. The study provided a set of algorithms by which one could size the blades and vanes of a machine with reasonable accuracy.

The attention then shifted to flow in cascades of radial or centrifugal blades. There, the base of empirical information was found to be less extensive and came to rely on a semi-theoretical treatment of flow deviation. Several methods were presented and compared with good agreement as to prediction of centrifugal slip. Slip, or slip coefficient, was found to depend strongly on the number of blades and the outflow angle requirements for the impeller. The concept of solidity in centrifugal cascades was found to be less important in design than was the case for axial flow machines.

Turbine blade cascades were examined in terms of flow deviation. A relatively simple method for flow prediction and design was presented for axial flow cascades that depend on the blade outlet geometry and the flow Mach number. A method was also provided to allow selection of blade solidity based on lift coefficient constraints.

Finally, the generation of viscous and end-wall or clearance losses was examined for cascades. The classical prediction techniques were presented with simple methods for estimating the efficiency of a given cascade layout. These loss predictions were based on the cascade geometry, the improved estimation of diffusion presented here, and the blade relative Mach numbers on the machine.

## 8.8 PROBLEMS

8.1. Repeat Problem 6.4, selecting appropriate solidity, camber, and pitch, using the diffusion factor limitation and Howell's modified rule for estimating deviation.

8.2. Air flows into an axial fan cascade through an inlet guide vane row, as sketched in Figure P8.2, with an absolute fluid outlet angle of -60°. The blade row is designed to generate a pure axial discharge flow.
(a) Select solidity and camber values for the inlet guide vanes.
(b) Select solidity and camber values for the blades.
(c) Estimate the ideal power.
Show all work and state all assumptions. (Hint: Howell's rule is for diffusing cascades only.)

8.3. A "simple-stage" single-width (SWSI) centrifugal fan has the following geometry: $w_1 = 5$ inches, $w_2 = 3$ inches, $d = 12$ inches, $D = 20$ inches, $N = 1800$ rpm, $\rho = 0.0233$ slugs/ft³, $\beta_1 = 25°$, and $\beta_2 = 35°$.
(a) Estimate the ideal performance as Q and $\Delta p_{Ti}$.
(b) Estimate the total efficiency for the fan, modify $\Delta p_{Ti}$ to $\Delta p_T$, and estimate the required power.

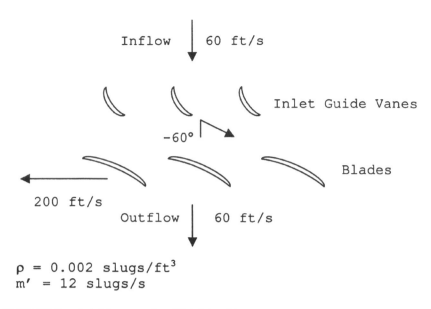

**FIGURE P.8.2**   Axial fan cascade with inlet guide vanes.

(c) Select a blade outlet angle and compatible blade number ($\beta_2$* and B) that will yield $\Delta p_T$ and Q. Use the full Stodola and the simplified Stodola forms for $\mu_E$ (Hint: Use enough blades to keep $\mu_E > 0.85$.)

8.4. For the fan of Problem 6.5, select a reasonable mean station solidity for the blade row, and determine the camber angle, $\phi_c$, required to generate $\Delta p_T$, including the effects of efficiency and deviation. Sketch the vector diagram and the blade shape and layout.

8.5. One can correct $\psi_{ideal}$ (see Problem 7.6) for nonviscous effects by including the Stodola formula for slip. Show that $\psi_T$, corrected for slip and viscous loss, becomes:

$$\psi_T = 2\left[1 - \frac{\dfrac{\pi \sin \beta_2}{N_B}}{\left(1 - \phi_e \cot \beta_2\right)}\right]\left[1 - \phi_e \cot \beta_2\right] - \left[\left(\frac{d}{D}\right)^2 + \phi_e^{\,2}\right]N_B C_{wb}$$

8.6. Use the result given in Problem 8.5 to show that an optimum number of blades for a given value of d/D, $\beta_2$, and $\phi_e$ is given by the following equation:

$$N_{Bopt} = \left[\frac{\left(2\pi \cos \beta_2\right)}{\left[\left(\dfrac{d}{D}\right)^2 + \phi_e^{\,2}\right]C_{wb}}\right]^{1/2}$$

8.7. Using the equation from Problem 8.6, explain how you would estimate $C_{WB}$.

8.8. For a blower with $\beta_2 = 60°$, $d/D = 0.7$, $\phi_e = 0.3$, and $C_{WB} = 0.025$, what is the "best" number of blades to use?

8.9. An axial water pump has a total pressure rise of 20 psig and a volume flow rate of 6800 gpm. Tip and hub diameters are 16 and 12 inches, respectively, and the rotational speed is 1100 rpm.
  (a) Estimate the total efficiency, and calculate the required power.
  (b) Construct a mean-station velocity diagram to achieve the pressure rise at the desired flow and the efficiency from part (a).
  (c) Lay out the blade shape at the mean station using the simple form of Howell's rule for deviation angle.

8.10. Redo the blade layout of Problem 8.9, part (c), using the more complex and more accurate "modified" Howell's rule. Compare the result to the "simple rule" result. What would the error in the pressure rise be using the simplified rule? Discuss the impact on pump design.

8.11. Use the Ainley-Mathieson method to detail the blade and vane shapes to achieve the performance of Problem 7.12 (mean radius only).

8.12. Rework the fan mean station of Section 8.1 (Chapter 8), using 12,000 cfm and 2.0 InWG. Analyze hub and tip stations using $d = 0.6\, D$.

8.13. In the chapter we discussed the formula for predicting deviation or slip in centrifugal blade cascades. The particular algorithm was the rather well-accepted Stodola formula: $\mu_E = 1 - (\pi/N_B)\sin\beta_2/(1 - \phi\cot\beta_2)$, $\phi = C_{r2}/U_2$. As $\beta_2$ approaches 90°, the formula reduces to $\mu_E = 1 - (\pi/N_B)\sin\beta_2$, or less flexibly, $\mu_E = 1 - (\pi/N_B)$.
  (a) Compare this result to the common expressions for radial-bladed impellers ($\beta_2 = 90°$) given by:

$$\text{Stanitz, 1951:} \qquad \mu_E = 1 - 0.63\left(\frac{\pi}{N_B}\right)$$

$$\text{Balje, 1952:} \qquad \mu_E = \cfrac{1}{\left[1 + \cfrac{6.2\left(\dfrac{d}{D}\right)^{2/3}}{N_B}\right]}$$

$$\text{Buseman, 1928:} \qquad \mu_E = \frac{\left(N_B - 2\right)}{N_B}$$

  (b) Identify the more conservative methods and discuss the design implications of using the methods producing higher values of $\mu_E$. When does the Balje method seem to offer an acceptable level of conservatism?

8.15. A centrifugal fan operating in air with $\rho = 0.00235$ slugs/ft$^3$ is running at 1785 rpm with a total efficiency of 85%. Fan diameter is $D = 30$ inches; inlet diameter is $d = 21$ inches. The blades have an outlet $w_2 = 7.5$ inches and an inlet width $w_1 = 9.0$ inches. If $\beta_1 = 32°$, $\beta_2 = 35°$, and $N_B = 12$ (number of blades),
   (a) Estimate the volume flow rate of the fan.
   (b) Estimate the ideal head rise or pressure rise of the fan.
   (c) Estimate the slip coefficient and the net total pressure rise (include effects of both slip and efficiency).

8.16. A blower is to be designed to supply 5 m$^3$/s of air at a pressure rise of 3000 Pa with an air density of 1.22 kg/m$^3$. Design a centrifugal fan impeller to achieve this specified performance:
   (a) Select $D$, $d$, $w_1$, $w_2$, and $N$.
   (b) Determine the efficiency using the modified Cordier value with Re and $\delta_0 = 2.0$.
   (c) Lay out the blade shapes ($\beta_1$, $\beta_2$, and $N_B$) accounting for slip.
   (d) Provide a table of all final dimensions with a dimensioned sketch or drawing.
   (e) Do an accurate representation of a blade mean camber line and channel to show blade angles and solidity.

8.17. A double-width centrifugal blower has the following geometry: $D = 0.65$ m; $w_2 = 0.15$ m; $w_1 = 0.20$ m; $d = 0.45$ m; $\beta_2 = 32°$. The fan has 15 blades and runs at 1185 rpm.
   (a) If the total flow rate is 8 m$^3$/s, determine the correct value of $\beta_1$.
   (b) With an inlet fluid density of $\rho_{01} = 1.19$ kg/m$^3$, estimate the pressure rise of the fan neglecting viscous losses. (Hint: Assume that the flow is incompressible for an initial solution. Then use $p/\rho^\gamma = $ constant to correct the pressure rise associated with an increased outlet density.)
   (c) Estimate the efficiency and use this value to correct the net pressure rise of the blower.

8.18. A vane axial fan is characterized by $D = 0.80$ m, $d = 0.49$ m, and $N = 1400$ rpm. The performance of the fan is $Q = 5.75$m$^3$/s, $\Delta p_T = 750$ Pa, with $\rho = 1.215$ kg/m$^3$ and $\eta_T = 0.85$. We need to select cascade variables to define the geometry for the blades and vanes. Allow the local diffusion factors, $D_L$, at hub, mean, and tip station to be 0.60, 0.45, and 0.30, respectively, for the blades and vanes.
   (a) Choose blade and vane solidity distributions that can satisfy these $D_L$ constraints.
   (b) Assume that the blades and vanes are very thin. With the solidities and flow angles determined, calculate deviation, camber, and incidence and stagger angles for the blade at hub, mean, and tip stations.
   (c) Repeat (b) for the vane row.

8.19. The performance requirements for a blower are stated as: flow rate, $Q = 9000$ cfm; total pressure rise, $\Delta p_T = 8.0$ InWG; ambient air density, $\rho = 0.0022$ slugs/ft$^3$.

(a) Determine a suitable speed and diameter for the blower impeller, and estimate the total efficiency. (Work with a wide centrifugal design, but justify this choice by your analysis).

(b) Correct the total efficiency for the effects of Reynolds Number and a generous running clearance of $2.5 \times$ Ideal (or $\delta_o = 2.5$) and the influence of a belt drive system (assume a drive efficiency of 95%).

8.20. Use the dimensionless performance from Figure 5.10 to select three candidates for the requirements of Problem 8.19 that closely match the results of your preliminary design decision in that problem.

8.21. Choose a suitable throat: diameter ratio, d/D, and width ratios, $w_1/D$ and $w_2/D$, for the fan of Problem 8.19. Define the velocity triangles in both the absolute and relative reference frames. Do so for both blade row inlet and outlet. Calculate the de Haller ratio for the blade.

8.22. For the centrifugal fan analyzed in Problem 8.21, select cascade variables to define the geometry for the blades. Select the number of blades needed according to Stodola's $\mu_E$. (Hint: Choose $\mu_E$ between 0.85 and 0.9 to maintain good control of the discharge vector.) Assume thin blades and use the flow angles already determined for the blade row inlet. On the basis of the chosen $\mu_E$, define the outlet angle for the blade metal, $\beta_2{}^*$, that will increase the previously calculated $C_{\theta 2}$ to $C_{\theta 2}/\mu_E$.

8.23. Develop a set of drawings that describe two views of the impeller of Problem 8.22. In the "front" view, show at least three blades as seen on the backplate to define camber shape and the blade-to-blade channel. For the "side" view, show the shape of the sideplate relative to blades and backplate, illustrating $w_1$, $w_2$, and d/D.

8.24. The performance requirements for an air mover are stated as: flow rate, $Q = 1$ m³/s; total pressure rise, $\Delta p_T = 500$ Pa; ambient air density, $\rho = 1.175$ kg/m³. The machine is constrained not to exceed 82 dB sound power level.

(a) Determine a suitable speed and diameter for the fan, and estimate the total efficiency. (Work with a vane axial design, but justify this choice with your analysis).

(b) Correct the total efficiency for the effects of Reynolds Number, a running clearance of $2.5 \times$ Ideal (or $\delta_o = 2.5$), and the influence of a belt drive system (assume a drive efficiency of 95%).

8.25. A small blower is required to supply 250 cfm with a static pressure rise of 8 InWG and air weight density of $\rho g = 0.053$ lbf/ft³. Develop a centrifugal flow path for a SWSI configuration. Use the optimal d/D ratio for minimum $W_1$, estimate the slip using the Stodola formula ($\mu_E \cong 0.9$), assume polytropic density change across the blade row (using $\eta_{T-C}$), and size the impeller outlet width by constraining $25° < \beta_2 < 35°$.

8.26. For the fan of Problem 8.24, let 500 Pa be the static pressure requirement, and:

(a) Estimate the change in power, total and static efficiency, and the sound power level, $L_w$.

(b) Choose a suitable hub–tip diameter ratio for the fan and repeat the calculations of part (a).

8.27. For the fan of Problem 8.24, with $\Delta p_T = 500$ Pa, define the velocity triangles in both the absolute and relative reference frames. Do so for both blade and vane rows at hub, mean, and tip stations. Calculate the degree of reaction for all stations.

8.28. For the vane axial fan analyzed in Problem 8.27, select cascade variables to define the geometry for the blades and vanes. Allow the local diffusion factors, $D_L$, at hub, mean, and tip stations to be 0.60, 0.45, and 0.30, respectively, for the blades and vanes.
   (a) Choose blade and vane solidity distributions that can satisfy these $D_L$ constraints.
   (b) Assume that the blades and vanes are very thin. With the solidities and flow angles already determined, calculate deviation, camber, and incidence and stagger angles for the blade at hub, mean, and tip stations.
   (c) Repeat (b) for the vane row.

8.29. Develop a set of curves for the blade geometric variables, versus radius, of the fan of Problem 8.28, using the above angles and the blade chord based on 17 blades. Develop curves for the vanes, too, using 9 vanes to determine the chord lengths.

8.30. For the turbine generator set of Problem 7.12, do a complete layout of the nozzles and vanes at the hub, mean, and tip radial stations. Account for the flow deviation using the Ainley-Mathieson method. Develop a plot of the nozzle and blade parameters as a function of radius.

8.31. For the centrifugal fan impeller of Problem 7.14, develop an exact shape for the blades.
   (a) Use the Euler form for $\mu_E = 0.9$ to select the blade number and the outlet metal angle for the blades.
   (b) Use the Stanitz formula with 11 blades to lay out the blade exit angles and compare to the impeller shape in (a).

8.32. For the fan in Problem 8.27, use $\sigma = 0.4$ at the blade tip to estimate the required camber. Use the NACA correlations and compare with the result using the low solidity corrections.

8.33. A heat exchanger requires 2500 cfm of air with the static pressure at 2.2 InWG above ambient pressure. The ambient density is $\rho = 0.00225$ slugs/ft$^3$. The installation for the fan requires that the sound power level be no greater than 85 dB.
   (a) Determine a suitable speed and diameter for a vane axial fan and estimate the total efficiency.
   (b) Modify the total efficiency for the influence of Reynolds Number and a running clearance of $\delta_0 = 2.0$.
   (c) Select a thin circular arc blade for the mean station of the blade and vane using the original Howell correlation, with a d/D ratio chosen from Figure 6.10. (Choose solidity on the basis of an equivalent diffusion of 0.5.)
   (d) Use the modified Howell correlation to select these sections and compare the result to the estimation of part (c).

(e) For a 10% thick NACA 65 Series cascade, define the blade and vane mean station properties using the full NACA correlation for blade angles and compare to the previous calculations.

# 9 Quasi-Three-Dimensional Flow

## 9.1 AXIAL FLOWPATH

In the analyses and calculations performed thus far, it was generally assumed that we were working with simple inlet and outlet conditions through a blade or vane row. Specifically, in axial flow machines, it was assumed that meridional flow surfaces through the machine are a simple set of concentric cylinders with a common axis—the axis of rotation. It was also assumed that the mass or volume flow rate is uniformly distributed in the annulus between the inner hub and outer casing. That is, one assumes that $C_x$ is a constant value across the inlet, across the outlet, and throughout the machine. This says, functionally,

$$C_x \neq f(r, \theta, x) \tag{9.1}$$

In many machines and their flowfields, these assumptions are well justified, although one naturally expects some distortion of the annular flow near the end walls (the hub and casing surfaces) due to the no-slip boundary conditions required for a real viscous fluid flow. However, in a more general treatment of throughflow, one needs to examine more carefully the flow along the curved surfaces associated with the axial motion and the rotating motion due to swirl (axial velocity and tangential velocity in the absolute frame of reference, $C_x$ and $C_\theta$). The rotating component of motion subjects the fluid to a centrifugal force so that if equilibrium of motion is to exist along any chosen streamline, the centrifugal force must be balanced by a gradient of pressure normal to the streamline, as illustrated in Figure 9.1.

## 9.2 RADIAL EQUILIBRIUM

To analyze the flow in detail, fully three-dimensional solutions of the potential flow field (ideal flow) or the viscous flow field have been developed as numerical solutions embodied in rather complex computer programs. These codes variously use the finite difference, finite volume, boundary element, or finite element formulations of the governing equations and accompanying discretization of the internal flow-path geometry of the machine being studied (Lakshminarayana, 1991; Baker, 1986; Anderson, 1996; Hirsch, 1988). An excellent review of these techniques has been presented in the ASME *Journal of Turbomachinery* (Lakshminarayana, 1991), and the subject will be addressed further in Chapter 10. These rigorous treatments of the turbomachinery flows are extensive and can require substantial effort in preparing the geometric inputs for analysis. In general, the detailed geometry of the machine

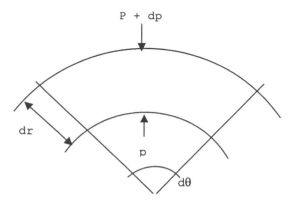

**FIGURE 9.1**   Radial equilibrium on the curved streamlines.

must be supplied by the user in order to seek a flowfield solution. An interesting exception to this requirement is the work developed by Krain and associates (Krain et al., 1984) where they systematically vary the geometry of a machine in a CAD-like process married to a flow solver. It can be argued that the greatest usefulness of these megacodes lies in performing studies requiring considerable detail and great accuracy of the flowfields for machines of known geometry. The techniques can be used to modify, improve, optimize, or verify the capability of an existing or proposed design.

## 9.3   TWO-SOLUTION METHODS

For use at a preliminary stage of design and layout, methods that do not require extensive details in geometries and machine boundaries can be more useful as a starting point. Less rigorous but simplified methods are frequently based on a concept called quasi-three-dimensional flow (Wu, 1952; Katsanis et al., 1969; Wright, 1982). These methods seek to determine a three-dimensional solution of the throughflow by the approximation of solving two fully compatible, two-dimensional flow problems on mutually perpendicular surfaces. One of these surfaces is the meridional surface of revolution on which one can examine the region between cross-sections of adjacent blades cut by this meridional surface shown in Figure 9.2. This part is called the blade-to-blade solution. The other surface is planar and passes through the axis of rotation. It is on this surface that one analyzes and determines the shapes of the meridional stream surfaces by solving for a circumferentially averaged velocity field. Such a field is illustrated in Figure 9.2. This solution drives the flow solution on the meridional surface of revolution by determining the shape and location where the blade-to-blade flow is analyzed. For best results, the two solutions are iterated with the meridional solution supplying tangential flow information to the circumferentially averaged flow on the planar surface until one solution fails to change the other. Such a solution, though not rigorously accurate, can provide an excellent approximation to the fully three-dimensional flow (Wright, 1984b). The method

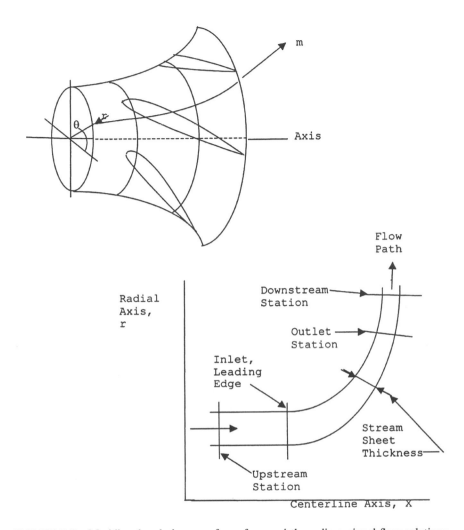

**FIGURE 9.2** Meridional and planar surfaces for quasi-three-dimensional flow solutions.

provides substantial reduction in effort and is most useful when the geometry is to be changed again and again.

## 9.4 SIMPLE RADIAL EQUILIBRIUM

For preliminary design work and establishing limiting parameters and seeking some fundamental insights into flow behavior, much can be learned by describing the throughflow in an axial machine using the concept of simple radial equilibrium. Here, the ability to establish a detailed description of the curving flow (with radial motion as well) through a blade or vane row is neglected in favor of establishing patterns of equilibrium flow at locations upstream and downstream of the blades.

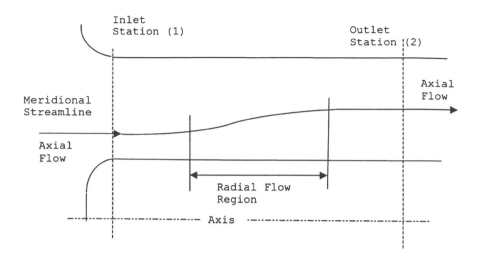

**FIGURE 9.3**  The simple radial equilibrium concept and geometry.

At these stations, we assume that radial velocity components are very small (negligible), but we require that axial and tangential components be compatibly matched through the simple balancing of the centrifugal force and a radial pressure gradient. True, a lot of detail is lost, but the flow in an annular cascade can be analyzed with reasonable accuracy based on the inlet and outlet conditions (Johnsen and Bullock, 1965). Figure 9.3 shows the geometry and illustrates the assumptions.

For steady, incompressible flow, the total pressure rise will be examined starting with the Euler equation from Chapter 8:

$$\Delta p_T = \eta_T \rho N r \left( C_{\theta 2} - C_{\theta 1} \right) \tag{9.2}$$

where

$$\Delta p_T = \left[ p + \left( \frac{\rho}{2} \right) C_x{}^2 + \left( \frac{\rho}{2} \right) C_\theta{}^2 \right]_2 - \left[ p + \left( \frac{\rho}{2} \right) C_x{}^2 + \left( \frac{\rho}{2} \right) C_\theta{}^2 \right]_1 \tag{9.3}$$

Differentiate this equation with respect to r in order to find the radial balancing pressure gradient to impose on the centrifugal force. Doing so and rearranging terms somewhat yields

$$\frac{dp_2}{dr} - \frac{dp_1}{dr} + \frac{\rho C_{x2} dC_{x2}}{dr} - \frac{\rho C_{x1} dC_{x1}}{dr} + \frac{\rho C_{\theta 2} dC_{\theta 2}}{dr} - \frac{\rho C_{\theta 1} dC_{12}}{dr}$$

$$= \rho N \left( \frac{d(\eta_T r C_{\theta 2})}{dr} - \frac{d(\eta_T r C_{\theta 1})}{dr} \right) \tag{9.4}$$

Now choose stations 1 and 2 so that radial equilibrium exists in the form $dp/dr = \rho \, C^2/r$ (with $C_r \cong 0$). Substituting yields

$$\frac{\left(\frac{1}{2}\right)dC_{x2}^{\ 2}}{dr} = N\left(\frac{d\left(\eta_T rC_{\theta2}\right)}{dr} - \frac{d\left(\eta_T rC_{\theta1}\right)}{dr}\right) - \frac{\left(\frac{1}{2}\right)dC_{x2}^{\ 2}}{dr} + \frac{\left(\frac{1}{2}\right)dC_{x1}^{\ 2}}{dr}$$

$$-\frac{\left(\frac{1}{2}\right)dC_{\theta2}^{\ 2}}{dr} + \frac{\left(\frac{1}{2}\right)dC_{\theta1}^{\ 2}}{dr} + \frac{\left(C_{\theta1}^{\ 2} - C_{\theta2}^{\ 2}\right)}{r} \quad\quad (9.5)$$

To achieve a more manageable form, assume that $C_{x1}$ is a constant (uniform inflow) and that $C_{\theta1}$ is zero (simple inflow). This yields

$$\frac{\left(\frac{1}{2}\right)dC_{x2}^{\ 2}}{dr} = \frac{N\eta_T d\left(rC_{\theta2}\right)}{dr} - \frac{\left(\frac{1}{2}\right)dC_{\theta2}^{\ 2}}{dr} - \frac{C_{\theta2}^{\ 2}}{r} \quad\quad (9.6)$$

Here, it is also assumed that $\eta_T$ is not a function of r. One can normalize the equation by dividing velocities by the tip speed, ND/2, and dividing length by D/2. Defining $\Psi$ and $\Phi$ as (see Chapter 7)

$$\Phi_2 = \frac{C_{x2}}{\left(\dfrac{ND}{2}\right)} \quad\text{and}\quad \Psi = \frac{C_{\theta2}}{\left(\dfrac{ND}{2}\right)} \quad\quad (9.7)$$

and

$$x = \frac{r}{\left(\dfrac{D}{2}\right)} \quad\quad (9.8)$$

the equilibrium equation becomes

$$\frac{d\Phi_2^{\ 2}}{2} = \eta_T \Psi dx + \eta_T x d\Psi - \Psi d\Psi - \left(\frac{\Psi^2}{x}\right)dx \quad\quad (9.9)$$

This form is a fairly nasty non-linear ordinary differential equation with arbitrary variation of $\Phi_2$ and $\Psi$ in the radial direction.

A fairly trivial solution exists if $\Psi = 0$. This is the case for no inlet swirl and no outlet swirl, and $\Phi_2 = $ constant satisfies the differential equation. That is, by conservation of mass, $\Phi_2 = \Phi_1$. Another simple but non-trivial solution exists when $\Psi = a/x$, where a is a constant. Substituting $\Psi = a/x$ into the equilibrium equation, one obtains

$$\frac{d\Phi_2{}^2}{2} = \eta_T\left(\frac{a}{x}\right)dx + \eta_T d\left(\frac{a}{x}\right) - \left(\frac{a}{x}\right)d\left(\frac{a}{x}\right) - \left[\frac{\left(\frac{a^2}{x^2}\right)}{x}\right]dx$$

(9.10)

$$= \left(\eta_T\left(\frac{a}{x}\right) - \eta_T\left(\frac{a}{x}\right) + \left(\frac{a^2}{x^3}\right) - \left(\frac{a^2}{x^3}\right)\right)dx = 0$$

Again, $\Phi_2 = \Phi_1$ satisfies the equation when $C_{\theta 2} = \text{Constant}/r$.

Recall from the study of fluid mechanics that the flowfield of a simple, potential vortex can also be described by this form for $C_\theta = a/r = \Gamma/2\pi r$, where $\Gamma$ is the strength of the line vortex. Such a flow is usually referred to as a free vortex swirl velocity distribution. In turbomachinery parlance, this is shortened to free vortex flow. If the $C_{\theta 2}$ values of an axial flow machine are to be distributed as a free vortex, one can use the Euler equation to describe the distribution of total pressure rise as

$$\Delta p_T = \eta_T \rho N r \left(C_{\theta 2} - C_{\theta 1}\right)$$

Using simple inflow with $C_{\theta 2} = a/r$ yields

$$\Delta p_T = \eta_T \rho N r \left(\frac{a}{r}\right) = a\eta_T \rho N = \text{constant}$$

(9.11)

This is a quantified restatement of the condition of uniform or constant work across the annulus of an axial machine. This condition was employed as a restriction for analyzing fans and pumps in the earlier chapters in order to treat the inflow and outflow values of $C_x$ as being identical. This makes the analysis very easy to deal with, but it enforces a strong restriction on the flexibility of the design approach. Since $C_{\theta 2}$ must increase as $a/r$ with decreasing radial position, a machine with a relatively small hub will require very large values of $C_\theta$ near the hub to achieve the free vortex conditions. It is known that when $C_\theta$ is large and U is small, the de Haller ratio for the blade is going to be unacceptably small. One can explore this more extensively, but for now there appears to be a need to relax the free vortex requirement on the blade loading distribution. However, to do so requires that one deal with the non-linearity of the equilibrium equation.

## 9.5   APPROXIMATE SOLUTIONS

Begin by allowing a more general form for $\Psi$ using a polynomial algebraic description. Retain the $a/r$ term as a baseline and use additional terms to permit a systematic deviation from the simplest case. We write

$$\Psi = \frac{a}{x} + b + cx + dx^2 + \ldots \tag{9.12}$$

The coefficients a, b, c, d, etc., are all simple constants that can be chosen to manipulate the swirl velocity or loading along the blade. $b = c = d = 0$ along with any remaining higher order terms reduces $\Psi$ to the original simple form. If one restricts to the first three terms (d and others being zero), one retains substantial generality in the swirl distribution. The equilibrium equation using $\Psi = a/x + b + cx$ becomes

$$\frac{d\Phi_2^2}{2} = \eta_T bdx + 2x\eta_T cdx - \frac{ab}{x^2} - \left(\frac{2ac}{x}\right)dx \tag{9.13}$$

The higher order terms in b and c (i.e., $b^2$, bc, and $c^2$) have been dropped from the equation. To do this, ensure that b and c are small compared to a, and that $b^2$ is small compared to b, etc. (e.g., $b \ll 1$, $c \ll 1$, and $a \approx 1$). This device for simplification of the ensuing mathematical manipulations is a common technique referred to as "order analysis" or perturbation theory. One can "perturb" the free vortex theory for equilibrium by including small deviations from a/r using small values of b and c. This implies an inherent restriction on the analysis but will yield a simple algebraic result that can then be examined in depth to gain some insight into the general flow behavior through axial machines.

In the simple form developed above, the differential equation can be integrated directly as

$$\frac{\Phi_2^2}{2} = \eta_T bx + \eta_T cx^2 - 2ac \ln(x) + \frac{ab}{x} + K_i \tag{9.14}$$

where $K_i$ is the constant of integration. One must use $K_i$ to enforce conservation of mass in the throughflow $\Phi$, so the constant is not at all arbitrary. Thus,

$$\Phi_2 = \left[2\left(\eta_T bx + \eta_T cx^2 - 2ac \ln(x) + \frac{ab}{x} + K_i\right)\right]^{1/2} \tag{9.15}$$

and we require

$$\int 2\pi x \Phi_1 dx = \int 2\pi x \Phi_2 dx$$

$$= 2^{3/2}\pi \int \left[2\left(\eta_T bx + \eta_T cx^2 - 2ac \ln(x) + \frac{ab}{x} + K_i\right)\right]^{1/2} xdx \tag{9.16}$$

$$= \pi \Phi_1\left(1 - x_h^2\right)$$

where the limits of integration are from $x_h$ to 1.0. Here, $x_h$ is the hub–tip ratio, d/D, and it is assumed that $\Phi_1$ is not a function of x. Integrating the square root of a polynomial with a logarithm added on is a rather non-trivial task that is left to the reader as an exercise. However, the equation remains a mathematically symbolic relation for the value of $K_i$ but remains unquantified for a real case. What to do?

What really needs to be done is to find an acceptable way to linearize the messy integrand above so that the calculus can be handled more readily. One can do so by introducing a perturbation variable form for $\Phi_2$ as

$$\Phi_2 = [1 + \varepsilon(x)]\Phi_1 \tag{9.17}$$

$\varepsilon(x)$ is a function of x that describes the deviation of the outlet axial velocity component from the uniformity of the inlet flow. That is, $\varepsilon(x) = 0$ for all x describes a uniform outflow, while $\varepsilon(x) = ex$ could describe a small linear redistribution of flow at the outlet if $e \ll 1$.

The troublesome $\Phi_2^2$ term becomes

$$\Phi_2^2 = [1 + \varepsilon(x)]^2 \Phi_1^2 = [1 + 2\varepsilon(x) + \varepsilon^2(x)]\Phi_1^2 \tag{9.18}$$

If willing to require that $\varepsilon(x) \ll 1$ so that $\varepsilon^2(x) \ll 2\varepsilon(x)$, then one can write

$$\Phi_2 = (1 + 2\varepsilon(x))^{1/2}\Phi_1 \tag{9.19}$$

Under the assumption that $\varepsilon^2(x) \ll 2\varepsilon(x)$, one can truncate a binomial expansion of $[1 + 2\varepsilon(x)]^{1/2}$ to $[1 + \varepsilon(x)]$ with no impairment in accuracy. This allows us to write

$$\int x\Phi_1 dx = \int x\Phi_2 dx = \int [1 + \varepsilon(x)]x\Phi_1 dx = \int x\Phi_1 dx + \int \varepsilon(x)x\Phi_1 dx \tag{9.20}$$

Through the perturbation device the requirement for conservation of mass has been reduced to the demand that

$$\int \varepsilon(x)x dx = 0 \tag{9.21}$$

where the integration is across the annulus, from $x_h$ to 1.0 (or from d/2 to D/2). From the relation $\Phi_2^2 = [1 + 2\varepsilon(x)]\Phi_1^2$, one can write

$$\varepsilon(x) = \left(\frac{1}{\Phi_1^2}\right)\left[\eta_T bx + \eta_T cx^2 - 2ac\ln(x) + \frac{ab}{x} + K_i - \frac{1}{2}\right] \tag{9.22}$$

so that $\int \varepsilon(x) \, x \, dx = 0$ is readily carried out and solved for $K_i$. The result is

$$\varepsilon(x) = \left(\frac{\eta_T}{\Phi_1^2}\right)\left[(x - x_1)b + (x^2 - x_2^2)c + \left(\frac{ab}{\eta_T}\right)\left(\frac{1}{x} - \frac{1}{x_3}\right) - \left(\frac{2ac}{\eta_T}\right)(\ln(x) + \ln(x_4))\right] \quad (9.23)$$

$x_1$, $x_2$, $x_3$, and $x_4$ are simple functions of $x_h = d/D$ given by

$$x_1 = \left(\frac{2}{3}\right)\left(\frac{(1 - x_h^3)}{(1 - x_h^2)}\right) = f_1(x_h)$$

$$x_2 = \left(\frac{(1 - x_h^2)}{2}\right)^{1/2} = f_2(x_h)$$

$$(9.24)$$

$$x_3 = \frac{(1 + x_h)}{2} = f_3(x_h)$$

$$\ln(x_4) = \frac{\left[1 + x_h^2(\ln(x_h^2) - 1)\right]}{\left[2(1 - x_h^2)\right]} = f_4(x_h)$$

This simple algebraic result allows calculation of $C_{x2}$ as a function of r for non-free-vortex blade loading distributions. Note that where $b = c = 0$, $C_{x2}$ reverts to $C_{x1} = $ constant, so one has a limit-case check.

Examination of the errors associated with the approximations of these results (Wright, 1988) shows that the errors in estimating $\Phi_2$ begin to exceed about 5% when the magnitude of $\varepsilon$ exceeds a maximum value of 0.25, in comparison to the result of "exact" numerical integration of the non-linear differential equation (Kahane, 1948; Ralston and Wright, 1987). Figure 9.4 shows the influence of a series of loading distributions on the non-uniformity of $\Phi_2$. The design study is constrained to equal pressure rise for every case for equal flow. The hub–tip ratio is $x_h = 0.5$, the flow rate is $\Phi_1 = 0.5$, and the efficiency is set at $\eta_T = 0.9$. Case 2 is the free-vortex baseline case and cases 1, 3, 4, and 5 represent increasing departure from free vortex loading. The figure also shows the corresponding variations of $\Phi_2$ with x. The free vortex shows a uniform outflow at $\Phi_2 = 0.5$, with the succeeding distributions showing greater and greater deviation from uniform outflow. The flow moves to the outside of the annulus while reducing the throughflow near the hub. Shifting the swirl generation load to the outboard region of the blade can lead to very low axial velocities near the hub and even to negative values, or flow reversal—a clearly unacceptable design condition.

Figure 9.5 shows a comparison for the predicted velocities associated with a blade design with constant swirl generation across the blade. Here, $\Psi = b = 0.2$ with $\Phi_1 = 0.4$, $x_h = 0.5$, $\eta_T = 0.9$, $c = 0.23$, $a = b = 0$, $a = c = 0$. The approximate solution

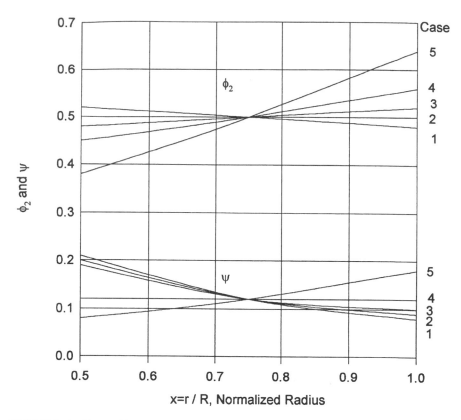

FIGURE 9.4   Non-uniform pressure rise across the annulus:
$\Phi_1 = 0.5$, $x_h = 0.5$, $\eta_T = 0.9$. (Wright, 1988.)

for $\Phi_2$ is shown along with an exact numerically integrated result (Ralston and Wright, 1986). A certain amount of error or difference is seen to be creeping into the simple solution as a result of the linearizing assumptions. The worst error appears to be about 4% in $\Phi_2$ at the blade tip, $x = 1.0$. The corresponding value of $\varepsilon$ is 0.12.

Figures 9.6 and 9.7 show similar results for linear swirl distributions with $c = 0.23$ and $c = 0.5$ ($\Psi = 0.23x$ and $\Psi = 0.50x$ with $a = b = 0$), respectively. The first case, Figure 9.6, shows a worst-case error of about 6%, while the second case, Figure 9.7, shows an 11% error, suggesting that the limits of the allowable perturbation have been exceeded.

The approximate solution is compared to experimental data as well using the test results of a study at NACA (Kahane, 1948) in which several fans with highly three-dimensional, tip-loaded flows were designed and tested (Wright, 1987). The results showed maximum errors generated in the linearized calculations are between 5 and 14% (near stall). To put these results into better perspective, overall performance prediction using the linearized and numerical solutions are compared to the experimental data from NACA in Figures 9.8a and 9.8b. Agreement with experiment is good for both methods of calculating the quasi-three-dimensional flow, suggesting

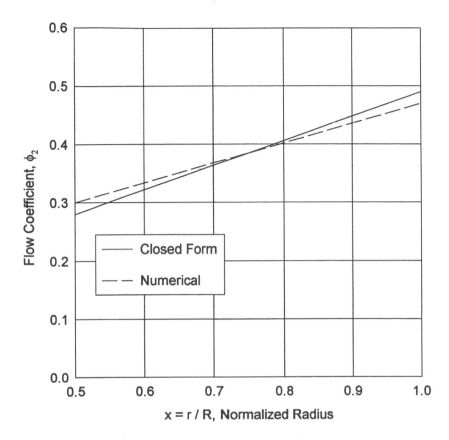

**FIGURE 9.5**  Non-uniform pressure rise across the annulus:
$\Phi_1 = 0.4$, $x_h = 0.5$, $\eta_T = 0.9$, $b = 0.2$, $a = c = 0$. (Wright, 1988.)

that local differences in the $\Phi_2$ distributions are submerged in the integrated performance results.

## 9.6  EXTENSION TO NON-UNIFORM INFLOW

To solve the flowfield through a stator vane row following a non-uniformly loaded rotor, the calculations for equilibrium must be able to account for the existence of a radially varying distribution of axial velocity. Jackson (1991) extended the earlier work on a closed form radial equilibrium flow to include the variation of axial velocity into the vane row. The fundamental assumptions employed were

1. The total pressure through the vane row is a constant (no energy addition and negligible losses) and density is constant so that the ideal Bernoulli equation applies.
2. The vane inlet flow is identical to the rotor outlet absolute velocities.
3. The vane row through-flow is governed by simple radial equilibrium.

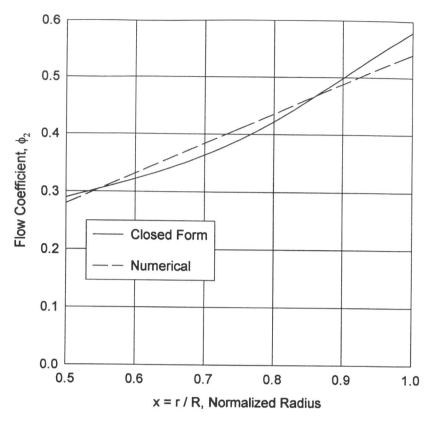

**FIGURE 9.6**  Non-uniform pressure rise across the annulus:
$\Phi = 0.4$, $x_h = 0.5$, $\eta_T = 0.9$, $c = 0.23$, $a = b = 0.0$. (Wright, 1988.)

Assigning the subscripts 3 to the vane inlet station and 4 to the vane outlet station, then

$$\Delta p_T = p_{T4} - p_{T3} = \left( p_4 - \frac{\rho C_{x4}^2}{2} + \frac{\rho C_{\theta4}^2}{2} \right) - \left( p_3 - \frac{\rho C_{x3}^2}{2} + \frac{\rho C_{\theta3}^2}{2} \right) \quad \textbf{(9.25)}$$

Differentiation with respect to r yields, on some rearrangement and assuming $C_{\theta4}$ is identically zero,

$$\frac{\left( \dfrac{dC_{x4}^2}{dr} \right)}{2} - C_{\theta3}^2 - \frac{\left( \dfrac{dC_{x3}^2}{dr} \right)}{r} - \frac{\left( \dfrac{dC_{\theta3}^2}{dr} \right)}{2} = 0 \quad \textbf{(9.26)}$$

If one normalizes the velocity and radius according to $\phi = C_x/U$, $\theta = C_\theta/U$, and $x = r/(D/2)$, then Equation (9.26) can be rewritten in the differential form as

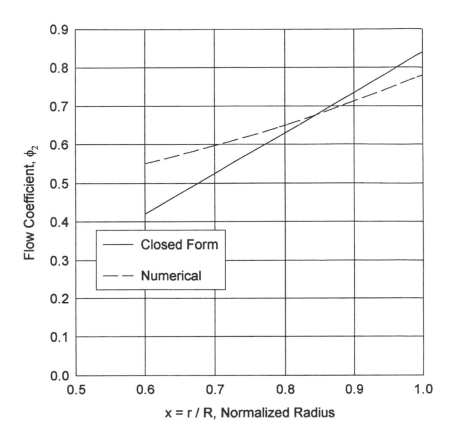

**FIGURE 9.7**   Non-uniform pressure rise across the annulus:
$\Phi_1 = 0.667$, $x_h = 0.6$, $\eta_T = 0.9$, $c = 0.5$, $a = b = 0$. (Wright, 1988.)

$$\frac{d\phi_4^{\,2}}{2} = \left(\frac{\theta_3^{\,2}}{x}\right)dx + \phi_3 d\phi_3 + \theta_3 d\theta_3 \tag{9.27}$$

It is assumed that $\theta_3 = \theta_2 = a/x + b + cx$ so that

$$\phi_4 d\phi_4 = \eta_T b dx + 2\eta_T cx dx \tag{9.28}$$

where the higher order terms have again been neglected. Integration yields the now familiar form

$$\frac{\phi_4^{\,2}}{2} = \eta_T bx + \eta_T cx^2 + K \tag{9.29}$$

where K is again the constant of integration to be used for conservation of mass. Using the perturbation form for $\phi_4$ where

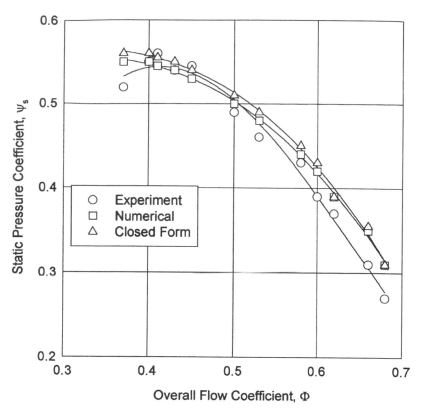

**FIGURE 9.8a** Comparison of the overall performance of the NACA constant swirl axial fan to linearized and numerical predictions. (Wright, 1988.)

$$\phi_4 = [1 + \epsilon(x)]\phi_3 \tag{9.30}$$

allows the vane outlet velocity to be expressed as

$$\phi_4 = \phi_3 + \left(\frac{\eta_T}{\phi_3}\right)\left[b(x - x_1) + c(x^2 - x_2^2)\right] \tag{9.31}$$

with

$$x_1 = \left(\frac{2}{3}\right)\frac{(1 - x_h^3)}{(1 - x_h^2)} \quad \text{and} \quad x_2^2 = \frac{(1 + x_h^2)}{2} \tag{9.32}$$

## 9.7   CENTRIFUGAL FLOWPATH

As in the case for axial throughflow, the flowfield in a mixed flow or radial flowpath has received considerable attention in the technical literature. The same finite

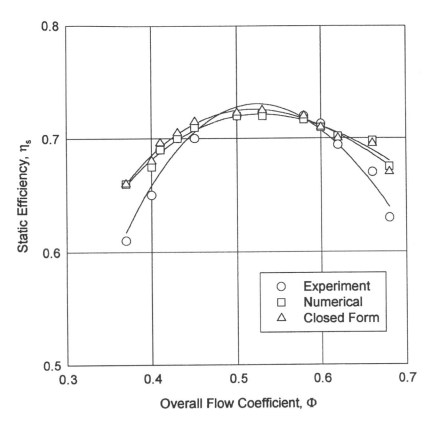

**FIGURE 9.8b**  Efficiency comparison for the NACA fan.

difference and finite element numerical solutions for inviscid, viscous, incompressible, and compressible flows have been applied to this flow problem as well. Review articles on the subject (Adler and Krimerman, 1980; Deconick and Hirsch, 1990; Lakshminarayana, 1991) provide an excellent background as well as a history of the development of these methods. Again, quasi-three-dimensional analysis provides a somewhat simpler, although less rigorous approach to the solution of these flows. Applications and comparison of the methods are available (e.g., Whirlow et al., 1981; Wright, 1982; Wright, 1984b) along with extensive comparisons to experimental data. In Wright's work of 1984, inviscid, incompressible calculations of the velocities and surface pressures on the blade and shroud surfaces of the impeller of a large centrifugal fan were computed using a finite element potential flow analysis and a finite difference quasi-three-dimensional analysis of the same flow (Katsanis, 1977). An instrumented impeller fitted with 191 internal pressure taps was used to provide the experimental evaluation of the computed results. The fan is a wide-bladed centrifugal fan with a specific speed at design point of $N_s$ = 1.48, $D_s$ = 2.14, and $\eta_T$ = 0.89. Experimental data and analytical prediction comparisons are shown in Figures 9.9 and 9.10. Data is presented here at one of the three stations along the span of the blade (midspan, near the backplate, and near

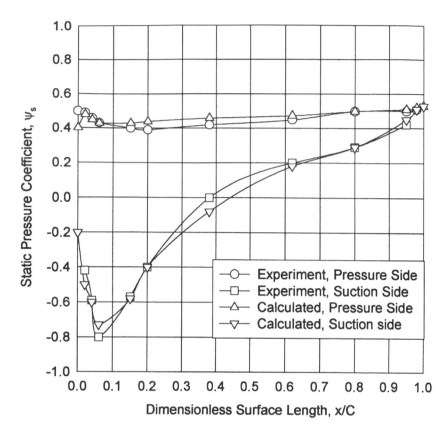

**FIGURE 9.9**  Comparison of the predicted blade surface pressures and experimental results (design point flow at midspan). (Wright, 1984b.)

the rotating shroud), the midspan station at design flow rate. Data is also provided along the shroud surface along a path midway between the camber lines of adjacent blades, along an extension of this path onto the stationary inlet bell, and at design flow. The blade and shroud pressures are given in coefficient form, where the value of $\psi_s$ was defined in the reference as

$$\psi_s = \frac{p_s}{\left(\dfrac{\rho U_2^{\,2}}{2}\right)} \qquad (9.33)$$

and $p_s$ is the gage static pressure.

In a follow-on effort, Shen (Shen, L.C., 1993) extended the study to include another, more readily available FEM code (Ansys, 1985) and the panel code developed at NASA (McFarland, 1982; McFarland, 1985), which relies on a surface singularity technique to develop surface and interior velocity distributions. Agreement with the other methods and the experimental data was very good. This may

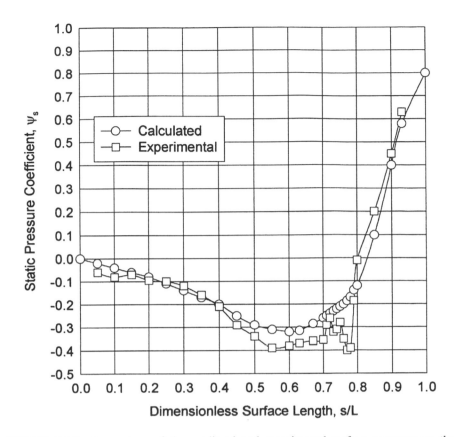

**FIGURE 9.10**  Comparison of the predicted and experimental surface pressures on the impeller shroud and inlet bell (design flow). (Wright, 1984b.)

be of particular importance concerning the panel method of McFarland. This computer code is compact, very easy to use, and can be run on personal computers of modest capability (e.g., with an Intel Corp. 80486 or Pentium chip, 8 MB of random access memory, and 80 MB of hard disk storage).

In general, the predictions of the more rigorous, fully three-dimensional finite element method yields predictions that are in no better agreement with the experimental data than the predictions of the quasi-three-dimensional method. Stated more positively, both methods provide good predictions of the surface velocities and pressures measured in the fan. As was true in the axial flow studies, both of these numerical methods required extensive input data preparation and refinements. For the purposes of preliminary design or systematic parametric analysis of design features, simpler and perhaps less rigorous, less accurate methods of analysis seemed desirable.

Simpler approaches to analysis of centrifugal flowfields include the sort of one-dimensional calculations used in earlier chapters, based on mean flow considerations. However, considering the rather large variations of the flow properties along the blade span seen in most studies, even at design point, it seems that an acceptable analysis must deal in some manner with the inherent three-dimensionality of the flow in centrifugal impellers.

Analysis and measurements of the flow in the more complex geometry of a centrifugal compressor impeller are also available (Krain, 1984). Krain's work employed quasi-three-dimensional predictions of velocities in the impeller and very detailed laser Doppler measurements of velocities in the blade-to-blade channels in the impeller. Analytical and experimental results for the velocity profiles in the blade channels showed mixed results. Velocity profiles taken near the inlet plane showed good analytical and experimental agreement. Near the outlet, complexity and three-dimensionality of the flow give messy results, and agreement between prediction and experiment is more difficult to achieve. Here, the flow appears to be dominated by viscous effects and may be developing the jet-wake characteristics discussed earlier. Elsewhere in the impeller, the analysis compares very well with the measured velocities. However, application of newer, fully viscous approaches being developed are the current and future hope for improvement. Uncertainty still exists in choosing the most appropriate flow transition and turbulence models for the turbomachinery environment (Mayle, 1991; Volino and Simon, 1995).

## 9.8   SIMPLER SOLUTIONS

The very simple approach we used for axial flowpath analysis—simple radial equilibrium—cannot be readily adapted directly for mixed or radial flow analysis. Difficulty arises from the fact that the meridional streamline generally has a very significant amount of curvature. The pressure gradient term must then involve both the $C_m$ value and its related radius of curvature, $r_{mc}$, and the $C_\theta$ value and its local radius of curvature, r. There have been relatively successful attempts to model equilibrium flows using a full streamline curvature model that accounts for both terms. These models are collectively called "streamline curvature" analyses, where dp/dr must be balanced by the two terms $C_m^2/r_{mc}$ and $C_\theta^2/r$ as

$$\frac{dp}{dr} = \frac{C_m^{\,2}}{r_{mc}} + \frac{C_\theta^{\,2}}{r} \qquad (9.34)$$

Here, radial velocities clearly must be accounted for, and the equilibrium equation coupled with the Euler equation and conservation of mass results in more complex differential equations. Recalling that the meridional stream surfaces interior to the machine must adjust their locations in terms of the governing equations, then $r_{mc}$ is known only at the shroud and hub surfaces prior to the solution of the flowfield. Therefore, an iterative approach with fairly complicated relations between flow and geometry is required to generate a converged solution. Convergence is of course strongly linked to the designer's choice of the distribution of the generation of swirl velocity, $C_\theta$, along the stream surface as well.

Numerical schemes have been developed (e.g., Novak, 1977) that work reasonably well. An example of application is found in a study of the influence of inlet swirl disturbance on centrifugal fan performance (Madhavan, 1985); this study employed a modified form of Novak's original numerical code for meridional flow analysis. As seen before, iterated numerical procedures can provide good information and design guidance, but they may become difficult or troublesome to work with because of their input requirements and the ever-present worries of convergence and accuracy of the solutions.

To illustrate general principles, to carry out parametric studies, or to investigate preliminary design layouts, a simple, reasonably accurate analysis is very useful. There have been many attempts to develop such an analysis in the past, and it is fair to say that most of these methods have been only partially successful. For example, Davis and Dussord (1970) developed a calculational procedure based on a modified mean flow analysis for radial flow machines by modeling the viscous flow in the impeller channels to provide "mean" velocities at an array of positions along the flowpath. Simplifying assumptions concerning the rate of work input along the flowpath allowed calculation of distributed diffusion and estimation of loss through the impeller. Meridional curvature was modeled only in terms of interpolation of end-wall geometry between shroud and hub. Thus, the results were valid primarily for very narrow impeller channels, in terms of blade height between shroud and hub. In spite of the approximations involved, Davis and Dussord were able to establish reasonable predictions for the performance of several compressor configurations. An example is shown in Figure 9.11, where predicted and experimental values are in reasonable agreement. The work, which was carried out in a competitive industrial environment, showed good promise but was reported without great detail on the method, its applications, and the details of the experimental results.

A more general approach was taken by Adler and Ilberg (1970), particularly in terms of the distribution of the flow in the region of the impeller inlet. They were concerned that in the presence of strong end-wall curvature in the inflow regions of pump or compressors, the frequently or usually used assumption of uniform inflow to the impeller was inadequate for designing the critical impeller inlet geometry. An attempt was carried out "to develop a simple method based on the flow equations, for the calculation of the flowfield in the entry to radial or mixed flow impellers." Their goal in part was to avoid the usual iterative procedures involved in satisfying mass conservation in the then currently available potential flow computations (Vavra, 1960).

The key to their solution lay in establishing distributions of streamline curvature along normals to the streamlines compatible with the physics of the flow and the end-wall geometries. Their results gave an expression for $C_m$ in terms of an equivalent uniform value $C_{mu}$ and the end-wall curvatures in the form

$$C_m = C_{mu} e^{-kz} \tag{9.35}$$

This form yields a decaying profile across the passage with the maximum magnitude occurring at the shroud side and the minimum value at the hub side of the flowpath. Examples of Adler's calculations are compared to the iterated potential flow solution on Figures 9.12a and 9.12b. The forms of these solutions strongly suggest the $z^{-n}$ type variation of $C_m$.

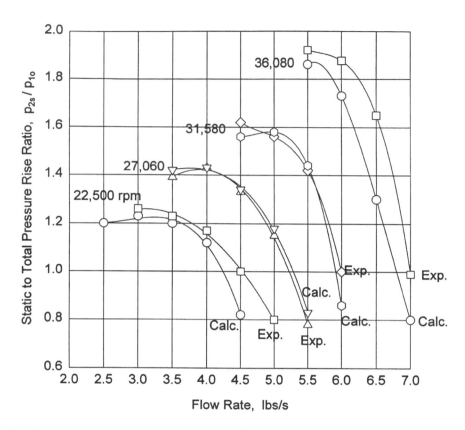

**FIGURE 9.11** Comparison of the Davis and Dussord model to measured compressor performance. (Davis and Dussord, 1970.)

Inlet velocity measurements made in the flow of a centrifugal fan (Gessner, 1967) show much the same behavior as the above predictions, possibly exaggerated by the small radius of curvature at the shroud.

Of particular interest in these results is the ratio of the maximum and minimum values of $C_m$ to the mean value. If one attempts to determine the inlet geometric angles of a blade in a centrifugal machine, then combining $C_m$ with U at a given meridional station sets the positioning of the blade leading edge. From the examples examined here, $C_m/C_{mu}$ varies from about 0.6 at the hub to 1.6 at the shroud. For a simple radial cascade, with $U = 2C_{mu}$, a mean setting angle for the blade leading edge would be $\beta_1 = \tan^{-1}C_{mu}/2C_{mu} = 26.6°$. At the shroud, the proper value would be $\beta_1 = \tan^{-1}1.6C_{mu}/2C_{mu} = 38.6°$. At the hub, the angle must be $\beta_1 = \tan^{-1}0.6C_{mu}/2C_{mu} = 14.0°$. This is quite a range of angles and suggests a twisting of the leading edge through an angle of about 24.6°.

Another way to view the result is to observe that if the blade leading edge is straight and set at 30°, a fairly severe range of angle of incidence, i, will occur over the axial span of the blade. This distribution is shown in Figure 9.13 as the leading edge incidence angle vs. span. Since a given airfoil shape can tolerate only a limited

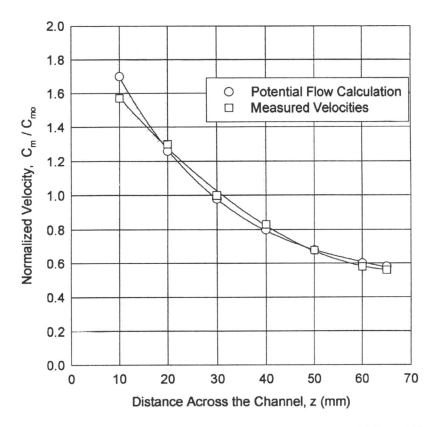

**FIGURE 9.12a**   Computations and experimental results for the flowpath of Adler and Ilberg, $\theta = 75°$. (Adler and Ilberg, 1970.)

amount of leading edge incidence to the oncoming relative vector without suffering some boundary layer separation or stalled flow (Koch et al,. 1978), it is important to be able to estimate the magnitude of i and to be able to control the magnitude of i through the radius of curvature of the end-walls. Blade twist and shroud curvature appear to be the critical parameters.

## 9.9   APPROXIMATE ANALYSIS

In order to arrive at a simple correlation for the distribution of approach velocity to blade leading edge in a mixed flow or centrifugal flow machine (Wright, 1984b), an earlier analysis used a streamline curvature approach (Novak, 1977) to analyze a systematically designed set of shroud configurations. The geometry of the inlet region was modeled as a hyperbolic shroud shape characterized by its minimum radius of curvature, as shown in Figure 9.14. Surface velocities were calculated along the hyperbolic contour to estimate the ratio of the maximum value of $C_m$ compared to the mean value $C_{mu}$. The mean value is based on $C_x = Q/(\pi d^2/4)$ and $U_1 = Nd/2$. It was hypothesized, using a simple ring vortex model (Gray et al., 1970),

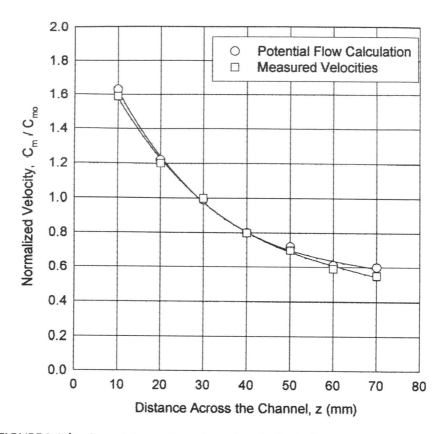

**FIGURE 9.12b**  Computations and experimental results for the flowpath of Adler and Ilberg, $\theta = 30°$.

that the maximum $C_m/C_{mu}$ value would occur at the minimum radius of curvature as indicated in Figure 9.14 and the magnitude would be greater than 1.0. Samples of these calculations are shown in Figures 9.15.

Based on these results and normalizing the velocity by $U_2 = ND/2$, $W_{max}$ could be expressed as

$$\frac{W_{max}}{U_2} = \left[ \left( \frac{d}{D} \right)^2 + F^2 \left( \frac{C_{mu}}{U_2} \right)^2 \right]^{1/2} \tag{9.36}$$

Based on these results and simple vortex modeling, F was expected to take the form

$$F = 1.0 + \frac{constant}{\left( \dfrac{r_c}{D} \right)^n} \tag{9.37}$$

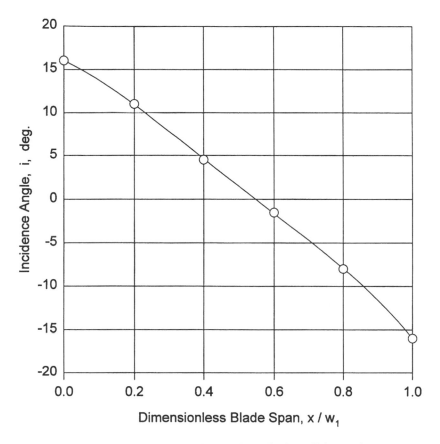

**FIGURE 9.13**    Distribution of leading edge incidence in the radial cascade.

In this form, F approaches 1.0 for very large values of $r_c/D$ (e.g., in a cylindrical duct). The function, F, is illustrated in Figure 9.16.

With these concepts in place, one is now in a position to begin constructing a heuristic model for the cross-channel distribution of $C_m$ in the region of a blade leading edge. Assuming that the distribution of $C_m$ from hub to shroud $(0 < z < w_1)$ varies according to $C_m/C_{mu} = a_1 + a_2/z^n$ and imposing conservation of mass while requiring $(C_m/C_{mu})_{max} = 1 + 0.3/(r_c/D)^{1/3}$ results in

$$\frac{C_m}{C_{mu}} = \frac{\left[1 + \dfrac{0.3}{\left(\dfrac{r_c}{D}\right)^{1/3}}\right]}{\left(1 + az^n\right)} \tag{9.38}$$

Mass conservation is satisfied by requiring

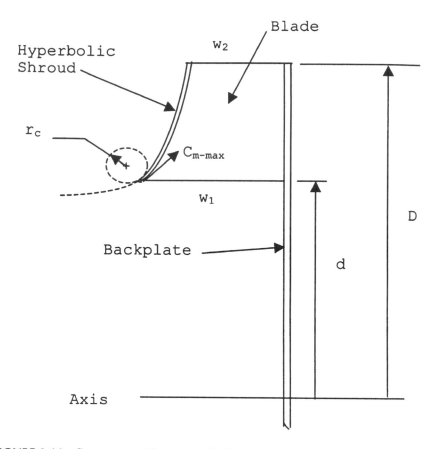

**FIGURE 9.14**　Geometry used in the hyperbolic shroud study.

$$\frac{\int C_m dA}{\int C_{mu} dA} = 1.0 = \frac{\int C_m dA}{C_{mu} A} \tag{9.39}$$

For the geometry defined by Figure 9.17, we can define $C_{mu}$ in terms of volume flow rate as

$$Q = C_{mu}\left(2\pi r_s^2\left(1 - \left(\frac{w'}{2}\right)\cos\theta\right)w'\right); \quad w' = \frac{w}{r_s}$$

$Q$ is also given as

$$Q = 2\pi r_s \int C_m dz - 2\pi\cos\theta \int C_m z dz \tag{9.40}$$

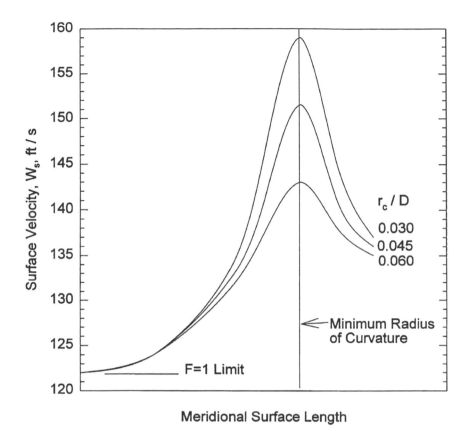

**FIGURE 9.15**   Sample surface velocity result on the hyperbolic shrouds.

with limits on z from 0 to w. Using dimensionless variable $w' = w'/r_s$ and $z' = z'/rs$ yields

$$Q = C_{mu} 2\pi r_s^2 \left( \int \left( \frac{C_m}{C_{mu}} \right) dz' - 2\pi \cos\theta \int \left( \frac{C_m}{C_{mu}} \right) z' dz' \right) \qquad (9.41)$$

We employ $C_m/C_{mu} = (C_{mo}/C_{mu})/(1 + (ar_s)z')$, where $n = 1$ and $C_{mo} = Q/A_o$. In terms of $r_c$ and $r_s$, and based on the calculations in Figure 9.15 (with $d/D = 0.7$), one obtains

$$C_{ou} = 1.0 + \frac{0.426}{\left( \dfrac{r_c}{r_s} \right)^{1/3}} \qquad (9.42)$$

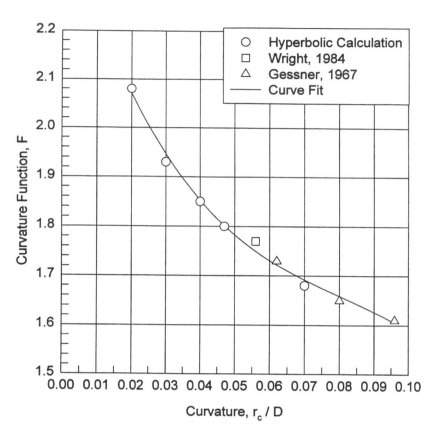

**FIGURE 9.16** Synthesis of the curvature function, F, based on the hyperbolic shroud model data.

Finally, defining for convenience that $b = ar_s$, the constraint on mass conservation reduces to

$$\left[\frac{w'\left(1+\dfrac{w'\cos\theta}{2}\right)}{\left(1+\dfrac{0.426}{\left(\dfrac{r_c}{r_s}\right)^{1/3}}\right)}\right] = \int\left[\frac{1}{(1+bz')}\right]dz' - \cos\theta\int\left(\frac{z'}{(1+bz')}\right)dz' \qquad (9.43)$$

with integration limits from 0 to $w'$. This relationship provides a solution for b.

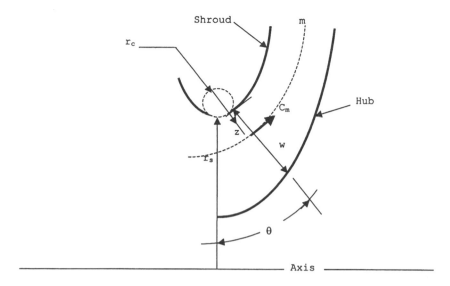

**FIGURE 9.17**  Geometry for the heuristic flowpath velocity distribution model.

The integrals evaluate readily to yield

$$\left[ \frac{w'\left(1 - \dfrac{w'\cos\theta}{2}\right)}{\left(1 + \dfrac{0.426}{\left(\dfrac{r_c}{r_s}\right)^{1/3}}\right)} \right] = \left(\frac{1}{b}\right)\left[ \ln(1 + bw')\left(1 + \frac{\cos\theta}{b}\right) - w'\cos\theta \right] \qquad (9.44)$$

For a given choice of geometry, the left side of this equation is simply a constant. The right side is badly transcendental in b and cannot be solved directly. Fortunately, it can be solved iteratively by successive approximations for a given set of geometric choices, yielding a parametric function of the three variables w', $\theta$, and $r_c/r_s$. A computer code was used to solve the transcendental problem for a wide range of the variables ($0 < w' < 1.25$, $0° < \theta < 90°$, $0.05 < (r_c/r_s) < 1.2$). A sample of results for b, with $\theta = 90°$, is shown in Figure 9.18. The results are seen to be rather nonlinear in both $r_c$ and w. Linear regression with least squares curve fitting yields a functional form for b of

$$b = \left[ 0.264(\cos\theta)^{1.18} + \frac{0.955}{w'} \right]\left( \frac{r_c}{r_s} \right)^{-0.38} \qquad (9.45)$$

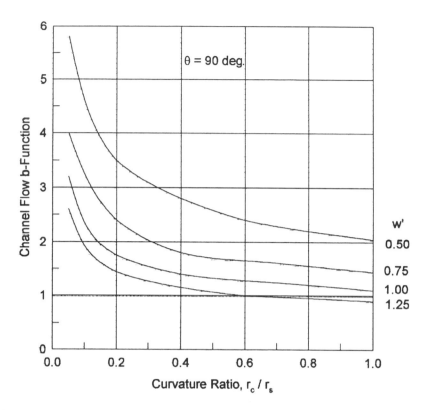

**FIGURE 9.18**   Sample results for the curve-fit form of the b-function.

with correlation coefficients in the range of 0.999. Thus, the heuristic model is closed in the form

$$\frac{C_m}{C_{mu}} = \frac{C_{ou}}{(1 + bz')} \tag{9.46}$$

with

$$C_{mu} = \frac{Q}{\left[ 2\pi r_s^2 w' \left( 1 - \frac{w' \cos\theta}{2} \right) \right]}$$

$$C_{ou} = 1.0 + \frac{0.426}{\left( \dfrac{r_c}{r_s} \right)^{1/3}} \tag{9.47}$$

$$z' = \frac{z}{r_s}; \quad w' = \frac{w}{r_s}$$

using b from above.

Using the geometry provided by Adler and Ilberg and by Gessner, this flow model was employed to estimate the $C_m/C_{mu}$ curves shown earlier. The results agree well with both the potential modeling and the experimental data given in the figures. This additional data, along with data taken from Wright (1984b), has been added to Figure 9.16. While the curvature function, F, was based only on the streamline curvature results represented in Figure 9.15, inclusion of the additional data in Figure 9.16 serves to qualify or validate the function. The model seems to be somewhat conservative. It is concluded that the simple flow model is useful for illustration and preliminary design layout in mixed flow geometries, at least for predicting the inflow velocity distribution near the blade leading edge. It is suspected that the increasing importance of viscous influences farther downstream in the blade channel will significantly reduce the relevance of the simple model as the flow approaches the channel exit. Also note that the approximate equation for $C_m/C_{mu}$ overpredicts the value at $z = 0$ in every case, although not greatly (about 7 or 8%).

## 9.10   EXAMPLES

The importance of defining the shape of these curves derives from its role in defining the inlet flow angles and blade diffusion levels. Consider an example for a centrifugal flow path along which $C_{mu}$ is a constant, implying a constant flow area along lines emanating from the center of the radius of curvature for values of $\theta$ between $0°$ and $90°$. That is, $C_{mu} = C_{mu}(\theta = 0) = C_{ou}$. If one defines $w_o$ as w at $\theta = 0$, this requires that

$$C_{mu} = C_{ou} = \left[\frac{Q}{\pi\left(r_s^2 - (r_s - w_o)^2\right)}\right] = \frac{\left(\dfrac{Q}{\pi r_s^2}\right)}{\left[1 - (1 - w_o')^2\right]} \tag{9.48}$$

so that

$$w' = \frac{\left[1 - (1 - C_1 \cos\theta)^{1/2}\right]}{\cos\theta} \tag{9.49}$$

where

$$C_1 = \left(1 - (1 - w_o')^2\right) \tag{9.50}$$

to ensure the constant average, meridional velocity.

For the example, select $w_0' = 0.8$ so that as a function of $\theta$, w' behaves as shown in Figure 9.18. Note that at $\theta = 90°$, the equation is indeterminate and must be evaluated in the limit. One can detail such a flowpath if values for $r_s$ and D are chosen. For $D = 0.5$ m, $r_s = 0.333$ m, and $Q = 4$ m³/s, one obtains $C_{mu} = 12.9$ m/s. The shape then depends on $r_c$, as sketched in Figure 9.19 and in Figure 9.20 for

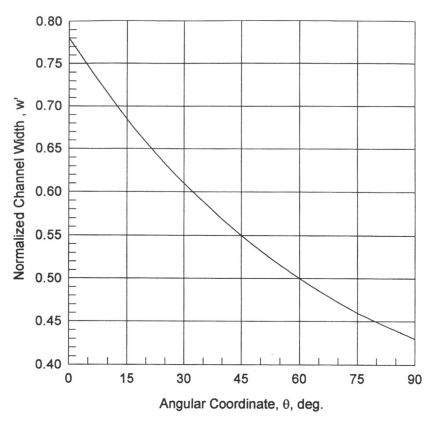

**FIGURE 9.19**  Normalized channel width distribution in the entrance region of the impeller.

$r_c = 0.05$ m, $0.10$ m, and $0.20$ m. From the viewpoint of mechanical design and cost, it is desirable to keep the axial length of the impeller as small as possible. As will be seen, a constraint on axial length can lead to negative effects on the velocity distribution in the blade channel. Choose the leading edge of the blade to lie along the $\theta = 45°$ line as shown in Figure 9.20 and use the previously developed equations to analyze the approach velocity, $C_m$, across the blade (from hub to shroud). Figure 9.21 shows the calculated results as $C_m$ vs. $z/w$ (fraction of channel width) for the configuration with $r_c = 0.05$ m (the impeller with the smallest axial length). $C_m$ ranges from 27 m/s at the shroud to 8 m/s at the hub. For the largest radius of curvature considered, $r_c = 0.40$ m, $C_m$ ranges from 21 m/s to 11 m/s. The ratio of minimum to maximum meridional velocity has been reduced from 3.4 to 1.9.

This relatively large change in $C_m$ is reflected in the variation of $\beta_1 = \tan^{-1}(C_{m1}/U_1)$. Given N = 1200 rpm, $\beta_1$ versus values of $z/w$ are shown in Figure 9.22. $r_c = 0.05$ m yields the range $18° < \beta_1 < 34°$ ($16°$ of blade twist), while $r_c = 0.40$ m gives $23° < \beta_1 < 27°$ ($4°$ of twist). This small amount of twist will lead to a much simpler blade, and, as one will soon see, lower levels of diffusion in the impeller. However,

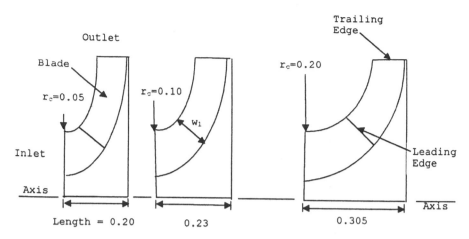

**FIGURE 9.20**   Three geometric channel layouts for the example study (dimensions in meters).

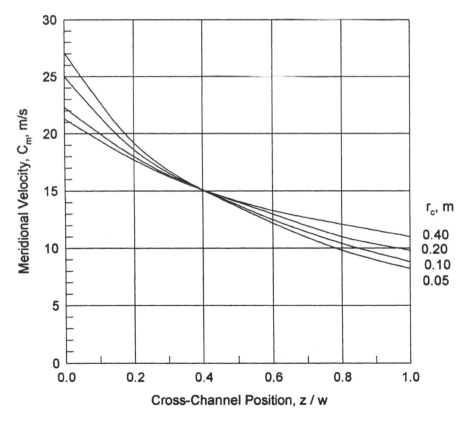

**FIGURE 9.21**   Velocity distribution across the channel for the example of Table 9.1.

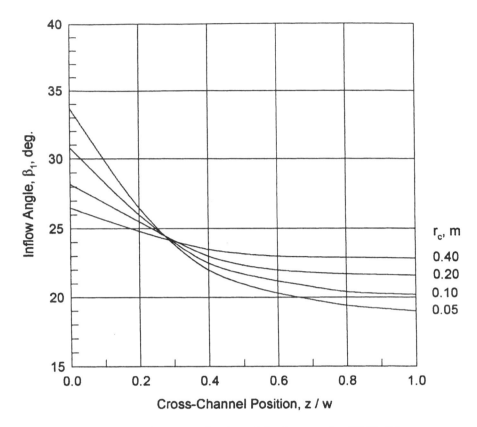

**FIGURE 9.22**   Inflow angles across the channel for the example of Table 9.1.

the impeller will be rather long in the axial direction, leading to higher costs and perhaps to excessive overhanging of rotating mass or large bearing spans.

If the impeller is designed to generate 60 m of head rise, one can calculate a relative outlet velocity of $W_2 = 18.8$ m/s. Using the mean value $C_{mu}$ at the blade leading edge gives a mean value of $W_1 = 22.4$ m/s, or a very mild mean de Haller ratio of 0.84. However, when examining the blade leading edge value of $W_1$, using $U_1 = 18.8$ m/s, one calculates de Haller ratios at shroud and hub locations as given in Table 9.1. Thus, in addition to inducing a highly twisted blade, a small radius of curvature can lead to high levels of diffusion, unstable flow along the shroud and perhaps reduced efficiencies as well.

As implied earlier, the larger radius of curvature allows use of an untwisted blade without the penalties associated with large leading edge incidence or increased de Haller ratios.

## 9.11   SUMMARY

This chapter introduced the concepts of fully three-dimensional flow in the blade and vane passages of a turbomachine. Because of the inherent great complexity of

**TABLE 9.1**
**De Haller Ratio Dependence**
**on Shroud Radius of Curvature**

| $r_c$ (m) | $W_2/W_1$ Shroud | $W_2/W_1$ Hub |
|---|---|---|
| 0.05 | 0.63 | 0.93 |
| 0.10 | 0.65 | 0.97 |
| 0.20 | 0.66 | 1.00 |
| 0.40 | 0.67 | 1.04 |

these flows, the idea was developed of using an approximate solution involving the superposition of two physically compatible flowfields in mutually perpendicular planes. This quasi or almost three-dimensional approach was pursued throughout the chapter to provide a tractable calculation scheme to aid in understanding the flows and in generating reasonable design geometry.

The classical results developed and applied here allow the relaxation of the rather simple but convenient assumption of uniform throughflow velocities, either purely axial or purely radial. Rather, one is able to estimate radial distributions of axial velocities in axial flow impellers and streamline-normal distributions of meridional velocities in centrifugal machines. In either case, the flow is constrained to maintain equilibrium between stream-normal pressure gradients and centrifugal force on the fluid along streamlines.

For axial flow machines, this concept is further simplified to Simple Radial Equilibrium and a non-linear integro-differential flow equation is developed. This rather intractable form is linearized using the concept of a perturbation solution around the known uniform solution. The result is an approximate, somewhat restricted, closed-form algebraic solution for non-uniform axial flow in a turbomachine. Comparison of the approximate result to numerical solutions and experimental results illustrates the utility of the method and its limitations.

Quasi-three-dimensional flow solutions are also available for centrifugal flowfields. Prior analytical and experimental results were used to illustrate some of the approximate but still rather intractable approaches to the problem. Simpler solutions are sought for use in broad studies of geometric influences on flow and for approximate layout of a machine in preliminary design. Such an approximate method was outlined and developed based on equilibrium considerations. Comparison of calculations from this simple method to experimental data on centrifugal and radial flow fans illustrates the reasonable estimates that can be generated with simple algorithms.

The approximate nature of these simpler prediction tools to estimate three-dimensional flow effects was emphasized throughout the chapter. Although they are capable of creating useful information for preliminary design layout, and yield insight into the complex flowfields, they will in practice be followed by rigorous, fully three-dimensional, fully viscous, and fully compressible computational flow analyses.

## 9.12   PROBLEMS

9.1. As outlined in the chapter, the pressure rise coefficient can be written as

$$\psi = \frac{a}{x} + b + cx$$

where a, b, and c are constants. Let $\psi = b$, with $a = c = 0$. Impose the streamline equilibrium condition of $d\phi_2{}^2$ with the conservation of mass constraint and show that

$$\phi_2 = \phi_1\left[1 - x_1\left(\frac{\eta_T\psi}{\phi_1{}^2}\right) + \left(\frac{\eta_T\psi}{\phi_1{}^2}\right)x\right]$$

where

$$x_1 = \frac{\left(\frac{2}{3}\right)\left(1 - x_h{}^3\right)}{\left(1 - x_h{}^2\right)}$$

9.2. With $\eta_T = 0.85$, $\psi = b = 0.35$, and $\phi_1 = 0.625$, use the results of Problem 9.1 to explore the influence of non-uniform loading and hub size ($x_h$) on the outlet velocity distribution. Calculate and graph the distribution of $\phi_2$ for these parameters for $x_h = 0.25, 0.375, 0.5, 0.625, 0.75$, and $0.875$.

9.3. Replicate the results shown in Figure 9.5 for the constant swirl solutions by the small perturbation method and the numerical. Use the results of Problem 9.1 to evaluate the other methods shown in the figure.

9.4. As was done in Problem 9.1, use the radial equilibrium streamline method to describe a special case outlet flow and swirl distribution. Here, use the single term $\psi = cx$ with c as a constant. Show that

$$\frac{\phi_2}{\phi_1} = \left[1 - \left(\frac{\eta_T c}{\phi_1{}^2}\right)x_2{}^2 + \left(\frac{\eta_T c}{\phi_1{}^2}\right)x^2\right]$$

where

$$x_2{}^2 = \frac{\left(1 + x_h{}^2\right)}{2}$$

9.5. Consider the example shown in Figure 9.6. With $x_h = 0.5$, $c = 0.23$, $\eta_T = 0.9$, and $\phi_1 = 0.4$, use the results of Problem 9.4 to evaluate the numerical and perturbation method solutions shown in the figure.

9.6. When one allows the outlet swirl velocity distribution to deviate from the free-vortex a/x form, one loses the simplicity of the two-dimensional cascade variables across the blade or vane span. Although the flow must now follow a path through the blade row, which has a varying radial position, one can still simplify the analysis by again considering inlet and out stations only. A common approximation for a fixed radial location on the blade is to consider the flow to be roughly two-dimensional, by using the *averaged* axial velocity entering and leaving, along with $U = U_1 = U_2$ and $C_{\theta 2}$ to form the velocity triangles. That is, the mean axial velocity,

$$C_{xm} = \frac{\left(C_{x1} + C_{x2}\right)}{2}$$

can be used with U to calculate

$$\beta_1 = \tan^{-1}\left(\frac{U}{C_{xm}}\right)$$

Similarly

$$\beta_2 = \tan^{-1}\left[\frac{\left(U - C_{\theta 2}\right)}{C_{xm}}\right]$$

One retains the character of the simple-stage assumptions by a minor adjustment in definition.

Use this definition and the results of Problem 9.2 to lay out the blade sections at hub, mean, and tip radial stations. Use a hub–tip ratio of 0.5.

9.7. Compare the camber and pitch distribution of the blade developed in Problem 9.6 with an equivalent blade for a fan meeting the same performance specifications with a free vortex swirl distribution. Lay out the blade with $\psi = a/x$ to give the same performance.

9.8. Determine the minimum hub size achievable for both the free vortex and the constant swirl fans of Problem 9.7. Use a de Haller ratio at the fan hub equal to 0.6 as the criterion for acceptability.

9.9. Repeat the study of Problem 9.8 using $D_L \le 0.6$ with $\sigma \le 1.5$.

9.10. Use the approximation for the mean axial velocity with the fan specified in Figure 9.6 to lay out a blade to meet the specified performance. Use the properties of Figure 9.6, including $x_h = 0.50$.

9.11. Explore the blade design of Problem 9.10 to find the smallest allowable hub size for the fan. Use the de Haller ratio criterion with $W_2/W_1 \ge 0.6$.

9.12. Design a vane row for the fan of Problem 9.5. Use the mean axial velocity across the blade at the entrance of the vane row. Assume that there is no further skewing of axial velocity component within the vane row.

9.13. Design a vane row for the fan of Problem 9.6. Use the same assumptions as were used in Problem 9.12.

9.14. Repeat the design of the blade row of Problem 9.12 using the non-uniform inflow from the rotor to calculate the non-uniform vane row outflow according to

$$\phi_4 = \phi_3 + \left(\frac{\eta_T}{\phi_3}\right)\left[b(x - x_1) + c\left(x^2 - x_2^2\right)\right]$$

with

$$x_1 = \frac{\left(\frac{2}{3}\right)\left(1 - x_h^3\right)}{\left(1 - x_h^2\right)} \qquad \text{and} \qquad x_2^2 = \frac{\left(1 + x_h^2\right)}{2}$$

9.15. Repeat the design of the blade row from Problem 9.13 using the ground rules of Problem 9.14.

9.16. Initiate the design of a vane axial fan to provide a flow rate of 0.5 m³/s with a total pressure rise of 600 Pa. Select the size speed and hub–tip ratio and do a preliminary free vortex swirl layout for the blades and vanes.

9.17. Extend the work of Problem 9.16 by preparing a swirl velocity distribution that is shaped as $C_{\theta 2} = ax^{1/2}$. (Hint: Although this is clearly not one of the terms in the series adopted for swirl distribution, one can approximate the curve by fitting three values—hub, mean and tip—of this function to $\psi = a/x + b + cx$.)

9.18. With the swirl of Problem 9.17, solve the outlet flow distributions for both blade and vane row and determine the vector triangles for blade and vane at the hub, mean, and tip stations. Set the hub size by requiring $W_2/W_1 \geq 0.6$

9.19. Lay out the cascade properties for both blade and vane for the fan design of Problem 9.17. Use the quasi-three-dimensional for the square-root swirl distribution, and compare the physical shape of these blades and vanes to those of the constant swirl fan of Problem 9.5.

9.20. The mass-averaged pressure rise for an arbitrary swirl distribution can be calculated from the Euler equation with weighted integration of $\psi$ across the blade span. Using $\psi = a/x + b + cx$, show that

$$\psi_{avg} = \frac{\int \psi \phi_1 x dx}{\int \phi_1 x dx}$$

and

$$\psi_{avg} = \frac{2a}{\left(1 + x_h\right)} + b + \frac{\left(\frac{2}{3}\right)c\left(1 - x_h^3\right)}{\left(1 - x_h^2\right)}$$

# 10 Advanced Topics In Performance and Design

## 10.1 INTRODUCTION

Previous chapters treated the flow inside the passages of a turbomachine as relatively simple flowfields. Even when we have tried to include the first-order effects of throughflow three-dimensionality, we have restricted our models and analyses to inviscid flows, and simply pointed out some of the outstanding differences between estimated and measured flow behavior. However, to try to achieve a fairly realistic view of turbomachinery flows, or to attempt significant improvement in flow modeling, one must begin to include some of the real complexities. The flows may be unsteady, three-dimensional, highly viscous, and frequently separated, and the machines are run at significantly off-design conditions with onset of stall and loss of stability.

## 10.2 FREESTREAM TURBULENCE INTENSITY

One of the more vexing characteristics of turbomachinery flows is the existence of strong levels of turbulence intensity imbedded in the core flow. This turbulence results from blade, vane, or end-wall boundary layer development, wake shedding, or the influence of less than ideal inflows with obstructions, bends, or other interferences. In terms of classical boundary layer considerations (Schlichting, 1979), the instability development length or distance between the first boundary layer instability and the point of completed transition to turbulent flow decreases remarkably in the presence of high turbulence levels (Granville, 1953). This behavior is illustrated here in a figure adapted from Schlichting (1979). The instability point is characterized by the magnitude of the Reynolds Number based on the momentum deficit thickness, $Re_\theta$. The fully turbulent condition is similarly set at the point where the transition has occurred. As seen in Figure 10.1, this length can be reduced by an order of magnitude with freestream turbulence levels as low as $T_u = 2\%$, where

$$T_u = 100\left(u'^{2}/U^2\right)^{1/2} \tag{10.1}$$

and $u'$ is the random, unsteady velocity component of the turbulence. Influence of this same phenomenon on the convective heat transfer and skin friction behavior is predictably important (Kestin et al., 1961 in Schlichting, 1979), and it is capable of causing orders of magnitude changes in heat transfer at values of freestream turbulence intensity from about 2.5%. To place these numbers in the context of prediction or analysis of flow in turbomachines, measurement of turbulence intensity in a

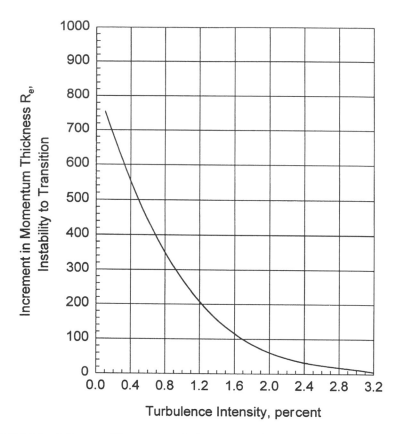

**FIGURE 10.1**   Influence of freestream turbulence on transition (after Schubauer and Scramstad as quoted in Schlichting, 1979).

compressor environment ranges typically from 2.5% at the inlet station to levels near 20% at stations downstream of blade and vane row elements (Wisler et al., 1987). This seems to imply that classical considerations of laminar flow instability and gradual transition may not provide an accurate assessment of what is going on inside the typical turbomachine.

An extensive review and critical study of laminar to turbulent transition in turbomachines was carried out by Robert Mayle as the 1990 Turbomachinery Fellow Award Paper for the American Society of Mechanical Engineers. Mayle ably traces some of the history of the earlier efforts in this field and summarizes the manner in which transition can be expected to occur. As concerns "natural transition," outlined above in terms of stability and development, little if any of these results are relevant to flow in a gas turbine engine or other type of turbomachine. Typically, transition in a turbomachine will occur at values of $Re_\theta$ an order of magnitude smaller than found in the classical smooth-flow results. In typically high levels of freestream turbulence, the process of instability, vortex formation, and growth can be replaced with rapid formation of turbulent spots, triggered or created directly by the freestream turbulence. For cases of practical interest here, this "bypass" of the natural or normal process of development has narrowed the range of phenomena to be considered to spot formation and subsequent development. Further, since blade and vane flows are frequently characterized by large pressure gradients and low Reynolds numbers, the occurrence of laminar separation can also lead to a short-cutting of the transition process. Typically, the transition may take place in a free shear layer above a "bubble" of separation, or the "bubble" may reattach the flow to the solid surface as a turbulent boundary layer. Recalling that the flow in the machine is usually unsteady as well, due to the motion of imbedded wakes shed by blade, vane, and inlet elements in the upstream flow, the boundary layer processes are clearly far from simple. Work continues in the field with Mayle (1990; 1991) and many other investigators active in the field (Addison et al., 1992; Wittig et al., 1988; and Dullenkoph, 1994).

## 10.3  SECONDARY AND THREE-DIMENSIONAL FLOW EFFECTS

As was discovered fairly early in the development of turbomachinery technology, the flow in and around the rotating elements of a machine is characterized by complex secondary influences. These effects are associated with blade-to-blade and radial pressure gradients, centrifugal force effects, spanwise flows, and flows through the seals and clearance gaps of the machines. Johnsen and Bullock (1965) present an early review of these messy effects and provide a clear discussion of the flows along with contemporary work by Smith (1954) and Hanson and Herzig (1953).

These early investigations confirmed the presence of "passage vortex" flows driven by the roll-up motion forced on the low momentum end-wall boundary layer fluid by the basic flow turning process and blade-to-blade pressure gradient. The presence of spanwise radial flow of vane boundary layers (inward flow due to pressure gradients) and blade boundary layer fluid (outward flow due to centrifugation) showed even higher levels of complexity. End-wall effects where a clearance exists between vane or blade and end-wall can lead to the roll-up of additional vortices in the flow passages. Figure 10.2a shows kinetic energy loss contours in a blade flow passage for an axial compressor downstream of the rotor. Build-up of centrifuged boundary layer fluid and tip flow effects as well as passage vortex formation near the inner end-wall can be seen clearly. Figure 10.2b shows a somewhat similar effect in a centrifugal blower passage (Wright et al., 1984c), which includes the influence of impeller inlet gap. In both cases, the basic flow is significantly perturbed by the "secondary effects."

Over the past 30 years, intensive investigation of these secondary flow phenomena has continued. Improved measurement and flow tracing techniques (Lakshminarayana and Horlock, 1965; Denton and Usui, 1981; Moore and Smith, 1984; Gallimore and Cumptsy, 1986) have contributed to the basic understanding of these flows. The concurrent modeling and calculational work had led until recently to two major concepts of how to handle the problem.

Adkins and Smith (1982), and more recently Wisler, Bauer, and Okiishi (1987), have concentrated on the convective models of secondary flow to provide explanation and modeling. On the other hand, Gallimore and Cumptsy (1986a, 1986b) had pursued equally successfully a model based on the random, turbulent diffusion process as the dominant mechanism of spanwise mixing. The two camps appeared

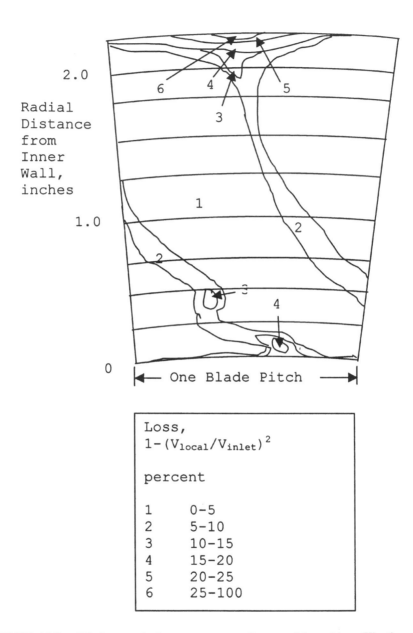

**FIGURE 10.2a**  Discharge velocity measurements for an axial machine. (Kinetic energy deficit contours, from Johnsen and Bullock, 1965.)

to be rather different and neither was willing to accept the notions of the other as to correctness. In a later paper, Wisler (Wisler, Bauer, and Okiishi, 1987) published something of a landmark paper, with commentaries by Gallimore and Cumptsy, L.H. Smith, Jr., G.J. Walker, B. Lakshninarayana, and others, which appears to have reconciled the two views as being components of the same problem and solution.

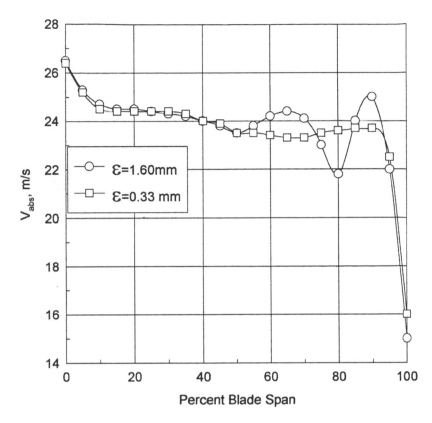

**FIGURE 10.2b**   Absolute discharge velocity from a centrifugal impeller with large and small inlet clearance. (Wright, 1984c).

It seems that either convective secondary flow or diffusive turbulent mixing can dominate the mixing processes, depending on the region of the passage. That is, an adequate calculational model will have to include both mechanisms.

   More recently, Leylek and Wisler (1992) have re-examined the physical measurements and flow-tracing data in terms of the results of extensive CFD work with detailed three-dimensional Navier-Stokes solution using high-order turbulence modeling. The results (and those of Li and Cumptsy, 1991) are extensive, and continue to support the view that both mechanisms of transport and mixing are important to a clear understanding of fully viscous, three-dimensional passage flows.

## 10.4   AXIAL CASCADE CORRECTIONS FOR LOW REYNOLDS NUMBER

A considerable amount of progress has been made in the development of procedures for correctly designing and manufacturing axial flow machines. One area of design procedures that has, until recently, been somewhat overlooked in the open literature is the effect that a very low Reynolds Number has on the performance of moderately loaded axial flow equipment. The term "low Reynolds Number" describes the conditions of the flow in which the Reynolds Number lies within the range of $10^3$ to $10^5$, where the Reynolds Number here is the chordwise value, $Re_c = W_1 C/\nu$. The effect of the low Reynolds Number on the flow is seen as the formation of a separation bubble in the boundary layer at a certain critical Reynolds Number. A separation bubble is a detachment of the boundary layer from the blade and is an area of swirling flow. Separation bubbles can form when the flow is in transition from laminar to turbulent flow. This effect is essentially undesirable in the design of axial flow equipment in that the separation bubble may cause performance to severely degrade or may lead to stall. These phenomena have been dealt with comprehensively in the Gas Turbine Scholar paper by Mayle (1990), who also provides an excellent review bibliography. Work in the area continues (Mayle, 1991; Gostelow, 1995).

Investigations have been made into this unique phenomenon by Roberts (1975, 1975a), Citavy and Jilek (1990), Cebeci (1989), Pfenninger and Vemura (1990), O'Meara and Mueller (1987), and Schmidt (1989). Although there have been methods presented for predicting the flow under these conditions, substantial disagreement on particular aspects of the flow remains. Also, many of these treatments require rather complex methods of integration and interpolation and are not very straightforward in their use. Although other methods are more straightforward in their use, they sacrifice accuracy in the end results. What is missing is a method that will yield a fairly accurate description of these low Reynolds Number flows without utilizing ponderous, complex routines. Wang (1993) studied such a model and used the results that have been presented by Roberts, and Citavy and Jilek. This information was used to create a less complicated procedure for determining the losses and deviation angles caused by low Reynolds Number separation bubbles, and it can be used in preliminary design routines. Figure 10.3, adapted from the results of Roberts (1975), illustrates the low Reynolds Number behavior of the flow, with fluid turning, $\theta_{fl}$, decreasing rapidly with decreasing $Re_c$ and the total pressure loss coefficient, $\omega_1$, increasing rapidly with decreasing $Re_c$.

The aim of the study of Wang was to develop a simple procedure for the prediction of the design characteristics of low Reynolds Number flows through axial flow cascades to be used in preliminary design studies. The study includes development of loss estimation algorithms for such cascades. The procedure provides a method in which complex computations are kept to a minimum at a reasonable cost to accuracy, and the results, which were checked against the results of available test data, can be incorporated into existing axial fan design procedures. The prediction of the losses involves many parameters, including camber angle, inlet flow angle, outlet flow angle, solidity, Reynolds Number, deviation angle, angle of incidence, angle of attack, and turbulence intensity. In order to simplify the process by which an approximation of the losses can be achieved, several of the values are held constant or disregarded for this method of prediction. The angle of incidence is designed to be near zero in most preliminary design routines and should not strongly affect the results. Turbulence intensity was considered negligible, since in most of the test data used it was less than 5%. Reynolds Number, inlet flow angle, outlet flow angle, camber angle, solidity, and deviation angle are either known values or can be determined in the algorithm.

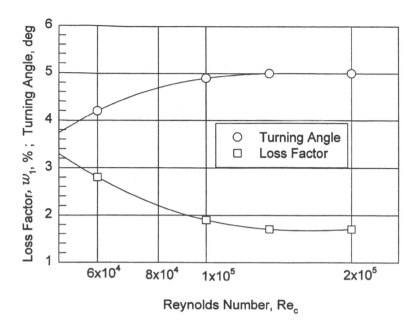

**FIGURE 10.3**  Flow turning and loss at low Reynolds numbers. (Roberts, 1975.)

To approximate the amount of losses incurred in an axial flow cascade operating in a low Reynolds Number regime, the methods provided by Lieblein (1959) and Roberts were used. As stated earlier, the method presented by Lieblein was not developed for low Reynolds Number loss prediction, but it can be used to determine the loss prior to laminar separation or the loss at the point of bursting of the laminar separation bubble. This can be done if it is assumed that the losses up to the point of bursting are not heavily dependent on the Reynolds Number. As seen in Figure 10.3, the losses depend weakly on the Reynolds Number until bursting occurs, and then there is a sharp increase in the loss curve at the point of bursting where the losses become very heavily dependent on the Reynolds Number.

The loss values up to the point of bursting can be approximated as constant, and the loss at the bursting point can be determined approximately from the method provided by Lieblein (1959). This loss is based on the momentum deficit thickness results as a function of the local diffusion factor as discussed in Chapter 8. The loss factor is calculated from a curve-fit equation

$$\omega_{1,b} = \omega_1 = \left(\frac{2\sigma}{\cos\beta_2}\right)\left(\frac{\cos\beta_1}{\cos\beta_2}\right)^2 \times \left[0.005 + 0.0049(D_L) + 0.2491(D_L)^5\right] \quad \textbf{(10.2)}$$

where

$$D_L = 1.0 - \left(\frac{\cos\beta_1}{\cos\beta_2}\right) + \left(\frac{\cos^2\beta_1}{2\sigma}\right)(\tan\beta_1 - \tan\beta_2) \quad \textbf{(10.3)}$$

To go beyond this point and begin to modify the loss values associated with a burst bubble, Wang relied on prior work (Roberts, 1975). The method provided by Roberts to be used here can be summarized in the following equations. He wrote the equation for prediction of losses beyond the burst as an increment to the value of loss just prior to burst, written as

$$\omega_{1sb} = K_1(\Phi TS)\Delta RX + \omega_{1,b} \tag{10.4}$$

where $K_1(\Phi TS)\Delta RX$ is the incremental function. $K_1$ is a constant determined by Roberts to be 0.016. $\Phi$ (a function of airfoil camber, $\phi_c$), T (a function of airfoil thickness, t/c), and S (a function of solidity, $\sigma$) are non-dimensional variables defined as follows:

$$\Phi = \frac{\phi_c}{\phi_{cref}}, \quad \phi_{cref} = 10° \quad (\phi_c \geq 5°)$$

$$T = \frac{\left(\dfrac{t}{c}\right)}{\left(\dfrac{t}{c}\right)_{ref}}, \quad \left(\frac{t}{c}\right)_{ref} = 0.1 \tag{10.5}$$

$$S = \frac{\sigma_{ref}}{\sigma}, \quad \sigma_{ref} = 1.0$$

$\Delta RX$ was given as the following equation in terms of the "deviation factor" DF

$$\Delta RX \cong 900\,\Delta DF \tag{10.6}$$

$\Delta DF$ is a change in the deviation factor defined by $DF_{sb}$ (sub-bursting or just prior to burst) and $DF_b$, (the value after bursting), so that

$$\Delta DF = DF_{sb} - DF_b \tag{10.7}$$

This equation was generated through a curve fit of Roberts' data using linear curve fitting, in keeping with earlier curve fits. $DF_{sb}$ and $DF_b$ are defined as follows

$$DF_{sb} = \frac{\delta_{sb}}{\left(\dfrac{\phi_c}{\sigma}\right)} \tag{10.8}$$

and

$$DF_b = \frac{\delta_b}{\left(\dfrac{\phi_c}{\sigma}\right)} \tag{10.9}$$

From all of the available data, $\delta$ becomes increasingly dependent on Re for values below about $10^5$. To express this effect in a simple function, $\Delta DF$ was rewritten as:

$$\Delta DF = DF_{sb} - DF_b = \frac{\delta_{sb}}{\left(\frac{\phi_c}{\sigma}\right)} - \frac{\delta_b}{\left(\frac{\phi_c}{\sigma}\right)} = \left(\frac{\delta_{sb}}{\delta_b} - 1\right)\left(\frac{\delta_b}{\left(\frac{\phi_c}{\sigma}\right)}\right) \qquad (10.10)$$

The term $\delta_b/(\phi_c/\sigma)$ can be rewritten using Howell's simple formulation as $m/\sigma^{b-1} + \delta_o\sigma/\phi_c$. For moderate camber and solidity, $\sigma^{b-1} \cong 1$ and $\delta_o\sigma/\phi_c \cong 0$. If a moderate inlet angle is chosen (around $\beta_1 = 50°$), then $m \cong 1/4$. That is, within the prevailing uncertainty, $\Delta DF$ may be approximated as

$$\Delta DF \cong \left(\frac{1}{4}\right)\left(\frac{\delta_{sb}}{\delta_b} - 1\right) \qquad (10.11)$$

Then, directing attention to $\omega_{1sb}$ and rewriting as $\Delta\omega_1 = \omega_{1sb} - \omega_{1,b}$,

$$\Delta\omega_1 = K_1 TS\Phi\left(\frac{900}{4}\right)\left(\frac{\delta_{sb}}{\delta_b} - 1\right) \qquad (10.12)$$

If $K_1(900/4)$ is replaced by a new constant $K_f$ and $\Phi$ is replaced by $C_{lo}$, T by t/c and S by $1/\sigma$, this result becomes

$$\Delta\omega_1 = K_f C_{lo}\left(\frac{t}{c}\right)\frac{\left(\frac{\delta_{sb}}{\delta_b} - 1\right)}{\sigma} \qquad (10.13)$$

$K_f$ will be provided empirically, and $\delta_{sb}/\delta_b$ will be curve fitted to data as well. Using the data of Citavy ($C_{lo} = 1.2$), $\delta_{sb}/\delta_b$ can be fit in simple form as

$$\frac{\delta_{sb}}{\delta_o} = \left(\frac{10^6}{Re_c}\right)^{0.2} \qquad (10.14)$$

as seen in Figure 10.4. Here, a curve fit is shown for the normalized deviation angle from the data of Citavy and Jilek (1990). Deviation angle data are also shown for the work of Roberts (1976) for comparison. The curve fit is shown with the measured deviation angles presented by Citavy and Jilek in Figure 10.4. Deviation angles from Roberts' data are included for comparison. The curve fit was constrained to conservatively estimate $\delta_{sb}/\delta_b$ from Citavy's data with a forced fit at Re = $10^6$.

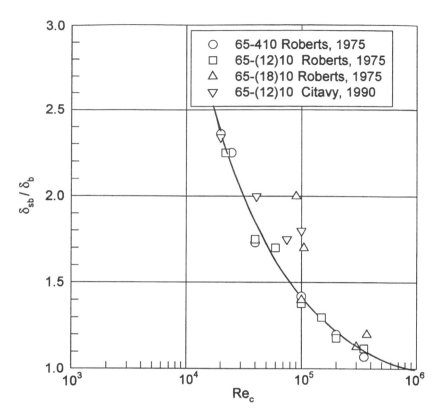

**FIGURE 10.4**  Comparison and correlation of deviation angles. (Data from Roberts (1975) and Citavy and Jilek (1990).)

A single low-camber data point from Roberts' data, seen in Figure 10.5, at $C_{lo} = 0.4$, is used to evaluate $K_f$ by matching $\Delta\omega_1$ at $Re = 10^5$. The resulting $K_f$ value is $K_f \cong 2$, so that

$$\Delta\omega_1 \cong 2 \; C_{lo}\left(\frac{t}{c}\right)\frac{\left(\left(\frac{10^6}{Re}\right)^{0.2} - 1\right)}{\sigma} \qquad (10.15)$$

This becomes the final approximation for loss for low Reynolds numbers, $Re_c \leq 10^6$. The previous equations form the basis of predicting both the losses and the deviation angles occurring in axial flow cascades at low Reynolds numbers. The algorithms can be tested against other information published by Roberts, Citavy and Jilek, and Johnsen and Bullock (NASA SP-36) to establish the consistency and accuracy of the results.

Figures 10.5, 10.6 and 10.7 are comparisons of the correlation embodied in the low Reynolds Number algorithms with the near-design loss data from Roberts' tests

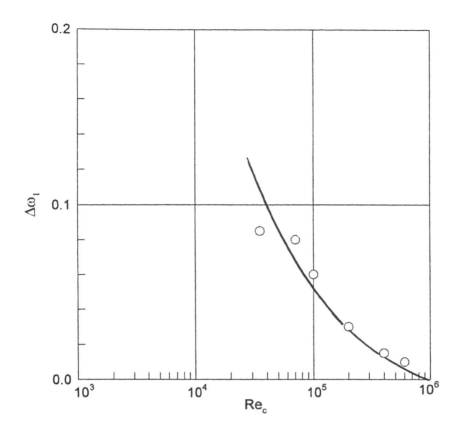

**FIGURE 10.5**   Comparison to Roberts' NACA 65-410 airfoil data (from Roberts, 1975).

with $C_{lo}$ = 0.4, 1.2, and 1.8. The results for $C_{lo}$ = 0.4 are excellent, of course, since the value of $\Delta\omega_1$ at Re = $10^5$ was used to size the constant $K_f$. Still, the complete data set is nicely simulated by the correlation. Figure 10.6, for $C_{lo}$ = 1.2, shows reasonable agreement over most of the data range. Roberts' data show a somewhat more complex form with perhaps an inflection with changing Reynolds Number compared to the very simple form chosen for the correlation (the "leveling off behavior" described by Johnsen and Bullock). Figure 10.7, for $C_{lo}$ = 1.8, shows a general overestimation for the correlation—as much as 30% at the moderate test value of Rec.

Figure 10.8 is a comparison of the correlation with the data of Citavy and Jilek at $C_{lo}$ = 1.2. Agreement is only fair with significant over-estimation of loss down to values of Re near $10^4$, accompanied by a nearly singular increase in $\Delta\omega_1$, at the lowest test value of Re. Finally, in Figure 10.9, the British C4 data, quoted from NASA SP-36 (Johnsen and Bullock, 1965), are plotted along with the correlation presented here. Overestimation prevails for all but the lowest Reynolds numbers with another sudden increase in $\Delta\omega_1$, at the lowest test values of $Re_c$.

Since Roberts' information did not explore losses at extremely low Reynolds numbers, and Citavy and Jilek show behavior that was unexpected, the results of

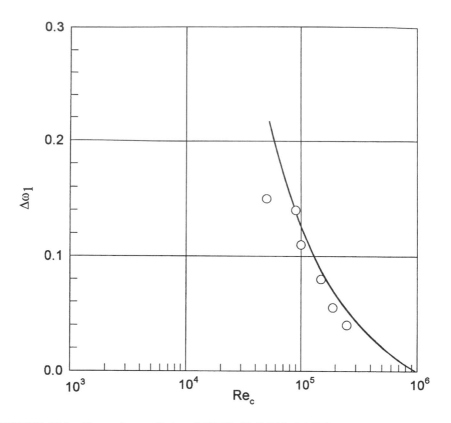

**FIGURE 10.6**    Comparison to Roberts' NACA 65-(12)10 airfoil data.

this algorithm must be taken only as a rough approximation at the lowest values of $Re_c$ ($Re_c \leq 10^4$). Although the loss estimation becomes doubtful at very low Reynolds numbers, the prediction of deviation angle is reasonably good, particularly for moderate blade loading.

Airfoils can be designed to handle flows that are within the low Reynolds Number regime. Studies performed by Pfenninger and Vemura (1990) address the problems of optimizing the design of low Reynolds Number airfoils. The studies consisted of low Reynolds Number airfoils ASM-LRN-003 and ASM-LRN-007, which were designed for high section lift-to-drag ratios. These investigations demonstrate that an airfoil can be designed to operate under low Reynolds Number conditions and still provide confidence in the performance of the airfoils with optimum control of transition.

These successful studies with the ASM airfoils employed a redesigned airfoil whose leading edge region was modified to significantly change the pressure distribution on the suction surface. Figure 10.10 qualitatively illustrates both the airfoil shape and the change in pressure. The airfoils of Pfenninger and Vemura were compared to the baseline performance and shape of the Eppler 387 design (Eppler and Somers, 1980). The fundamental result is that the camber line in the nose region

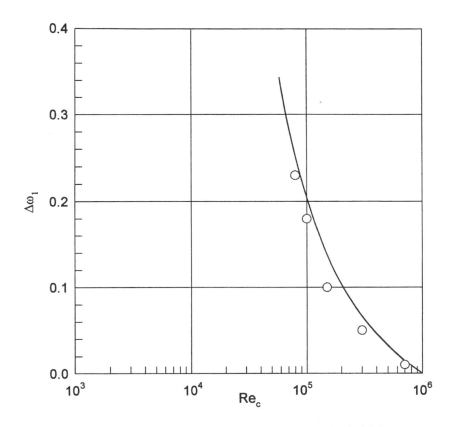

**FIGURE 10.7**   Comparison to Roberts' data for the NACA 65-(18)10 airfoil.

is modified to create a locally negative leading edge incidence of a few degrees. This adequately reduces the adverse pressure gradient on the suction surface responsible for the formation and subsequent bursting of the laminar separation bubble.

These results can be incorporated into design procedures using conventional cascade algorithms by simply adding a few degrees of camber to an existing airfoil selection and reducing the blade pitch angle slightly to avoid excess fluid turning. The resulting configuration should achieve a negative leading edge incidence sufficient to suppress bubble formation or bursting. This concept has not been verified experimentally.

Recommendations for future algorithms created to predict the increased losses and deviation angles caused by low Reynolds Number separation bubbles should include some correlation that deals with the angle of attack in the algorithm. Also, if more information at low Reynolds Number becomes available, a better correlation for the deviation angle and loss in the very low Reynolds Number regime should be possible.

The concepts of Pfenninger and Vemura, as embodied in the suggestion for using extra camber at the leading edge, accompanied by an equivalent reduction in blade pitch, should be investigated. The question of whether a simple change in conven-

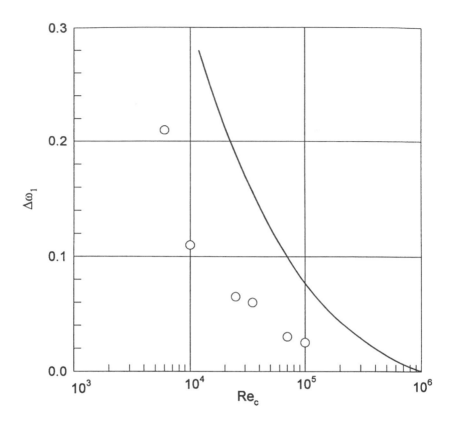

**FIGURE 10.8** Comparison to Citavy's data for the NACA 65-(12)10 airfoil.

tional cascade parameters can be used to alleviate the damaging effects of low Reynolds number on performance would be answered.

## 10.5 ONSET OF STALL AND LOSS OF STABILITY

As viewed for the operations of pumping machines in earlier chapters, the steady-state functioning of the machine is determined by the matching of the fan, pump, blower, or compressor characteristic to the corresponding point on the system resistance curve. The flow through the pumping machine is equal to the flow through the resistance system, and the pressure rise of the pump is equal to the pressure drop in the system. Such a system is in equilibrium and is assumed to be stable, as considered for the simple picture presented in Chapter 1. However, if one is dealing with the limitations of real pumping systems, then one needs to consider the influence of a small disturbance imposed on the fan or other machine.

If the disturbance is a small increase in flow rate and the fan and system return to the original unperturbed flow point, the system and fan are in stable equilibrium. If the flow rate, when disturbed, continues to change, the system is said to be statically unstable. This is illustrated in Figure 10.11 (adapted from Greitzer's work

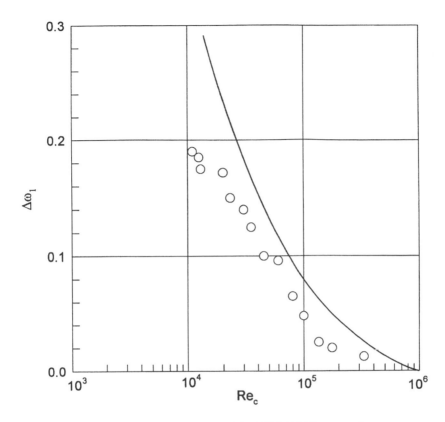

**FIGURE 10.9**   Comparison to the NASA SP-36 10C4/40P40 Data at –4°

(Greitzer, 1981)). Here, the point of operation prior to the disturbance is at point A. The increase of flow from A will cause the pressure drop in the system to be greater than it was, leading to a slowing down of the flow in the system and a subsequent decrease in flow. That is, the classic return toward the unperturbed condition defines the requirement for stability. If the system were operating at point B on the pump characteristic with a higher resistance system and was again disturbed with a small increase in flow, the pump would generate an increase in pressure rise greater than the increase in system resistance, leading to a further increase in flow. The system does not tend to return to its unperturbed equilibrium and, by definition, is said to be unstable. At point B, the slope of the pump characteristic curve is steeper, or more positive, than the slope of the system resistance curve (unstable flow), while at A, the slope of the pump curve is much less than that of the system resistance (stable). Thus, we can establish the simplest criterion for pumping stability or static stability of the pumping system. Unfortunately, one must also be very wary of conditions that can lead to dynamic instability or the growth of oscillation of the flow rate around the initial setting. In Figure 10.11, this condition can prevail when the pump is trying to operate between points A and B or very nearly at the peak pressure rise of the pump. A simple pumping system involving a closed volume of

Basic Airfoil

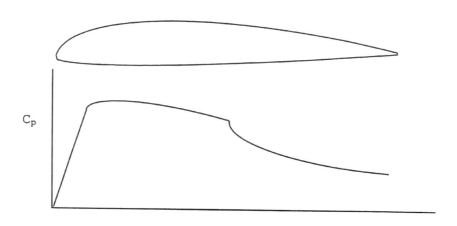

Recambered Nose Region

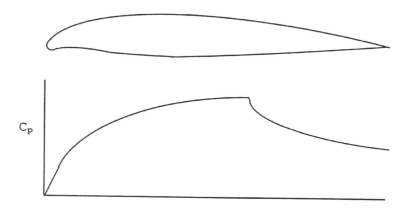

**FIGURE 10.10**    Airfoil shapes and pressure distributions for low Reynolds Number design.

air, $V_a$, a flow length, $L$ and a flow area $A$, can be analyzed for both static and dynamic stability criteria (Greitzer, 1981) to show that stability requires that

$$\left(\frac{dp_p}{dm'}\right) < \left(\frac{dp_{sys}}{dm'}\right) \qquad \text{(static)} \qquad \textbf{(10.16)}$$

and

$$\left(\frac{dp_p}{dm'}\right) < \frac{K_s}{\left(\dfrac{dp_{sys}}{dm'}\right)} \qquad \text{(dynamic)} \qquad \textbf{(10.17)}$$

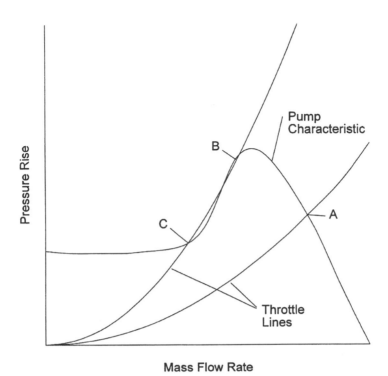

**FIGURE 10.11** Illustration of static and dynamic instability conditions for a pumping machine. (Greitzer, 1996.)

$K_s$ is a system property defined as

$$K_s = \left( \frac{\gamma p_1 L}{\rho V_a A} \right) \tag{10.18}$$

where $\gamma$ is the ratio of specific heats (see Chapter 1). $K_s$ is positive definite and may be quite small for systems of large volume, $V_a$, or short flowpath length, L. This implies that $(dp_p/dm')$ will be slightly positive but can be and often is near the zero-slope condition at the onset of dynamic instability. This condition is shown as point C in Figure 10.11. It is clearly less restrictive than the criterion for static instability and more readily encountered in a system with increased or increasing resistance. Greitzer (1996) provides a very simple and clear definition of the underlying process leading to dynamic instablity in terms of energy or power required to sustain oscillation. He examines the product of the increments to performance, $\Delta m'$ and $\Delta p$, the perturbations in mass flow and pressure rise. There are, as stated, two possible conditions of interest: the case of positive pump curve slope and negative pump curve slope. Assuming quasi-steady behavior in the pump, for the first case the change in m' is positive when the change in $\Delta p$ is positive so that the product is always positive. Energy is being added by the pump to the flow to drive the oscillation. For the negative pump slope, the change in m′ is always of opposite sign to the change in Δp, the product is always negative, and the energy is being removed from the flow by a dissipation process, damping out the oscillation.

The required positive, or at least zero, slope condition on the pump characteristic is normally accompanied by a stalled flow condition in the flow passages of the machine, as discussed in terms of loading limits in Chapter 6. Very frequently, this stall or flow separation is of the rotating type of stall where one or more flow passages or groups of flow passages in the turbomachine experience separation. The flow leaves the blade surfaces and/or the passage end-walls, and these stall "cells" move or rotate relative to the impeller itself. Rotating stalls have been observed and documented in the turbomachinery technical literature (Greitzer, 1980; O'Brien et al., 1980; Laguier, 1980; Wormley et al., 1981; Goldschmied et al., 1981; Madhavan and Wright, 1985) in both axial and centrifugal pumping machines. If the initial flow separation on a blade surface occurs on the suction surface, as is the case for flow rates below the design flow, then the stalled cell has a natural tendency to move to the next suction surface in a direction opposite to the direction of rotation of the impeller. Figure 10.12 shows what is happening as flow is diverted to the passage behind the first-stalled passage and leading edge incidence is increased on the blade ahead and decreased on the blade behind. This forward motion of the stall cell or cells is the form taken by classical rotating stall. Cells can propagate in the opposite direction when flow is excessively above the design flow rate, or when the flow has been deliberately pre-swirled in the direction of motion of the impeller (as in the use of pre-spin control vanes to reduce blower performance) as shown by Madhavan and Wright (1985).

The speed of movement or the frequency observed in the stationary frame of reference differs from the rotating speed or frequency by the relative motion of the cells. When cells of rotating stall are fully established, the pressure pulsation observed in the stationary frame occurs at two-thirds of running speed for the classical suction surface stall and at four-thirds of running speed for the pressure surface stall at high flow or pre-swirled inflow. Figure 10.13, from the centrifugal fan experiments, shows the variation of pressure pulsation frequency as a function of flow rate variation about the mean flow.

The existence of these rotating stall cells is not necessarily a great hazard to operation of the machine by themselves. Pressure pulsations are more nearly a low-frequency acoustical problem than a structural danger to the fan and its system, with pressure amplitudes of the pulsation ranging to about 20 to 40% of static pressure rise. If the pulsation frequency coincides with a natural acoustic frequency of the system, a resonance condition may arise. Problems can arise anyway, because the occurrence of rotating stall may trigger an overall unstable system response known as surge. Based on the diffuser data of Sovran and Klomp (1975), as discussed in Chapter 7, one can estimate the maximum static pressure increase in a cascade as a function of geometry. One can use $\psi_{s\text{-max}}$ as the stalling value of the cascade and $L/w_1$ as the geometric parameter (see Figure 7.36), where $\psi$ is defined as

$$\psi_{s-max} = \frac{\Delta p_s}{\left(\rho U_2{}^2\right)} \tag{10.19}$$

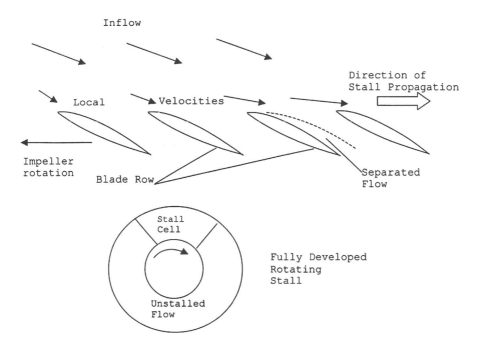

**FIGURE 10.12**  Stall cell formation and propagation.

This concept is reinforced in an earlier correlation by Koch (1981) where he estimates $\psi_{s\text{-max}}$ as a function of the length ratio $L/g_2$. L is essentially the blade chord, and $g_2$ is the "staggered pitch," or $g_2 = s \cos\beta_2$ (we had earlier used simply s). Koch's correlation is shown in Figure 10.14 (without the supporting experimental data) and shows that one can expect to achieve, at most, a pressure rise coefficient of perhaps 0.6 or less before stall. As pointed out by Greitzer (1981), Wright and Madhavan (1984a), and Longley and Greitzer (1992), the onset of stall in the impellers of both axial and centrifugal machines can be significantly influenced by the existence of flow distortion or non-uniformity in the impeller inlet. Most distortions encountered in practice include radially varying or circumferentially varying steady-state inflows and unsteadiness in the fixed coordinates as well. An example, for a centrifugal impeller tested by Wright, Madhavan, and DiRe (1984), shows the level of reduction in stall margin that can result from poor installation or inattention to inlet conditions. As seen in Figure 10.15, the stall pressure rise value can be reduced by 5 to 10% with a significant variation of velocity magnitude into the fan inlet. Here, $V_{rms}$ is based on a normalized standard deviation of the inlet velocity profile, $V_i$, compared to the mean value, $V_m$, as

$$V_{rms} = \sum \frac{\left[\dfrac{\left(V_i - V_m\right)^2}{n}\right]^{1/2}}{V_m} \tag{10.20}$$

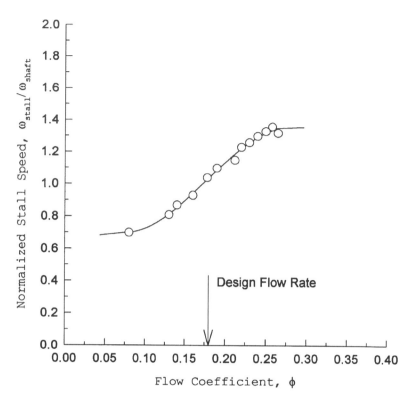

**FIGURE 10.13** Stall cell speed and pressure pulsation frequency. (Madhavan and Wright, 1984a.)

The pressure is normalized by the stall pressure rise with a uniform inlet flow as $\psi' = \psi_s/\psi_{s\text{-}o}$.

The great risk to performance and noise, or even the structural integrity of the turbomachine or its system, is the possibility that operation in stalled flow or with rotating stall cells can trigger the major excursions in pressure rise and flow rate known as surge. The range of pressure generated in surge can be as high as 50% of design values, and flow excursions may include significant levels of total flow reversal through the machine. Very high levels of stress in both machine and system can readily lead to severe damage or destruction of the equipment. In an analysis of resonator compression system models, Greitzer (1976) (employing modeling techniques used by Emmons (1955)) defined a parameter on which the proclivity of the entry into surge conditions depends. Defining

$$B = \left(\frac{U}{2a}\right)\left(\frac{V_c}{A_c L_c}\right)^{1/2} \qquad (10.21)$$

where $V_c$, $A_c$, and $L_c$ describe the compression system geometry in terms of an external plenum volume, and flow cross-sectional area and length. U is the impeller

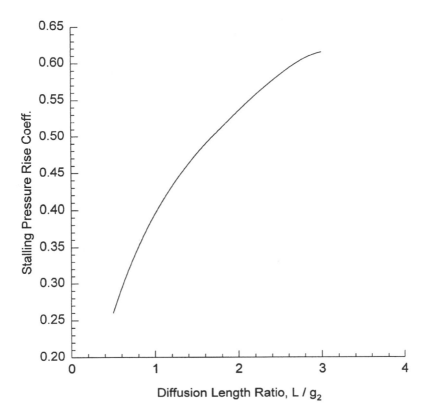

**FIGURE 10.14**   Koch's stalling pressure rise correlation.

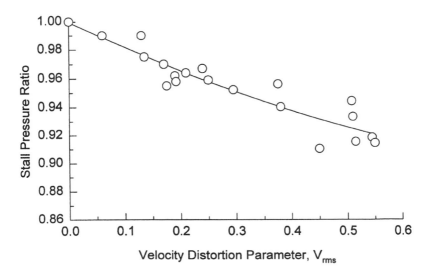

**FIGURE 10.15**   Influence of inlet flow distortion on stall margin.

speed, and a is the speed of sound. For a given system layout, there is a critical value of B, $B_{crit}$, above which a compression system will exhibit large, oscillating excursions in flow rate and pressure rise—surge. Below the critical value, the system will experience a transient change to a steady rotating stall operating condition with poor performance and high noise levels, but not the wild behavior of surge.

Determination of the critical value of B can be done analytically or experimentally (see Greitzer, 1981) for a particular system layout. Clearly, when an existing compression system is experiencing surge problems and strong oscillation in performance, B can be used to determine an approach to fix the problem. Knowing $B_{crit}$, one can modify the geometry of the system or the speed of the turbomachine to reduce B sufficiently to pull the system out of surge.

## 10.6  COMPUTATIONAL FLUID DYNAMICS IN TURBOMACHINERY

The ultimate aim for the work on computational fluid dynamics (CFD) for turbomachinery has been to develop comprehensive numerical systems to simulate the detailed flowfield in a turbomachine. The work and results of the very many investigators involved cover the range from improved preliminary design and layout to fully developed schemes that rigorously solve the discretized fully three-dimensional, viscous, compressible, and often unsteady flow. In the preliminary design area, the work of McFarland (1993) and Miller, Oliver, and Miller (1996) are fairly typical of such efforts. McFarland developed an overall framework to size the geometric layout of multi-staged machines. The method retains a linearized formulation of the governing equations to gain compactness and speed for this initial design step. The work of Miller et al. is an excellent example of the geometrically oriented design approach that provides, as a final result, a complete physical surface description for use in design, analysis, and manufacturing applications. Conceptually, the work is an extension of the earlier work of Krain (1984) mentioned in Chapter 9. These methods, and others like them, provide the starting point for the very detailed flowfield analysis that can be used to verify or modify the selected geometry. The design layout is thus improved or optimized in an iterative process to increase performance, boost efficiency, or achieve better off-design performance and stability.

The overall schemes needed to approach a full numerical simulation of the flowfields in turbomachines or their components—from pumps to aircraft engines—will generally employ available, well-established CFD codes (e.g., Dawes, 1988) as an engineering tool integrated within the overall framework (Casey, 1994). The more rigorous treatment of these flowfields through CFD allows an experienced designer to develop "better engineered and more clearly understood designs" at low engineering cost. The approach outlined by Casey covers selection of machine types, overall sizing, mean-line analysis, experimental correlations, and other preliminary design steps as a precurser to assessment by CFD analysis and subsequent iterative refinement of a design. One such scheme is described by Topp, Myers, and Delaney (1995), with emphasis on development of graphical user interface software to simplify user input and overall handling of results. Blech et al. (1992) outline a philosophy for constructing a simulation system that is capable of "zooming" from an overall analysis using relatively simplistic lumped parameter analysis to successive levels of two- or three-dimensional analysis on an interactive basis. Although these concepts have been limited in application because of computational speed capabilities, continued improvement in hardware and application of parallel processing will steadily alleviate the difficulties.

The task of the turbomachinery designer or analyst is to successfully employ the maturing methodology and related computer codes available for the rigorous flowfield analysis. An excellent review and technical background for these techniques is given in Chapters 18 and 19 in the *Handbook of Fluid Dynamics and Fluid Mechanics* (Shetz and Fuhs, Eds., 1996); in Chapter 18, the hardware appropriate to CFD analysis is discussed in terms of architectural evolution, power and size, and graphics and networking. The requirements for fluid flow computation are discussed, and a historical perspective of the development of numerical flowfield solutions is presented. As pointed out by the authors (Stevens, Peterson, and Kutler, 1996), the factors still limiting computational capability, and thus defining the areas of required improvement and research, include improved physical modeling (such as for flow transition and turbulence), solution techniques (improved algorithms), increased computer speed and size, and improved data for the validation or qualification of computed results. Ultimately, the improvements needed will require significant progress in grid generation techniques and benchmark testing, and improvements in output data analysis and graphical display of results (Stevens, Peterson, and Kutler, 1996). The handbook (Shetz and Fuhs, Eds., 1996) supplies a detailed review of the key topics of CFD in Chapter 19, well beyond the scope intended for this chapter. Topics include finite difference formulations, finite element and finite volume formulations, grid generation (structured, unstructured and transformed), and the key topics of accuracy, convergence, and validation.

A very thorough and extensive treatment of the theory and application of CFD to turbomachinery is given in the book written by Lakshminarayana (1996) (see also the earlier work, Lakshminarayana, 1991). The material presented there ranges from a detailed review of the basic fluid mechanics through a presentation of the numerical techniques involved, and it considers those problems that are particular to the application of CFD to turbomachinery flowfields. Since such a thorough treatment of the subject is beyond the intended scope of both this book and this chapter, Lakshiminarayana's book is strongly recommended to the reader who expects to be using the capabilities of CFD in turbomachinery design and analysis.

The "essential ingredients" for good analysis are listed as (Lakshiminarayana, 1996) the proper use of the governing equations, the imposition of the necessary boundary conditions, adequate modeling of the discretized domain, relevant modeling of the flow turbulence, and the use of an appropriate numerical technique. As expected, clear emphasis is placed on the need for analytical calibration and experimental validation of the calculational process. The turbulence modeling must be reasonably capable of including the influence of end-wall flows, tip leakages, shed wakes, and other features typical of the flow fields in a turbomachine (see Section 10.3).

Lakshiminarayana (1966) also reviews the history and development of calculational techniques for turbomachinery flows over the past few decades. The reader is reminded that many of the methods developed earlier may find good application in advanced design and analysis. For example, the use of inviscid flow analysis, perhaps coupled with boundary layer calculations, can be particularly appropriate in the early design phases or in regions where the flow is substantially inviscid in behavior. Although these flows may be revisited in a later, more rigorous study using three-dimensional full Navier-Stokes methods, great reductions in time and expense can be achieved through the use of the more approximate technique at the appropriate time.

Finally, in his concluding remarks on computation of turbomachinery flows, Lakshiminarayana states that "...there is a need for calibration and validation" of any codes to be used for "production runs" in analysis and design. He states that, for the purpose of validation of computations, "Benchmark quality data are scarce in the area of turbomachinery. This issue should be addressed before complete confidence in these codes can be achieved."

As an example of application of turbomachinery CFD codes, one can examine the work reported by McFarland (1993) in a little more detail. McFarland was able to modify his original work on blade-to-blade solutions (McFarland, 1984) to include all of the blade rows of a multi-stage machine in a single flow solution. Calculations from the multi-stage, integral equation solution method were verified by comparing the CFD results to experimental data from the Large Rotating Rig at the United Technologies Research Center (Dring, Joslyn, and Blair, 1987). McFarland describes the geometry of the turbine test case as "a one and one half stage axial turbine" whose "vane and blade shapes were typical of 1980 designs." The geometry analyzed was the IGV, rotor blade, and OGV at the mean radial station ($r = r_m$). Typical results of the analysis are presented in Figure 10.16 for the IGV and Figure 10.17 for the turbine rotor blade, as an averaged pressure coefficient along the normalized surface length (x/C) of the pressure and suction surfaces. The averaging is performed over each blade in a given row. Here, the pressure coefficient is defined as $C_p = (p_{01} - p)/(\rho U_m^2/2)$. The comparison of the CFD results with the experimental data is quite good and was obtained with very low computer time and cost. This quickness and economy can allow the investigator to examine a wide range of design configurations in a specific problem, as well as justify the approximations of McFarland's flow model. However, he points out that "the lack of unsteady and viscous terms in the governing equations cause the method to inaccurately calculate variations in unsteady blade loading" (non-averaged results). The lack of a loss model also leads to an inaccuracy in total pressure, growing row by row. A user-specified adiabatic efficiency was to be added to the method to alleviate this shortcoming (McFarland, 1993). Although the trade-off for low run time has been at the expense of rigor in the solution, the method supplies a very useful tool for design.

Another example of CFD application in a complex turbomachine is furnished by the paper of Hathaway, Chriss, Wood and Strazisar (1993) on work done at the NASA Lewis Research Center. This very interesting example includes detailed measurements of the flow in the impeller passages of a low-speed centrifugal compressor (LSCC) and accompanying hot-wire probe, five port pressure probe,

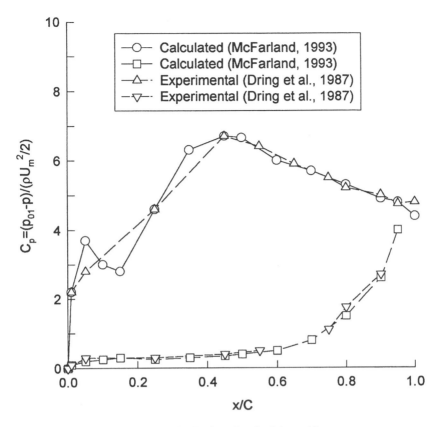

**FIGURE 10.16**   Surface pressure distributions for the inlet guide vane.

and laser velocity probe data of the compressor's internal flow field, in addition to a complete analysis of the flow using CFD.

The purpose of their studies was to provide detailed experimental data on the primary and secondary flow development in the internal passages of an unshrouded centrifugal compressor impeller. The complex curvature of these channels, coupled with strong rotational forces and the effects of clearance between the impeller and the stationary shroud, was expected to generate strong secondary flows capable of major disruption of the core flow in the channels. The goal was to achieve an understanding of these complicated phenomena that could lead to performance and efficiency improvements in the compressor through impeller redesign.

Earlier studies of these complex flows include Eckardt's laser velocimetry data in a high-speed shrouded impeller (Eckardt, 1976), and Krain's work (1988), and Krain and Hoffman's studies (1990) with laser anemometry data from backswept impellers. All of these studies showed strong secondary flows with low-momentum fluid migrations through the impeller channels. These earlier efforts provided focus for the work of Hathaway et al. for their investigation.

Examples of the level of detail seen in the flowfields, both experimentally and computationally, are shown in Figure 10.18 (adapted from Hathaway et al., 1993). The

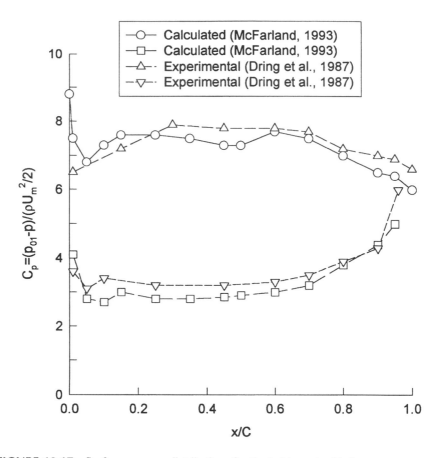

**FIGURE 10.17**  Surface pressure distributions for the turbine rotor blade.

flow near the blade channel inlet illustrates the nearly potential nature of the early flow development region. The flow near the exit of the impeller channel shows the dominance of viscous effects on the flow, with high velocities near the suction surface and a region of low-momentum flow nearer the pressure surface. In both cases, the agreement between computed and measured velocities is good, although the low velocities in the low-momentum region near the exit are somewhat more pronounced in the measured flow than in the computed flow. This well-executed simulation of the "real" flow provides the basis for running "calculational experiments" on the layout geometry. The objective of such an exercise is to try to modify the low-momentum regions in the blade channels to reduce loss and increase efficiency.

## 10.7   SUMMARY

The models and analytical methods used in earlier chapters have all, to some degree, sidestepped many of the difficulties that arise in real turbomachinery flowfields. Although the simpler methods seek to model the dominant physical phenomena in

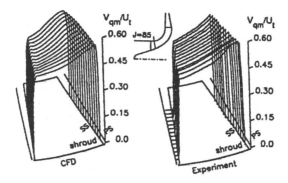

**(a) Flow Near the Channel Inlet**

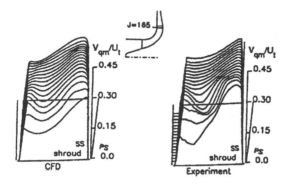

**(b) Flow Near the Channel Exit**

**FIGURE 10.18** Computed and measured velocities in the centrifugal compressor impeller. (Adapted from Hathaway et al., 1993.)

the flow, a search for additional rigor or more in-depth understanding will require that some of these problem areas be addressed. Several such topics are briefly introduced in this chapter and discussed.

The inherently high levels of freestream turbulence intensity can dominate the development and behavior of the boundary layers and other strongly viscous regions of flow. Transition from laminar to turbulent flow depends very strongly on the turbulence intensity, and the behavior in a turbomachine is usually quite different from that in a typical external flowfield. Further, development and growth of separated flows, stalls, and flow mixing in the flow passages are also strongly dependent on the level and scale of turbulence imbedded in the flow.

The basic concepts for flow in the blade channels must reflect the influences of the channel secondary flows. These develop as a result of strong gradients of pressure from blade to blade (circumferentially) and from hub to shroud (radially or normally). The flow is additionally complicated by the growth of passage vortices that may be triggered from the movement of flow through the end-wall or tip gaps in machines without integral shrouds around the blades. These end-wall gap or leakage flows can lead to a strong rotational motion superposed on the basic core flow in the passage. The dichotomy of convective and diffusive mixing of the flow in a channel was discussed briefly, and the need to include both mechanisms in realistic flow modeling was noted.

One of the more difficult flow problems that one must work with is the influence of very low Reynolds numbers on the flow turning and losses in blade rows. The experimental information on these flows in the open literature is limited, so that effective modeling is hampered by some uncertainty. The problem is discussed in the chapter, and a rough predictive model is developed based on some extension of the original model of Roberts (1974) and additional experimental data. The result can provide adequate estimation of the effect of low Reynolds Number on the deviation angles of blade flow, but it yields more uncertain results for the prediction of total pressure losses.

A brief introduction of the problems of static and dynamic stability was given in terms of the models developed over the years by Greitzer (1981, 1996). Operation of a machine at the "zero slope or positive slope" condition on the pressure-flow characteristic of the machine will lead to instability of the flow. Development of rotating stall cells can, for certain system conditions, trigger a potentially destructive onset of surge in the system. The models presented in the chapter can help the user or designer of turbomachines avoid these unacceptable regions of operation.

Finally, the basic concepts of the application of CFD methods in turbomachinery design and flow analysis were introduced. The literature in this field was briefly discussed, with some review of the history. Examples were presented that are representative of both the preliminary design layout and the iterative approach to design refinement and improvement.

# Appendix A
# Fluid Properties

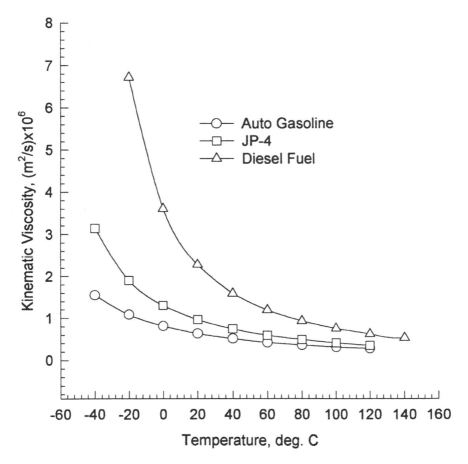

**FIGURE A.1** Variation of kinematic viscosity of fuels with temperature. (Data from Schetz and Fuhs, 1996.)

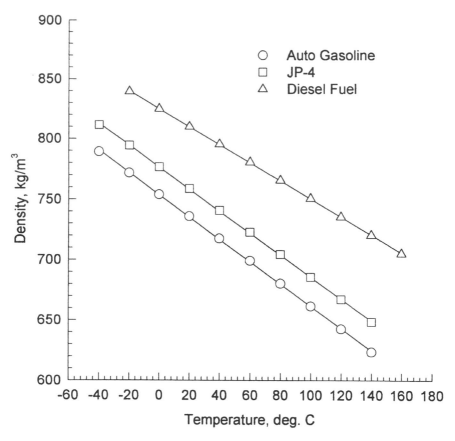

**FIGURE A.2**   Variation of density of fuels with temperature. (Data from Schetz and Fuhs, 1996.)

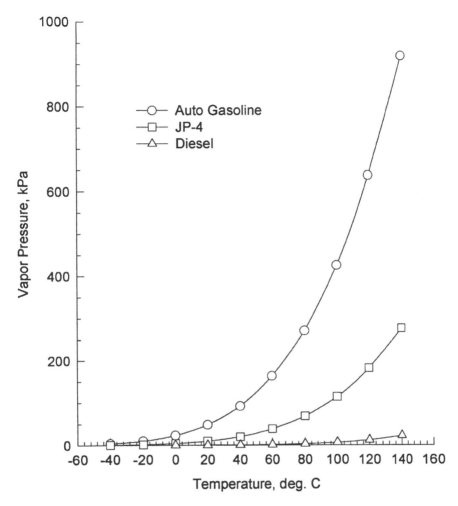

**FIGURE A.3**  Variation of fuel vapor pressure with temperature. (Data from Schetz and Fuhs, 1996.)

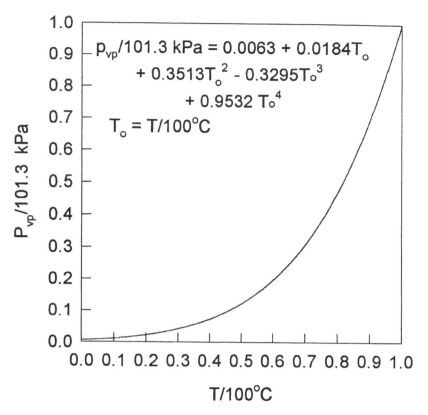

**FIGURE A.4** Normalized vapor pressure for water with a fourth order curve fit (data from Baumeister et al., 1978).

## TABLE A.1
## Approximate Gas Properties at 20°C, 101.3 kPa (from Marks' Handbook, Baumeister et al., 1978)

| Gas | M | $\gamma$ | $\rho$ (kg/m³) |
|---|---|---|---|
| Air | 29.0 | 1.40 | 1.206 |
| Carbon dioxide | 44.0 | 1.30 | 1.831 |
| Helium | 4.0 | 1.66 | 0.166 |
| Hydrogen | 2.0 | 1.41 | 0.083 |
| Methane | 16.0 | 1.32 | 0.666 |
| Natural gas | 19.5 | 1.27 | 0.811 |
| Propane | 44.1 | 1.14 | 1.835 |

## TABLE A.2
## Approximate Values for Dynamic Viscosity Using $\mu = \mu_0 (T/T_0)^n$ (based on data from Marks' Handbook, Baumeister et al., 1978, with $T_0 = 273$ K)

| Gas | $\mu_0$ (kg/ms) $\times 10^5$ | n |
|---|---|---|
| Air | 1.71 | 0.7 |
| $CO_2$ | 1.39 | 0.8 |
| $H_2$ | 0.84 | 0.7 |
| $N_2$ | 1.66 | 0.7 |
| Methane | 1.03 | 0.9 |

## TABLE A.3
## Approximate Liquid Densities at 20°C and 1 Atmosphere (from Marks' Handbook, Baumeister et al., 1978)

| Liquid | $\rho$ (kg/m³) | $\mu$, (kg/ms) $\times 10^5$ |
|---|---|---|
| Ethyl alcohol | 790 | 1.20 |
| Gasoline (S.G. = 0.68) | 675 | 0.29 |
| Glycerine | 1261 | 1412 |
| Kerosene | 806 | 1.82 |
| Mercury | 13,546 | 1.55 |
| Machine oil (heavy) | 905 | 453 |
| Machine oil (light) | 905 | 86.7 |
| Fresh water | 998 | 1.00 |
| Sea water | 1022 | 1.08 |

# Appendix B
# References

Addison, J. S. and Hodson, H. P. (1992), "Modeling of Unsteady Transitional Layers," *ASME J. of Turbomachinery*, Vol. 114, July.

Adenubi, S. O. (1975), "Performance and Flow Regimes of Annular Diffusers with Axial-Flow Turbomachinery Discharge Inlet Conditions," ASME Paper No. 75-WA/FE-5.

Alder, D. and Ilberg, H. (1970), "A Simplified Method for Calculation of the Flow Field at the Entrance of a Radial or Mixed-Flow Impeller," ASME Paper No. 70-FE-36.

Adler, D. and Krimerman, Y. (1980), "Comparison between the Calculated Subsonic Inviscid Three Dimensional Flow in a Centrifugal Impeller and Measurements," *Proc. 22nd Annu. Fluids Engineering Conf.*, New Orleans, p. 19-26.

Adkins, R. C. (1975), "A Short Diffuser with Low Pressure Loss," *ASME J. Fluids Eng.*, p. 297-302, September.

Adkins, G. G. and Smith, L. H., Jr. (1982), "Spanwise Mixing in Axial-Flow Turbomachines," *ASME J. Eng. for Power*, Vol. 104.

Ainley, D. G. and Mathieson, G. C. R. (1951), "A Method of Performance Estimation of Axial-Flow Turbines," ACR R&M 2974, 1951.

Allen, C. H. (1957), *Noise Control*, Vol. 3, (in Richards, E. J., and Mead, D. J., 1968).

Allis-Chalmers (1986), "Catalog of General Purpose Pumps," Industrial Pump Division.

AMCA Standards (1985), "Test Standards for Fans and Blowers," Bulletin 210.

AMCA (1982), "Establishing Performance Using Laboratory Models," Publication 802-82.

ANSYS (1985), "Engineering Analysis System Users' Manual," Vol. 1 and 2, Swanson Analysis Systems.

ASME (1965), Power Test Codes, "Test Code for Pumps," PTC-8.

ASME (1985), Power Test Codes, "Test Code for Compressors and Exhausters," PTC-10.

Anderson, J. D. (1984), *Modern Compressible Flow with Historical Perspectives*, McGraw-Hill, New York.

Anderson, J. D. (1996), *Computational Fluid Dynamics: The Basics with Applications*, McGraw-Hill, New York.

Balje, O. E. (1981), *Turbomachines*, John Wiley & Sons, New York.

Balje, O. E. (1968), "Axial Cascade Technology and Application to Flow Path Designs," *ASME J. Eng. for Power*, Vol. 81.

Balje, O. E. (1962), "A Study on Design Criteria and Matching of Turbomachines," *ASME J. Eng. for Power*, Vol. 84, No. 1, p. 83.

Bathie, W. W. (1996), *Fundamentals of Gas Turbines*, 2nd ed., John Wiley & Sons, New York.

Baumeister, T., Avallone, E. A., and Baumeister, T., III (1978), *Marks' Standard Handbook for Mechanical Engineering*, McGraw-Hill, New York.

Beckwith, T. G., Marangoni, R. D., and Lienhard, J. H. (1993), *Mechanical Measurements*, Addison-Wesley, New York.

Beranek, L. L., Kampermann, G. W., and Allen, C. H. (1953), of the *Journal of the Acoustical Society of America*, Vol. 25. (In Richards, E. J., and Mead, D. J., 1968).

Beranek, L. L., and Ver, I. (1992), *Noise and Vibration Control Engineering Principles and Applications*, John Wiley & Sons, New York.

Blech, R. A., Milner, E. J., Quealy, A., and Townsend, S. E. (1992), "Turbomachinery CFD on Parallel Computers," NASA TM-105932, NASA Lewis Research Center.

Bleier, F. P. (1998), *Fan Handbook: Selection, Application and Design*, McGraw-Hill, New York.

Broach, J. (1980), *Acoustic Noise Measurement*, Bruel & Kjaer, New York.

Brown, L. E., "Axial Flow Compressor and Turbine and Loss Coefficients: A Comparison of Several Parameters," *ASME J. Eng. for Power*, Vol. 94.

Burdsall, E. A., and Urban, R. H. (1973), "Fan-Compressor Noise: Prediction, Research and Reduction Studies," Final Report FAA-RD-71-73.

Busemann, A. (1928), "The Head Ratio of Centrifugal Pumps with Logarithmic Spiral Blades" ("Das Forderhohenverhaltnis Radialer Krielselpumpen mit Logarithmisch-Spiraligen Schaufeln"), *Z. Agnew Math. Mech.*, Vol. 8. (in Balje, (1981).

Casey, M. V. (1994), *The Industrial Use of CFD in the Design of Turbomachinery*, AGARD 1994 (also NASA N95-14127-03-34).

Casey, M. V. (1985), "The Effects of Reynolds Number on the Efficiency of Centrifugal Compressor Stages," *ASME J. Gas Turbines and Power*, Vol. 107.

Cebeci, T. (1988), "Essential Ingredients of a Method for Low Reynolds Number Airfoils," *AIAA J.*, Vol. 27, No. 12, p. 1680-1688.

Chapman, S. (1990), *Electric Machinery Fundamentals*, 2nd ed., McGraw-Hill, New York.

Citavy, J., and Jilek, J. (1990), "The Effect of Low Reynolds Number on Straight Compressor Cascades," ASME Paper No. 90-GT-221.

Cohen, H., Rogers, G. F. C., and Saravanamuttoo, H. I. H. (1973), *Gas Turbine Theory*, Longman Group, Ltd., Toronto.

Compte, A., Ohayon, G., and Papailiou, K.D. (1972), "A Method for the Calculation of the Wall Layers Inside the Passage of a Compressor Cascade With and Without Tip Clearance," *ASME J. Eng. for Power*, Vol. 104.

Cordier, O., (1955), "Similarity Considerations in Turbomachines," VDI Reports, Vol. 3.

Cumpsty, N. A. (1977), "A Critical Review of Turbomachinery Noise," *ASME J. Fluids Eng.*

Csanady, G. T. (1964), *Theory of Turbomachines*, McGraw-Hill, New York.

Davis, R. C. and Dussourd, J. L. (1970), "A Unified Procedure for Calculation of Off-Design Performance of Radial Turbomachinery," ASME Paper No. 70-GT-64.

Dawes, W. N. (1988), "Development of a 3-D Navier-Stokes Solver for Application to All Types of Turbomachinery," ASME Paper No. 88-GT-70.

de Haller, P. (1952), "Das Verhalten von Traflugol gittern im Axialverdictern und im Windkanal," Brennstoff-Warme-Kraft, Vol. 5.

Denton, J. D. and Usui, S. (1981), "Use of a Tracer Gas Technique to Study Mixing in a Low Speed Turbine," ASME Paper No. 81-GT-86.

Dring, R. P., Joslyn, H. D., and Blair, M. P. (1987), "The Effects of Inlet Turbulence and Rotor/Stator Interactions on the Aerodynamics and Heat Transfer of a Large-Scale Rotating Turbine Model, IV—Aerodynamic Data Tabulation," NASA CR-179469.

Dullenkoph, K. and Mayle, R. E. (1984), "The Effect of Incident Turbulence and Moving Wakes on Laminar Heat Transfer in Gas Turbines," *ASME J. of Turbomachinery*, Vol. 116.

Eckardt, D. (1976), "Detailed Flow Investigations Within a High-Speed Centrifugal Compressor Impeller," *ASME J. Fluids Eng.*, Vol. 98, p. 390-402.

Emery, J. C., Herrig, L. J., Erwin, J. R., and Felix, A. R. (1958), "Systematic Two-Dimensional Cascade Tests of NACA 65-Series Compressor Blades at Low Speed," NACA-TR-1368.

Eppler, R. and Somers, D. M. (1980), "A Computer Program for the Design and Analysis of Low Speed Airfoils," NASA TM 80210.

Emmons, H. W., Pearson, C. E., Grant, H. P. (1955), "Compressor Surge and Stall Propagation," *Trans. ASME*.

Evans, B. J. (1971), "Effects of Free Stream Turbulence on Blade Performance in a Compressor Cascade," Cambridge University Engineering Department Report Turbo/TR 26.

Evans, B. J. (1978), "Boundary Layer Development on an Axial-Flow Compressor Stator Blade," *ASME J. Eng. for Power*, Vol. 100, 1978.

Farn, C. L. S., Whirlow, D. K., and Chen, S. (1991), "Analysis and Prediction of Transonic Turbine Blade Losses," ASME Paper No. 91-GT-183.

Fox, R. W. and McDonald, A. T. (1992), *Introduction to Fluid Mechanics*, John Wiley & Sons, New York.

Gallimore, S. J. and Cumptsy, N. A. (1986), "Spanwise Mixing in Multi-Stage Axial Flow Compressors: Part I. Experimental Investigations," *ASME J. Turbomachinery*, Vol. 8.

Gallimore, S. J., and Cumptsy, N. A. (1986), "Spanwise Mixing in Multi-Stage Axial Flow Compressors. Part II. Throughflow Calculations Including Mixing," *ASME J. Turbomachinery*, Vol. 8.

Gessner, F. B. (1967), "An Experimental Study of Centrifugal Fan Inlet Flow and Its Influence on Fan Performance," ASME Paper No. 67-FE-21.

Goulds (1982), *Pump Manual: Technical Data Section*, 4th ed., Goulds Pump, Inc., Seneca Falls, NY.

Graham, J. B. (1972), "How to Estimate Fan Noise," *J. of Sound and Vibration*, May 1972.

Graham, J. B. (1991), "Prediction of Fan Sound Power," *ASHRAE Handbook, HVAC Applications*.

Gray, R. B. and Wright, T. (1970), "A Vortex Wake Model for Optimum Heavily Loaded Ducted Fans," *AIAA J. Aircraft*, Vol. 7, No. 2.

Granville, P. S. (1953), "The Calculations of Viscous Drag on Bodies of Revolution," David Taylor Model Basin Report No. 849.

Granger, R. A., Ed. (1988), *Experiments in Fluid Mechanics*, Holt, Reinehart & Winston, New York.

Gostelow, J. P. and Blunden, A. R. (1988), "Investigations of Boundary Layer Transition in an Adverse Pressure Gradient," ASME Paper No. 88-GT-298.

Goldschmeid, F. R., Wormley, D. M., and Rowell, D. (1982), "Air/Gas System Dynamics of Fossil Fuel Power Plants—System Excitation," EPRI Research Project 1651, Vol. 5, CS-2006.

Greitzer, E. M. (1980), "Review—Axial Compressor Stall Phenomena," *ASME J. of Fluids Eng.*, Vol. 102, pp. 134–151.

Greitzer, E. M. (1981), "The Stability of Pumping Systems—The 1980 Freeman Scholar Lecture," *ASME J. Fluids Eng.*, Vol. 103, p. 193-242.

Greitzer, E. M. (1996), "Stability in Pumps and Compressors," Section 27.8, *Handbook of Fluid Dynamics and Turbomachinery*, (Ed. Joseph Shetz and Allen Fuhs), John Wiley and Sons.

Haaland, S. E. (1983), "Simple and Explicit Formulas for the Friction Factor in Turbulent Pipe Flow," *ASME J. Fluids Eng.*, March.

Hansen, A. G. and Herzig, H. Z. (1955), "Secondary Flows and Three-Dimensional Boundary Layer Effects," (in NASA SP-36, Johnson and Bullock, 1965).

Hanson, D. B. (1974), "Spectrum of Rotor Noise Caused by Atmospheric Turbulence," *J. Acoust. Soc. of Am.*, Vol. 56.

Hathaway, M. D., Chriss, R. M., Wood, J. R., and Strazisar, A. J. (1993), "Experimental and Computational Investigation of the NASA Low-Speed Centrifugal Compressor Flow Field," NASA Technical Memorandum 4481.

Hirsch, C. (1974), "End-Wall Boundary Layers in Axial Compressors," *ASME J. Eng. for Power*, Vol. 96.

Hirsch, C. (1990), *Numerical Computations of Internal and External Flows*, Vols. 1 and 2, John Wiley & Sons, New York.

Holman, J. P. and Gadja, W. J., Jr. (1989), *Experimental Methods for Engineers*, 7th ed., McGraw-Hill, New York.

Horlock, J. H. (1958), *Axial Flow Compressors, Fluid Mechanics and Thermodynamics*, Butterworth Scientific, London.

Howard, J. H. G., Henseller, H. J., and Thornton, A. B. (1967), Performance and Flow Regimes for Annular Diffusers," ASME Paper No. 67-WA/FE-21.

Howden Industries (1996), Axial Fan Performances Curves.

Howell, A. R. (1942), "The Present Basis of Axial Flow Compressor Design. Part I. Theory and Performance," ARC R&M 2095.

Howell, A. R. (1945), "Fluid Mechanics of Axial Compressors," *Proc. Institute of Mechanical Engineers*, Vol. 153.

Huebner, K. H., Thornton, E. A., and Byrom, T. G. (1995), *The Finite Element Method for Engineers*, 3rd ed., John Wiley & Sons, New York.

Hughes, W. F. and Brighton, J. A. (1991), *Theory and Problems of Fluid Dynamics*, 2nd ed. (Schaum's Outline Series), McGraw-Hill, New York.

Hunter, I. H. and Cumptsy, N. A. (1982), "Casing Wall Boundary Layer Development Through an Isolated Compressor Rotor," *ASME J. Eng. for Power*, Vol. 104.

Hydraulic Institute (1983), *Pump Test Standards*.

Industrial Air (1986), *General Fan Catalog*, Industrial Air, Inc.

ISO-ASME (1981), "Calculation of Orifice Discharge Coefficients," *Flow Test Standards*.

Idel'Chik, I. E. (1966), *Handbook of Hydraulic Resistance*, AEC-TR-6630 (also Hemisphere, New York, 1986).

Jackson, D. G., Jr., and Wright, T. (1991), "An Intelligent Learning Axial Fan Design System," ASME Paper No. 91-GT-27.

Japikse, D., Marscher, W. D., Furst, R. B. (1997), *Centrifugal Pump Design and Performance*, Concepts ETI, White River Junction, VT.

Johnsen, I. A. and Bullock, R. O. (1965), "Aerodynamic Design of Axial Flow Compressors," NASA SP-36.

Johnson, M. W. and Moore, J. (1980), "The Development of Wake Flow in a Centrifugal Impeller," *ASME J. Turbomachinery*, Vol. 116.

Jorgenson, R., Ed. (1983), *Fan Engineering—An Engineer's Handbook on Fans and Their Applications*, Buffalo Forge Co., New York.

Kahane, A. (1948), "Investigation of Axial-Flow Fan and Compressor Rotor Designed for Three-Dimensional Flow," NACA TN- 1652.

Karassik, I. J., Krutzch, W. C., Fraser, W. H., and Messina, J. P. (1986), *Pump Handbook*, 2nd ed., McGraw-Hill, New York.

Katsanis, T. (1969), "Fortran Program for Calculating Transonic Velocities on a Blade-to-Blade Surface of a Turbomachine," NASA TN D-5427.

Kestin, J., Meader, P. F., and Wang, W. E. (1961), "On Boundary Layers Associated with Oscillating Streams," *Appl. Sci. Res.*, Vol. A 10.

Kittredge, C. P. (1967), "Estimating the Efficiency of Prototype Pumps from Model Tests," ASME Paper No. 67-WA/FE-6.

Krain, H., and Hoffman, W. (1989), "Verification of an Impeller Design by Laser Measurement and 3-D Viscous Flow Calculations," ASME Paper No. 89-GT-159.

Krain. H. (1988), "Swirling Impeller Flow," *ASME J. Turbomachinery*, Vol. 110, No. 1, p. 122-128.

Krain, H. (1984), "A CAD Method for Centrifugal Compressor Impellers," *J. Eng. for Gas Turbines and Power*, Vol. 106, p. 482-488.

Koch, C. C., and Smith, L. H., Jr. (1976), "Loss Sources and Magnitudes in Axial-Flow Compressors," *ASME J. Eng. for Power*, Vol. 98, No. 3, p. 411-424.

Koch, C. C. (1981), "Stalling Pressure Capability of Axial-Flow Compressors," *ASME J. Eng. for Power*, Vol. 103, No. 4, p. 645-656.

Laguier, R., (1980), "Experimental Analysis Methods for Unsteady Flow in Turbomachinery," *Measurement Methods in Rotating Components of Turbomachinery*, ASME, New York, pp. 71–81.

Lakshminarayana, B. (1996), *Fluid Dynamics and Heat Transfer of Turbomachinery*, John Wiley & Sons, New York.

Lakshminarayana, B. (1991), "An Assessment of Computational Fluid Dynamic Techniques in the Analysis and Design of Turbomachinery—The 1990 Freeman Scholar Lecture," *ASME J. Fluids Eng.*, Vol. 113.

Lakshminarayana, B. (1979), "Methods of Predicting the Tip Clearance Effects in Axial Flow Turbomachinery," *ASME J. Basic Eng.*, Vol. 92.

Lewis, R. I. (1996), *Turbomachinery Performance and Analysis*, John Wiley & Sons, New York.

Leylek, J. H. and Wisler, D. C. (1991), "Mixing in Axial-Flow Compressors: Conclusions Drawn from Three-Dimensional Navier-Stokes Analyses and Experiments," *ASME J. Turbomachinery*, Vol. 113.

Li, Y. S. and Cumptsy, N. A. (1991), "Mixing in Axial Flow Compressors. Part 1. Test Facilities and Measurements in a Four-Stage Compressor," *ASME J. Turbomachinery*, Vol. 113.

Li, Y. S. and Cumptsy, N. A. (1991), "Mixing in Axial Flow Compressors. Part 2. Measurements in a Single-Stage Compressor and Duct," *ASME J. Turbomachinery*, Vol. 113.

Lieblein, S. (1956), "Loss and Stall Analysis of Compressor Cascades," *ASME J. Basic Eng.*, September.

Lieblein, S. and Roudebush, R. G. (1956), "Low Speed Wake Characteristics of Two-Dimensional Cascade and Isolated Airfoil Sections," NACA TN 3662.

Lieblein, S. (1957), "Analysis of Experimental Low-Speed Loss and Stall Characteristics of Two-Dimensional Blade Cascades," NACA RM E57A28.

Longley, J. P. and Greitzer, E. M. (1992), "Inlet Distortion Effects in Aircraft Propulsion Systems," AGARD Lecture Series LS-183.

Mayle, R. E. and Dullenkopf, K. (1990), "A Theory for Wake Induced Transition," *ASME J. Turbomachinery*, Vol. 112.

Mayle, R. E. (1991), "The Role of Laminar-Turbulent Transition in Gas Turbine Engines," ASME Paper No. 91-GT-261.

Madhavan, S. and Wright, T. (1985), "Rotating Stall Caused by Pressure Surface Flow Separation in Centrifugal Fans," *ASME J. for Gas Turbines and Power*, Vol. 107, pp. 775–781.

Madhavan, S., DiRe, J., and Wright, T. (1984), "Inlet Flow Distortion in Centrifugal Fans," ASME Paper No. 84-JPGC/GT-4.

Mattingly, J. (1996), *Elements of Gas Turbine Propulsion*, McGraw-Hill, New York.

McDonald, A. T. and Fox, R. W. (1965), "Incompressible Flow in Conical Diffusers," ASME Paper No. 65-FE-25.

McDougall, N. M., Cumptsy, N. A., and Hynes, T. P. (1989), "Stall Inception in Axial Compressors," ASME Gas Turbine and Aeroengine Conference, Toronto, June 1989.

McFarland, E. R. (1985), "A FORTRAN Computer Code for Calculating Flows in Multiple-Blade-Element Cascades," NASA TM 87104.

McFarland, E. R. (1984), "A Rapid Blade-to-Blade Solution for Use in Turbomachinery Design," *ASME J. Eng. for Power*, Vol. 100, p. 376-382.

McFarland, E. R. (1983), "A Rapid Blade-to-Blade Solution for Use in Turbomachinery Design," ASME Paper No. 83-GT-67.

Mellor, G. L. (1959), "An Analysis of Compressor Cascade Aerodynamics. Part I. Potential Flow Analysis with Complete Solutions for Symmetrically Cambered Airfoil Families," *ASME J. Basic Eng.*, September.

Mellor, G. L. (1959), "An Analysis of Compressor Cascade Aerodynamics. Part II. Comparison of Potential Flow Results with Experimental Data," *ASME J. Basic Eng.*, September.

Mellor, G. L., and Wood, G. M. (1971), "An Axial Compressor End-Wall Boundary Layer Theory," *ASME J. Basic Eng.*, Vol. 93.

Miller, P. L., Oliver, J. H., Miller, D. P., and Tweedt, D. L. (1996), "BladeCAD: An Interactive Geometric Design Tool for Turbomachinery Blades," NASA TM-107262, NASA Lewis Research Center.

Moody, L. F. (1925), "The Propeller Type Turbine," *Proc. Am. Soc. of Civil Eng.*, Vol. 51.

Moore, J. and Smith, B. L. (1984), "Flow in a Turbine Cascade. Part 2. Measurement of Flow Trajectories by Ethylene Detection," *ASME J. Eng. for Gas Turbines and Power*, Vol. 106.

Moreland, J. B. (1989), "Outdoor Propagation of Fan Noise," ASME Paper No. 89-WA/NCA-9.

Morfey, C. L. and Fisher, M. J. (1970), "Shock Wave Radiation from a Supersonic Ducted Rotor," *Aeronaut. J.*, Vol. 74, No. 715.

Myers, J. G., Jr., and Wright, T. (1994), "An Inviscid Low Solidity Cascade Design Routine," ASME Paper No. 93-GT-162.

Nasar, S. A. (1987), *Handbook of Electric Machines*, McGraw-Hill, New York.

Neise, W. (1976), "Noise Reduction in Centrifugal Fans: A Literature Survey," *J. Sound and Vibration*, Vol. 45, No. 3.

Novak, R. A. (1967), "Streamline Curvature Computing Procedure for Fluid Flow Problems," *ASME J. Eng. for Power*, October.

New York Blower (1986), *Catalog on Fans*, New York Blower Co.

Oates, G. C., Ed. (1985), *Aerothermodynamics of Aircraft Engine Components*, American Institute of Aeronautics and Astronautics, New York.

Oates, G. C., Ed. (1984), *Aerothermodynamics of Gas Turbine and Rocket Propulsion*, American Institute of Aeronautics and Astronautics, New York.

Obert, E. F. (1973), *Internal Combustion Engines*, 3rd ed., Harper & Row, New York.

O'Brien, W. F., Jr., Cousins, W. T., and Sexton, M. R. (1980), "Unsteady Pressure Measurements and Data Analysis Techniques in Axial Flow Compressors," *Measurement Methods in Rotating Components of Turbomachinery*, ASME, New York, pp. 195–202.

O'Meara, M. M. and Mueller, T. J. (1987), "Laminar Separation Bubble Characteristics on an Airfoil at Low Reynolds Numbers," *AIAA J.*, Vol. 25, No. 8, August.

Papailiou, K. D. (1975), "Correlations Concerning the Process of Flow Decelerations, *ASME J. Eng. for Power*, April.

Pfenninger, W. and Vemura, C. S. (1990), "Design of Low Reynolds Number Airfoils," *AIAA J. Aircraft*, Vol. 27, No. 3, p. 204-210, March.

Ralston, S. A. and Wright, T. (1987), "Computer-Aided Design of Axial Fans Using Small Computers," *ASHRAE Trans.*, Vol. 93, Pt. 2, Paper No. 3072.

Rao, S. S. (1990), *Mechanical Vibrations*, 2nd ed., Addison-Wesley, New York.

Reneau, L. R., Johnson, J. P., and Kline, S. J. (1967), "Performance and Design of Straight Two-Dimensional Diffusers," *J. Basic Eng.*, Vol. 89, No. 1.

Richards, E. J. and Mead, D. J.(1968), *Noise and Acoustic Fatigue in Aeronautics*, John Wiley & Sons, New York.

Rhoden, H. G. (1952), "Effects of Reynolds Number on Air Flow Through a Cascade of Compressor Blades," ARC R&M 2919.

Roberts, W. B. (1975), "The Effects of Reynolds Number and Laminar Separation on Cascade Performance," *ASME J. Eng. for Power*, Vol. 97, p. 261-274.

Roberts, W. B. (1975), "The Experimental Cascade Performance of NACA Compressor Profiles at Low Reynolds Number," *ASME J. Eng. for Power*, July.

Rodgers, C. (1980), "Efficiency of Centrifugal Impellers," AGARD CP282, Paper No. 2.

Schetz, J. A. and Fuhs, A. E., Eds. (1996), *Handbook of Fluid Dynamics and Turbomachinery*, John Wiley & Sons, New York.

Schlichting, H. (1979), *Boundary Layer Theory*, 7th ed., McGraw-Hill, New York.

Schmidt, G. S. and Mueller, T. J. (1989), "Analysis of Low Reynolds Number Separation Bubbles Using Semi-Empirical Methods," *AIAA J.*, Vol. 27, No. 8, August 1989.

Schubauer, G. B. and Skramstad, H. K. (1947), "Laminar Boundary Layer Oscillations and Stability of Laminar Flow, National Bureau of Standards Research Paper 1772, and *J. Aeronaut. Sci.*, Vol. 14.

Sharland, C. J. (1964), "Sources of Noise in Axial Fans," *J. Sound and Vibration*, Vol. 1, No. 3.

Shigley, J. S. and Mischke, C. K. (1989), *Mechanical Engineering Design*, McGraw-Hill, New York.

Shepherd, D. G. (1956), *Principles of Turbomachinery*, Macmillan and Company, New York.

Shen, L. C. (1990), "Computer Codes to Analyze a Centrifugal Fan Impeller—A Comparison Between the Panel Method, Quasi-Three Dimensional Finite Difference Solutions, and Experimental Results," Masters Thesis in Mechanical Engineering, The University of Alabama at Birmingham, Birmingham, AL.

Smith, A. M. O. (1956), "Transition, Pressure Gradient and Stability Theory," *Proc. IX Int. Conf. Appl. Mechanics*, Vol. 4.

Smith, A. M. O. (1975), "High-Lift Aerodynamics," *AIAA J. Aircraft*, Vol. 12, No. 6.

Smith, L. H., Jr. (1955), "Secondary Flows in Axial-Flow Turbomachinery," *Trans. ASME*, Vol. 177.

Smith, L. H., Jr. (1969), "Casing Boundary Layers in Multi-Stage Axial Flow Compressors," *Fluid Mechanics of Internal Flow*, Elsevier Press, New York.

Smith, L. C. (1976), "A Note on Diffuser Generated Unsteadiness," *ASME J. Fluids Eng.*, September 1975.

Sovran, G., and Klomp, E. D. (1967), "Experimentally Determined Optimum Geometries for Diffusers with Rectilinear, Conical, or Annular Cross-Sections," *Fluid Mechanics of Internal Flow*, Elsevier, Amsterdam, The Netherlands.

Stanitz, J. D. and Prian, V. D. (1951), "A Rapid Approximate Method for Determining Velocity Distributions on Impeller Blades of Centrifugal Compressors," NACA TN 2421.

Stepanoff, A. J. (1948), *Centrifugal and Axial Flow Pumps*, John Wiley & Sons, New York.

Steendahl, J. D. (1988), "Optimizing Turbine Efficiency Using Index Testing," *Mech. Eng. Mag.*, p. 74-77, October.

Stevens, K. G., Peterson, V. L., and Kutler, P. (1996), "Introduction to Computational Fluid Dynamics," Chapter 18 of *The Handbook of Fluid Dynamics and Fluid Machinery*, (J. A. Shetz and A. E. Fuhs, Eds.), John Wiley & Sons, New York.

Strub, R. A., Bonciana, L., Borer, C. J., Casey, M. V., Cole, S. L., Cook, B. B., Kotzur, J., Simon, H., and Strite, M. A., (1984), "Influence of Reynolds Number on the Performance of Centrifugal Compressors," *ASME J. Turbomachinery*, Vol. 106, No. 2.

Stodola, A. (1924), Steam and Gas Turbines, Springer, New York.

Thumann, A. and Miller, R. K. (1986), Fundamentals of Noise Control Engineering, Fairmont Press, Atlanta.

Thoma, D. and Fischer, K. (1932), "Investigation of the Flow, Conditions in a Centrifugal Pump," *Trans. ASME*, Vol. 58.

Topp, D. A., Myers, R. A. and Delaney, R. A., "TADS: A CFD-Based Turbomachinery and Analysis Design System with GUI, Volume 1: Method and Results," NASA CR-198440, NASA Lewis Research Center.

Van Wylen, G. J., and Sonntag, R. E. (1986), Fundamentals of Classical Thermodynamics, 3rd ed., John Wiley & Sons, New York.

Vavra, M. H. (1960), Aerothermodynamics and Flow in Turbomachines, John Wiley & Sons, New York.

Verdon, J. M. (1993), "Review of Unsteady Aerodynamics Methods for Turbomachinery Aeroelastic and Aeroacoustic Applications," *AIAA J.*, Vol. 31, No. 2.

Volino, R.J., and Simon, T. W. (1995), "Bypass Transition in Boundary Layers Including Curvature and Favorable Pressure Gradient Effects," *ASME J. Turbomachinery*, Vol. 117, No. 1.

Walker, G. J. and Gostelow, J. P. (1990), "Effects of Adverse Pressure Gradient on the Nature and Length of Boundary Layer Transition," *ASME J. Turbomachinery*, Vol. 112.

Wang, R. E. (1993), "Predicting the Losses and Deviation Angles with Low Reynolds Separation on Axial Fan Blades," Masters Thesis in Mechanical Engineering, The University of Alabama at Birmingham.

Weber, H. E. (1978), "Boundary Layer Calculations for Analysis and Design," *ASME J. Fluids Eng.*, Vol. 100.

Wennerstrom, A. J. (1991), "A Review of Predictive Efforts for Transport Phenomena in Axial Flow Compressors," *ASME J. Turbomachinery*, Vol. 113.

Weisner, F. J. (1967), "A Review of Slip Factors for Centrifugal Impellers," *ASME J. Eng. for Power*, Vol. 89.

White, F. M. (1994), *Fluid Mechanics*, McGraw-Hill, New York.

Wilson, D. G. (1984), *The Design of High-Efficiency Turbomachinery and Gas Turbine Engines*, MIT Press, Cambridge, MA.

Wilson, K. C., Addie, G. R., Sellgren, A., and Clift, R. (1997), *Slurry Transport Using Centrifugal Pumps*, 2nd ed., Chapman and Hall, New York.

Wisler, D. C., Bauer, R. C., and Okiishi, T. H. (1987), "Secondary Flow, Turbulent Diffusion, and Mixing in Axial-Flow Compressors," *ASME J. Turbomachinery*, Vol., 109.

Wittig, S., Shulz, A., Dullenkopf, K., and Fairbanks, J. (1988), "Effects of Free-Stream Turbulence and Wake Characteristics on the Heat Transfer Along a Cooled Gas Turbine Blade," ASME Paper No. 88-GT-179.

Wormley, D. M., Rowell, D., and Goldschmied, F. R. (1982), "Air/Gas System Dynamics of Fossil Fuel Power Plants—Pulsations," EPRI Research Project 1651, Vol. 5, CS-2006.

Wright, S. E. (1976), "The Acoustic Spectrum of Axial Fans," *J. Sound and Vibration*, Vol. 85.

Wright, T. (1974), "Efficiency Prediction for Axial Fans," *Proc. Conf. Improving Efficiency in HVAC Equipment for Residential and Small Commercial Buildings*, Purdue University.

Wright, T., Tzou, K. T. S., Madhavan, S., and Greaves, K. W. (1982), "The Internal Flow Field and Overall Performance of a Centrifugal Fan Impeller—Experiment and Prediction," ASME Paper No. 82-JPGC-GT-16.

Wright, T. (1982), "A Velocity Parameter for the Correlation of Axial Fan Noise," *Noise Control Eng.*, p. 17-26, July-August.

Wright, T., Madhavan, A., and DiRe, J. (1984a), "Centrifugal Fan Performance with Distorted Inflows," *ASME J. Gas Turbines and Power*, Vol. 106, No. 4.

Wright, T., Tzou, K. T. S., and Madhavan, S. (1984b), "Flow in a Centrifugal Fan Impeller at Off-Design Conditions," *ASME J. Gas Turbines and Power*, Vol. 106, No. 4.

Wright, T. (1984c), "Centrifugal Fan Performance with Inlet Clearance," *ASME J. Gas Turbines and Power*, Vol. 106, No. 4.

Wright, T. (1984d), "Optimal Fan Selection Based on Fan-Diffuser Interactions," ASME Paper No. 84-JPGC/GT-9.

Wright, T., Baladi, J. Y., and Hackworth, D. T. (1985), "Quiet Cooling System Development for a Traction Motor," ASME Paper No. 85-DET-137.

Wright, T., and Ralston, S. (1987), "Computer Aided Design of Axial Fans Using Small Computers," *ASHRAE Trans.*, Vol. 93, Part 2.

Wright, T. (1988), "A Closed-Form Algebraic Approximation for Quasi-Three-Dimensional Flow in Axial Fans," ASME Paper No. 88-GT-15.

Wright, T. (1989), "Comments on Compressor Efficiency Scaling with Reynolds Number and Relative Roughness," ASME Paper No. 89-GT-31.

Wright, T. (1995), "In Search of Simple Models for the Prediction of Fan Performance and Noise," *Noise Control Engineering J.*, Vol. 43-4, p. 85-89.

Wright, T. (1996), "Low Pressure Axial Fans," Section 27.6, *Handbook of Fluid Dynamics and Turbomachinery*, (Ed. Joseph Shetz and Allen Fuhs), John Wiley & Sons.

Wu, C. H. (1952), "A General Theory of Three-Dimensional Flow in Subsonic and Supersonic Turbomachine in Radial, Axial and Mixed Flow Types," NACA TR 2604.

Yang, T. and El-Nasher, A. M. (1975), "Slot Suction Requirements for Two-Dimensional Griffith Diffusers," *ASME J. Fluids Eng.*, June.

Zurn (1981), *General Fan Catalogs*.

Zweifel, O. (1945), "The Spacing of Turbomachine Blading, Especially with Large Angular Deflection," *The Brown Boveri Review*, Vol. 32, December (in Balje, 1981).

# Index